INTERNATIONAL TECHNOLOGICAL UNIVERSITY
This Book is Donated by:
PROF. WAI-KAI CHEN

Date:

A SYMPOSIUM TO HONOR
C.C. LIN

APPLIED MATHEMATICS, FLUID MECHANICS, ASTROPHYSICS

A SYMPOSIUM TO HONOR
C.C. LIN

22-24 June 1987
Massachusetts Institute of Technology,
Cambridge, USA

Editors
David J. Benney (M.I.T.)
Frank H. Shu (UC. Berkeley)
Chi Yuan (CCNY)

World Scientific
Singapore • New Jersey • Hong Kong

Published by

World Scientific Publishing Co. Pte. Ltd.
P. O. Box 128, Farrer Road, Singapore 9128

U. S. A. office: World Scientific Publishing Co., Inc.
687 Hartwell Street, Teaneck NJ 07666, USA

Applied Mathematics, Fluid Mechanics, Astrophysics
SYMPOSIUM TO HONOR C C LIN

Copyright © 1988 by World Scientific Publishing Co Pte Ltd.

All rights reserved. This book, or parts thereof, may not be reproduced in any form or by any means, electronic or mechanical, including photocopying, recording or any information storage and retrieval system now known or to be invented, without written permission from the Publisher.

ISBN 9971-50-245-3

Printed in Singapore by General Printing Services Pte. Ltd.

EDITOR's PREFACE

This volume records the papers presented at a symposium held during 22-24 June 1987 at the Massachusetts Institute of Technology to honor the career of Professor C. C. Lin on the occasion of his retirement from active service as Institute Professor. The papers are arranged in six broad subject areas that overlap Professor Lin's own research interests: applied mathematics, the theory of hydrodynamic stability and turbulence, general fluid mechanics, plasma physics, galactic astrophysics, and general astrophysics. The individual contributions have great diversity, yet a few common themes emerge when one reads these pages: the enormous influence which C. C. Lin has exerted on many fields, the permanence and utility of the concepts that he has helped to foster in these subjects, and the tremendous respect and fondness he has engendered among colleagues and students. In a rich life devoted to science, he has served as an inspiration and an unerring guide for us all, and it is with deep gratitude and appreciation that we dedicate this book to him on his seventy-first birthday.

David J. Benney, Associate Editor

Frank H. Shu, Associate Editor

Chi Yuan, Chief Editor

CONTENTS

Preface   v

I. APPLIED MATHEMATICS

Space Guidance Evolution at MIT — The Early Days   3
   Richard H. Battin

Theoretical Biology as a Branch of Applied Mathematics: Some Examples   15
   Lee A. Segel

Looking at a Computational Physics Environment   29
   Robert H. Berman

Sound Waves in Fluids: Mathematical Models and Physical Reality   35
   Garret Birkhoff

A Class of Exact Solutions in Viscous Incompressible Magneto-hydrodynamics   46
   A. D. D. Craik

Stress Singularities at a Rim of Circular Cylinders   49
   Yihan Lin and Frederic Y. M. Wan

II. STABILITY & TURBULENCE

On the Theory of Turbulence for Incompressible Fluids   59
   Pei-Yuan Zhou(Chou) and Shi-Yi Chen

Nonlinear Euler Partial Differential Equations: Singularities in their Solution   81
   J. T. Stuart

Nonlinear Stability of Rotating Pipe Flow   96
   T. R. Akylas and N. Toplosky

Bistable Cellular Flames   108
   A. Bayliss, B. Matkowsky and M. Minkoff

Whirling of a Flexible Cylinder Partially Filled with Liquid 116
   S. H. Crandall and J. W. Mroszczyk

Linear Stability Theory for a Subsonic Mixing Layer in
the Viscous Critical Layer Regime 124
   V. Djordjevic and L. G. Redekopp

Perturbations of Jeffery-Hamel Flows 129
   Philip G. Drazin

Instability Driven Boundary Layers on Curved Walls 134
   P. Hall and W. D. Lakin

Kelvin-Helmholtz Stability and Two-Phase Flow 144
   D. Y. Hsieh

High Taylor Number Couette Flow between Concentric
Cylinders and Long Eccentric Cylinders 150
   Martin Lessen

Nonlinear Stability of a Reversed Flow over a Flat Plate
with Suction 164
   S. P. Lin and H. B. Chen

Barotropic Instability of the Bickley Jet 168
   S. A. Maslowe

Instability of Stratified Shear Flow with $Ri > 0.25$ Everyplace 174
   Richard S Lindzen

Nonlinear Development of Görtler Vortices and the Generation
of High Shear Layers in the Boundary Layer 175
   A. S. Sabry and J. T. C. Liu

III. GENERAL FLUID MECHANICS

Some Hele Shaw Cell Flows and Their Visualization 187
   Louis N. Howard

Forced Generation of Solitary Waves 198
   T. Yaotsu Wu

Finite Groups of Gravity Waves   213
   Chia-Shun Yih

Solitons and Other Macrostructures in Fluid Dynamics   230
   Ru Ling Chou and C. K. Chu

Wave Propagation on an Ice-Covered Ocean   237
   Erik Mollo-Christensen and Antony K. Liu

Diffusion-Controlled Reaction in a Vortex Field   241
   Ronald G. Rehm, Howard R. Baum and Daniel W. Lozier

Amundson's Model of the Combustion of a Carbon Ball   249
   S. I. Rosencrans

Inertia Effects of a Localized Force near a Wall
in a Slow Shear Flow   253
   M. Shibata and C. C. Mei

Some Model Equations for Surface Waves on a Fluid   256
   M. C. Shen

IV. PLASMA PHYSICS

Physics of Space and Laboratory Plasmas   265
   Bruno Coppi

Stability of Rotating Electron Beams   267
   Y. Y. Lau

Ion Beam Inertial Fusion: Some Issues of
Interest for Applied Mathematics   275
   James. W-K. Mark

Plasma Processes Relevant to Extraterrestrial Phenomena   291
   C. S. Wu

V. GALACTIC ASTROPHYSICS

Observations of Spiral Structure in Galaxies   299
   R. J. Allen

Modal Theory of Galactic Spiral Structure   319
   Giuseppe Bertin

Observational Evidence on the Density Wave Theory —
An Intensive Study of M81   332
   Frank Bash

Color Imaging of Spiral Galaxies   338
   *Debra Meloy Elmegreen, Bruce G. Elmegreen,
and Philip E. Seiden*

Radial Amplitude Variations in Spiral Arms   345
   Preben Grosbol

Distribution Functions for Triaxial Galaxies   350
   C. Hunter

Shapes of Star-Gas Waves in Spiral Galaxies   358
   Stephen H. Lubow

Goings-On at the Center of a Galaxy   366
   R. H. Miller and B. F. Smith

N-Body Simulations of the Cloudy Interstellar Medium
in Density Wave-Dominated Galaxies:
Orbit Trapping, Sloshing, Self Gravity, and Spiral Structure   373
   William W. Roberts, Jr., David S. Adler, and Glen R. Stewart

The Dispersion Relation and the Masses of Galactic Disks   381
   W. W. Shane

VI.   GENERAL ASTROPHYSICS

Soliton Stars   387
   T. D. Lee

The Interaction of Acoustic Radiation with Turbulence   399
   Peter Goldreich

Supernova, Neutrinos and Cosmology   405
   Hong-Yee Chiu

Forced Precession of the Comet Halley Nucleus   422
   W. H. Julian

Gravitational Instabilities in the Thin Dust Disk
of the Solar Nebula   428
   Myron Lecar

A Probable Mechanism for Bipolar Outflows
Near Young Stellar Objects   434
   Z. Y. Yue, B. Zhang, and G. Winnewisser

I. APPLIED MATHEMATICS

Space Guidance Evolution at MIT—The Early Days

Richard H. Battin

The Charles Stark Draper Laboratory, Inc.
Adjunct Professor of Aeronautics and Astronautics
Massachusetts Institute of Technology

> This is a story that begins in the early 1950's when the author completed his doctoral program with C. C. Lin and joined the MIT Instrumentation Laboratory. The new and exciting programs then were self-contained guidance systems. How this early activity of designing a backup system for the Atlas intercontinental ballistic missile led to MIT's receiving the first prime contract for the Apollo lunar landing navigation, guidance, and control system is told from the author's own perspective. The fundamental analytic activities for Apollo were the work of several of MIT's Applied Mathematics graduates who remember with great fondness the influence on their careers of C. C. Lin and his contemporaries.

What a great idea this is! I am delighted to be included in this celebration. C. C. Lin played a major role in my life. He was my mentor, my doctoral thesis advisor, and he guided my first feeble attempts at independent research. He was coauthor of my first paper "On the Stability of the Boundary Layer over a Cone." Later we both went into outer space—I to the Moon and he to the Galaxies.

Recently, I scanned the roster of the MIT Math Department and found C. C. Lin to be the only remaining active Professor of those I knew as a student. Over there in the Emeritus list were the familiar names: Fran Hildebrand, Ted Martin, Dirk Struik, George Thomas, Warren Ambrose. These were my professors. Next year C. C. Lin will join them and he will be sorely missed.

I also noticed that the number of Professors of Mathematics almost exactly matches the number of Professors of *Applied* Mathematics. This was not always the case. When C. C. Lin transferred from Brown University to MIT as an Associate Professor in 1947, the effect was to double the number of professors in Applied Mathematics. Before him was Eric Reissner who is now Professor Emeritus of Applied Mechanics at U. C. San Diego. His specialty was Elasticity.

Let's face it. In those days Mathematics meant *Pure* Mathematics. If you could apply it, it wasn't Mathematics. As a student, I took no courses in Advanced Calculus or Applied Mathematics. I taught those courses—but only to *Engineering* students. Math majors were above such things. Some of us even thought MIT was an acronym for the

"Massachusetts Institute of Topology."

1. BACKGROUND

I started life at MIT in Electrical Engineering and transferred to the Math Department for my graduate work. The motivation was an innate desire to teach. Had I stayed in Engineering, I could only be a Lab Assistant. In Mathematics I could *actually* do classroom teaching.

The double life has provided a great cover. It is convenient to counter a penetrating question in mathematics with the explanation: "I am really an engineer." Likewise, colleagues seem satisfied when I attribute deficiencies in engineering to the excuse: "I am actually a mathematician."

Jobs were scarce in 1951 for PhD's in Mathematics so I was fortunate to be hired by the MIT Instrumentation Laboratory at a salary of $425 per month. Others from the Math Department had preceded me—Hal Laning, Elmer Frey, and Bill Root.

During the month following graduation, I had my first contact with numerical integration. It was not a pleasant one. Using a mechanical desk calculator, I attempted, with little success, to solve an equation derived in my thesis. It was an ordinary differential equation with variable coefficients. Later at the Instrumentation Lab I learned to use an Analog Computer for such purposes. Wires were glued to drums to represent the variable coefficients and the solution of my thesis equation was accomplished in a matter of minutes. I was really impressed.

For our numerical work at the Instrumentation Lab, there was a room filled with young women who operated those ubiquitous desk calculators. Friden soon introduced a mechanical marvel which could actually extract square roots albeit with a great deal of noise and commotion. But it cost $1800 and was usually in the custody of the repair man when anyone needed it.

But the big innovation in computing was the IBM Card Programmed Calculator (or CPC for short) which was acquired in 1952. We could now do arithmetic operations at the fantastic rate of 100 per minute. But read-write memory was at a premium—27 mechanical counters each holding a ten decimal digit number with sign and housed in bulky units known as "ice boxes."

Development of an experimental electronic digital computer was underway at MIT in the early 1950's. The Whirlwind computer, as it was called, completely filled a large building off-campus and had a memory of 1024 sixteen-bit words electrostatically stored on cathode-ray tubes. (Today, of course, much more capability can be had on a single silicon chip.)

Hal Laning was a student of Eric Reissner and headed the Math group at the Instrumentation Lab when I joined up. He had received his doctorate in 1947—the year that C. C. Lin arrived at MIT. In the summer of 1952, after six months of experience as a user of Whirlwind, Hal became obsessed with the idea that computers should be programmable using the ordinary language of mathematics.

Bill Root and I had great fun at Hal's expense.

"What a dumb idea! How could you teach a computer to read equations?"

But, in the end, the laugh was on us. In just a few months he had done it. Hal Laning had developed the very first algebraic compiler. He called it "George"—from the old saw

"Let George do it."

In June of 1954, almost two years later, John Backus and a team of experts from IBM came to MIT for a demonstration of George. As a result of their visit, algebraic expressions found their way into the Fortran language.

2. THE Q-SYSTEM

In the early fifties the MIT Instrumentation Lab was picked by the Air Force to provide a self-contained guidance system as a backup to Convair for the new super-secret Atlas Intercontinental Ballistic Missile.

Initially, Hal Laning and I were the only ones at the Lab involved in the analytic work for the Atlas system. There were no "standard" methods for guiding ballistic missiles. We had to invent one.

In those days, guidance systems had to be relatively simple since only analog hardware could be used for on-board mechanization. We developed a method based on linear perturbations called the "Delta guidance system." It was not easy to implement since considerable reference data required storage and a complete navigation system was needed.

But we were determined to make the Delta system work, despite its deficiencies, until I made my first trip to Convair San Diego in the summer of 1955.

The system proposed by the Convair engineers was no less complicated than Delta guidance. But there were some new concepts that I brought back to Cambridge: "correlated velocity" and "velocity-to-be-gained."

The immediate result of my trip was not what I expected. Hal Laning totally abandoned the Delta system. From that moment, Delta guidance was my millstone to bear alone. Hal seemed no longer interested in guiding the Atlas missile, or anything else for that matter. But after several weeks, he reappeared with an idea and needed a sympathetic ear. It had to do with a redefinition of the Convair concept of correlated velocity, and a simple differential equation for velocity-to-be-gained. In a few days, Delta guidance was to become an orphan.

Whenever I reminisce about the early days, one thing stands out vividly. It is the simplicity of the Fundamental Equation for the velocity-to-be-gained vector $\mathbf{v}_g$:

$$\frac{d\mathbf{v}_g}{dt} + \mathbf{Q}\mathbf{v}_g + \mathbf{a}_T = 0 \qquad \text{where} \qquad \mathbf{Q} = \frac{\partial \mathbf{v}_c}{\partial \mathbf{r}}$$

Here $\mathbf{a}_T$ is the missile thrust acceleration. The Q matrix is the gradient, with respect to the current position vector $\mathbf{r}$, of the free-fall velocity vector $\mathbf{v}_c$ required to reach the target at a specified time. For the partial derivative, the target vector $\mathbf{r}_T$ as well as the remaining flight time are held constant.

This was the basis of the Q-guidance system. One of its principal virtues was the absence of the Earth's gravity vector since gravity is not simple to compute with analog hardware.

Another virtue was that the elements of the Q matrix could be regarded as slowly varying functions of time and approximated by simple polynomials. For intermediate range applications (1500 miles or less) they could even be constants. The attainable accuracy of the system was one mile which was acceptable for that period in our military history.

The computation of velocity-to-be-gained is only one element of the system. Of equal importance is a method to control the missile in pitch and yaw, so that the thrust acceleration will cause all three components of the velocity-to-be-gained vector to vanish simultaneously.

It is important to note that the system was not controlling the thrust magnitude—just its direction.

The elegant solution to this control problem came in a brilliant flash of insight. It was all so simple and obvious! If you want to drive a vector to zero, then just align the time rate of change of the vector with the vector itself.

In this way, the components of the vector cross product

$$\mathbf{v}_g \times \frac{d\mathbf{v}_g}{dt}$$

could be used as the basic autopilot rate signals—a technique that came to be known as "cross-product steering."

A report on the Q-system was presented at the first Technical Symposium on Ballistic Missiles held in 1956 at the Ramo-Wooldridge Corporation in Los Angeles. There was only one session on Inertial Guidance, and all papers except ours dealt with inertial instruments—the Q-system had no competition! We were elated to say the very least.

The Q-system was first implemented on the Thor IRBM and then on the Polaris fleet ballistic missile, but not the Atlas for which it had been designed. Ironically, the system eventually used for the Atlas was some form of Delta guidance.

3. THE MARS PROBE

On October 4, 1957 the Russians launched their Sputnik. I had been away from MIT for a year exploring opportunities in industry. A few months later I learned that Hal Laning had a simulation of the solar system running on the IBM 650 and was "flying" round trips to Mars.

I could hardly wait to get back to MIT. A report by Hal Laning, Elmer Frey, and Milt Trageser on the feasibility of an unmanned photographic reconnaissance flight to Mars had just been published under the auspices of the Ballistic Missile Division of the U. S. Air Force. They were promoting a crash program and predicted the launching of such a vehicle within the next five to seven years.

A small group formed to flesh out the system proposal for the Mars mission. Hal and I were responsible for trajectory determination, as well as suitable navigation and guidance techniques. The project culminated a year or so later in a three volume report, and a full-scale model of the spacecraft.

Today, that model hangs in the lobby of the Draper Laboratory's Hill Building.

To my surprise, it soon became evident that we did not know how to solve the orbital boundary-value problem. That seemed ludicrous after all our previous work with ballistic missile trajectories.

Hal had been calculating round-trip Martian trajectories by "trial and error"—adjusting and readjusting the spacecraft initial conditions and determining the orbit by numerically solving the equations of motion. There had to be a better way!

And, of course, there was. But it was not easy to come by. One did not study Celestial Mechanics unless he planned to be an astronomer and astronomers were not designing orbits for missions to Mars. Probably no one at MIT was even interested in the subject.

So it was with some trepidation, that I presented my method of orbit determination at the annual meeting of the Institute of the Aeronautical Sciences in 1959.† My lack of background in Celestial Mechanics did little to inspire self-confidence in the novelty of the technique.

But, as it happened, the method eventually became the basis of the major orbit-determination programs of the Jet Propulsion Laboratory for

---

† "The Determination of Round-Trip Planetary Reconnaissance Trajectories," *Journal of the Aerospace Sciences*, Vol. 26, September 1959, pp. 545–567.

its series of unmanned interplanetary probes, and of the Navy and Air Force for targeting ballistic missiles.

To support the Mars project, we developed trajectories with flight times of roughly three years, and launch dates in 1962–1963. In this case, the spacecraft makes two orbits of the Sun while the Earth does three. Later we investigated round-trip missions to Venus, having flight times of only a year and a quarter.

One day, when plotting these flybys of Venus, I was impressed by the proximity of the spacecraft orbit and the Martian orbit on the return trip to Earth. This was the result of the increased velocity induced during the close pass of Venus. The interesting possibility of a dual contact with both planets seemed feasible—a celestial billiard game if you will. If such orbits were possible, they must certainly be rare.

After much effort, I found that ideal circumstances would exist on June 9, 1972. A spacecraft in a parking orbit launched from Cape Canaveral could be inserted into just such an interplanetary orbit with a modest injection velocity. Each leg of the journey would require roughly $\frac{4}{10}$ of a year. The spacecraft would pass some 4400 miles above the surface of Venus and receive a gravity assist sending it in the direction of Mars. It would then pass Mars with an altitude just over 1500 miles and return to Earth on September 13, 1973. This remarkable orbit (the first of its kind) is illustrated in Fig. 1.

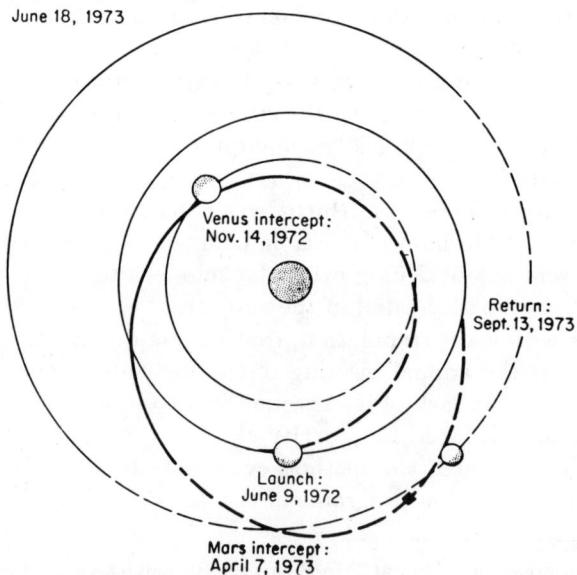

Fig. 1. Double-reconnaissance trajectory.

That was 1960 and the launch date seemed incredibly far off—twelve whole years! But the day finally came and passed without even a comment. Today the Voyager spacecraft on its Grand Tour of the solar system is a spectacular demonstration of such missions.

The Mars probe preliminary design was ready for customer review in the summer of 1959. Although the Air Force had been our sponsor, a new government agency—the "National Aeronautics and Space Administration" would control our destiny.

With view-graphs, reports, and the wooden spacecraft model, we arrived in Washington on the same day as Nikita Khrushchev. Our presentation was well received. But the high-level NASA audience we had expected was busy entertaining the Russians. We were sent home with a pat on the head and the promise of some future study money. Our dreams of instant glory in interplanetary space began to fade.

4. PRELUDE TO APOLLO

The NASA study contract kept our small team together but the absence of a specific goal diminished much of the enthusiasm. Now we were simply doing "interplanetary navigation system studies." There was no reason to suspect that a new goal lay just ahead to challenge us beyond our wildest imaginations.

Navigation data for the Mars probe was to be gathered by an onboard sextant and processed by a spacecraft digital computer. The observation data would be used to determine spacecraft position and a correction to the spacecraft clock. Periodically, small changes in velocity would be made by a small propulsion system as directed by the computer.

The terminology was not yet common, but we were in fact dealing with an "estimator" of a four-dimensional "state vector." (Years later, the state space would be nine-dimensional and we would be estimating things like radar biases and landmark locations.) We linearized the measurements about a reference point, and developed a weighted least-squares process to obtain the celestial fix.

I struggled alone on some annoying problems with our navigation algorithm. The matrix inversion required for the method of least squares had numerical troubles. My efforts to remedy this resulted in a recursive form of the estimator which allowed the separate incorporation of redundant measurements.

The "Interplanetary Navigation System Study" for NASA was published in April 1960. It was classified because it quoted some confidential Centaur missile data. The recursive form of the estimator was in the unclassified Appendix.† Today, those familiar with the Kalman filter would recognize the similarity between the two.

As luck would have it, Rudolf Kalman published his paper in the *open literature* just one month earlier.‡

Meanwhile, Hal Laning joined forces with Ramon Alonzo to design a small control computer with unique characteristics for space applications:

1) variable speed to save power,
2) relatively few transistors,
3) parallel word transfer,
4) automatic counter incrementing, and
5) automatic interrupt.

The program and constants were wired in a "core rope"—a memory with unusually high bit densities which could not be altered electronically. This would have been ideal for the Mars probe.

Little did Hal and Ray suspect that their computer would one day take man to the Moon.

5. THE RACE TO THE MOON BEGINS

After a nine-month hiatus in our work for NASA, came another six-month contract—this time for a preliminary design study of a guidance and navigation system sponsored by the NASA Space Task Group.

Later that year on May 25, 1961, President John F. Kennedy in his Special Message to Congress on Urgent National Needs said:

"I believe that this nation should commit itself to achieve the goal, before this decade is out, of landing a man on the Moon and returning him safely to Earth."

Jim Webb, the NASA Administrator, asked Doc Draper to develop the Apollo guidance and navigation system.

"When will it be ready?" asked Webb.
"When you need it," said Draper.
"How do I know it will work?" Webb persisted.
"I'll go along and operate it for you."

And he most certainly would have done it, had they only let him.

---

† "Computational Procedures for the Navigational Fix," Appendix B of *Interplanetary Navigation System Study*, Report R-273, MIT Instrumentation Laboratory, Cambridge, Mass., April 1960.

‡ "A New Approach to Linear Filtering and Prediction Problems," *Journal of Basic Engineering, Transactions of the American Society of Mechanical Engineers*, Vol. 82D, March 1960, pp. 35–45.

But the fact remains that

The first major Apollo contract awarded by the space agency was to the MIT Instrumentation Laboratory!

At first it was to be a completely self-contained system—absolutely no ground communications with the Apollo vehicle would be allowed. We assumed this was to prevent Russian interference with an Apollo flight. Gradually that requirement eroded away—but not before we had designed and implemented computer algorithms to permit completely autonomous missions.

Several characteristics of the Apollo guidance computer, or AGC as it was called, made self-contained algorithms a challenge to implement:

1) a 16 bit word, necessitating mostly double precision calculations;
2) a modest memory—36,864 read-only, and 2048 read-write; and
3) moderate speed—23.4 μsec add time.

Small as that was, it was a big improvement over what was available in the fall of 1961. Then the AGC had 4096 words of fixed memory, 256 words of erasable memory, and operated only half as fast. After all it was designed for the far simpler Mars mission.

Technology advances over the years allowed us to expand the capacity of the AGC without changing its physical size. The volume was one cubic foot—a dimension which could change only at great cost. That was all the room the spacecraft designer had provided.

The first mission programming we did for the AGC was to implement the recursive navigation algorithm. Of course, the computer program changed many times during the years but not the concept. A diagram we used countless times for NASA briefings is shown in Fig. 2.

Note that the reference trajectory has been replaced by the integrated vehicle state. The necessity for this important change was obvious when we first addressed the implementation problem. The modification is generally referred to now as the "extended" Kalman filter.

The idea of using the original Q-system to guide the Apollo vehicle during its many and varied powered maneuvers was soon rejected. Its principal virtue had been the ease of mechanization on board the vehicle with analog hardware. With the AGC we had at our disposal for the first time ever a powerful general purpose digital computer as the key ingredient of a vehicle-borne guidance system. Why not use it?

In fact, the Q matrix could be avoided altogether. We had only to write

$$\frac{d\mathbf{v}_g}{dt} = \frac{d\mathbf{v}_r}{dt} - \mathbf{g} - \mathbf{a}_T$$

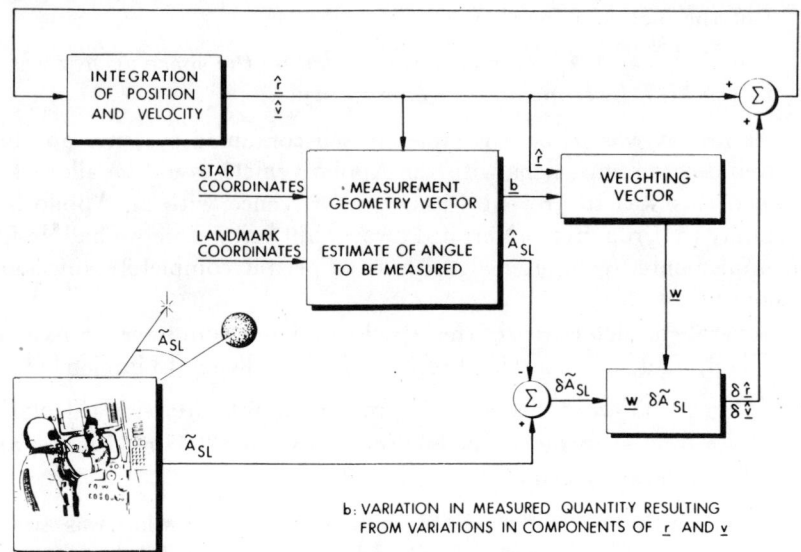

Fig. 2. Apollo coasting-flight navigation.

as the differential equation for velocity-to-be-gained provided that we had an analytic expression for the required velocity $\mathbf{v}_r$. We were no longer worried about computing gravity—it posed no problem for the AGC.

We had a convenient expression for velocity when the target and flight time are specified. It remained to be seen how many of the major orbital maneuvers could be accomplished conceptually by a single velocity impulse, and if simple formulas could be found for the required velocities.

One by one, we accumulated suitable expressions for a variety of possible Apollo maneuvers. For example, when the Apollo command module returned to Earth, it had to impact the atmosphere at a specified angle—otherwise it would either skip out of the atmosphere or be destroyed by overheating. A simple formula for the required velocity was obtained which later appeared as an exercise in my book "Astronautical Guidance."

For the first unmanned guided Apollo flight in August 1966, the required velocity was defined to achieve an orbit of given eccentricity and energy. The list goes on, but does not include the lunar landing since this maneuver cannot be performed with a single impulsive burn.

The computation of the error signal required for control is shown in Fig. 3. Numerical differentiation of the required velocity was simpler than programming the analytic derivative. Near the end of the maneuver, when v_g is small, cross-product steering is terminated, the vehicle holds a constant attitude, and engine cut-off is made on the basis of the magnitude of

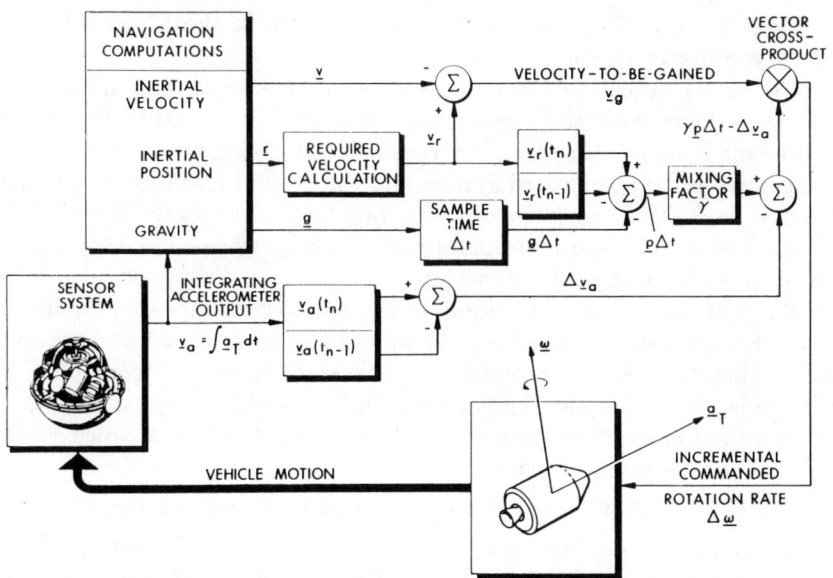

Fig. 3. Apollo cross-product steering.

the vector $\mathbf{v}_g$.

Steering to intercept a given target at a specified time came to be known as Lambert guidance after Johann Heinrich Lambert, the famous eighteenth century Alsatian scholar. Lambert guidance placed one of the heaviest burdens on the AGC. An iterative solution of Lambert's time equation to obtain the required velocity had to be calculated cyclically in real time.

To complete all the calculations in the time available was a programmer's nightmare. Ever since, the problem has fascinated me. My latest solution was published in 1984.†

THE FLIGHT OF APOLLO 8

Apollo 8 was the first manned spaceflight beyond Earth orbit and the first to demonstrate the feasibility of onboard, self-contained space navigation. For many of us it was the most exciting of all. The 5 minute burn of the S–IVB engine over Hawaii, to boost the speed of the Apollo spacecraft to the 24,200 mph necessary to escape the Earth, was certainly the longest and most thrilling 5 minutes of my life.

---

† "An Elegant Lambert Algorithm," *Journal of Guidance, Control, and Dynamics*, Vol. 7, November-December 1984, pp. 662–670. Coauthored with Robin M. Vaughan.

"You are on your way," said Chris Kraft, from the Mission Control room, "you are *really* on your way."

Jim Lovell, the Navigator Command Module Pilot, had spent many hours practicing with the Apollo sextant simulator at MIT. During the flight he made many star-elevation measurements using both the Earth and Lunar horizons. The onboard system and the ground tracking system each provided navigation results which were practically identical. The evidence was conclusive that the astronauts could have performed successfully on their own without ground contact.

Early in the morning of December 24, Apollo 8 disappeared behind the Moon. For 34 minutes there was no way of knowing what had happened. During that time a 247-second burn took place under the control of the MIT guidance system and Apollo 8 was in lunar orbit. The astronauts announced their orbital parameters provided by the AGC when voice contact was resumed. Mission Control asked

"How do they know? We haven't had time to track them."

They were not yet experienced with self-contained inertial systems.

At 8:40 p.m. on Christmas Eve in 1968, the Apollo 8 astronauts were on television broadcasting from lunar orbit. "For all the people on Earth," said Bill Anders, "the crew of Apollo 8 has a message we would like to send you." He paused a moment and then began reading:

In the beginning God created the Heaven and the Earth ...

... and God saw that it was good.

The commander Frank Borman added: "And from the crew of Apollo 8, we close with good night, good luck, a Merry Christmas, and God bless all of you—all of you on the good Earth."

Later, in the Washington Post, there appeared an editorial

"At some point in the history of the world someone may have read the first ten verses of the Book of Genesis under conditions that gave them greater meaning than they had on Christmas Eve. But it seems unlikely ... This Christmas will always be remembered as the lunar one."

THEORETICAL BIOLOGY
AS A BRANCH OF APPLIED MATHEMATICS:
SOME EXAMPLES*

Lee A. Segel
Department of Applied Mathematics
Weizmann Institute, Rehovot, ISRAEL

Some examples drawn from the author's research will be presented wherein theoretical investigations of biological phenomena have given rise to partially novel methodologies or points of view that may be of some general value in other branches of applied mathematics. (1) The so-called quasi-steady state assumption is a fundamental example of frequent instances in theoretical biology where advantage can be taken of the simultaneous existence of widely separated time scales. It is shown that appropriate scaling can markedly sharpen estimates of the range of validity of this approximation. (2) A new methodology for modelling dynamically changing networks is applied to simple morphogenesis in fungal colonies. A brief discussion is given of possible application to the complex polymerization and cross-linking that characterize cytoplasmic fibers. (3) It is shown that problems in ecology and immunology can be illuminated by considering nonlocal interactions (and hence integro-differential equations) that generate patterns in "aspect space" or "shape space." This extends familiar ideas from applications of stability theory to spatial pattern formation.

1. INTRODUCTION

Almost twenty years ago a discussion with C. C. Lin engendered a switch in my field of research from fluid mechanics to theoretical biology. In one of our frequent conversations concerning the nature of applied mathematics, C. C. raised a question that the logician Paul Cohen had asked him: "Why should one direct a good student into applied mathematics?" In answer, we reasoned that as science grows it is increasingly important to transfer a theoretical idea from one field to another and to reveal underlying unities in seemingly disparate investigations. If such arguments are taken seriously, they encourage an applied mathematician to work in more

---

* A paper similar to this one will appear in the proceedings of a meeting on Nonlinear Problems in Biology and Medicine, Los Alamos, 1987.

than one field. I decided to do just this. Metallurgy and economics both tempted me, but biology seemed more exciting.

My training as a graduate student under C. C.'s direction had already instilled in me the view that one's initial attempts at research could profitably be viewed as a case study in applied mathematics. With approaches to a scientific problem understood and with a toolkit of concepts, one needn't be afraid of entering a new area of application. Armed with this faith and a tenth grade course in biology, I began reading the biology articles in the *Scientific American* for background, studying selected books, and visiting laboratories of friends in the biological sciences. The key event was arranging a sabbatical at Cornell Medical School in the theoretical biology group of the late Sol Rubinow.

For this volume dedicated to C.C., I thought it appropriate to sketch very briefly certain of my (and colleagues') ideas in theoretical biology that I believe may have some general impact on applied mathematics. Although my major research goal has been to "impress biologists" without regard to mathematical sophistication, I have welcomed opportunities to add a little bit to the methodological or conceptual toolkits of fellow applied mathematicians.

A secondary goal in the discourse to follow is to provide to investigators from other fields something of the flavor of modern theoretical biology. I hope that such readers will begin to agree with my view that biology is now an area just as fertile for the applied mathematician as more traditional areas such as elasticity and fluid dynamics. It must be stressed that the presentation here is perforce extremely brief; the interested reader should turn to the original articles for a fuller exposition.

2. IMPROVING THE QUASI-STEADY STATE ASSUMPTION

Biochemical kinetics is an area of biology in which theory has played an important role for many years. Mathematical representations for the temporal development of chemical concentrations have provided a major vehicle for the formulation and testing of hypotheses concerning the molecular nature of the self-regulating chemical reactions that lie at the core of life.

Every respectable biochemistry textbook contains a discussion of the basic reaction in which an enzyme (concentration E) interacts with another molecule called the reaction *substrate* S to form an enzyme-substrate *complex* C. The complex may break apart into its original constituents E and

S, or (upon utilization of the catalytic capability of the enzyme) into an altered substrate, called the *product* P, together with the original enzyme. The reaction is schematized by

$$E + S \underset{k_{-1}}{\overset{k_1}{\rightleftharpoons}} C \overset{k_2}{\to} E + P \ . \tag{1}$$

According to the natural assumptions embodied in the *law of mass action*, the following equations govern the reaction:

$$dE/dt = -k_1 ES + k_{-1}C + k_2 C = -dC/dt \ , \tag{2a,b}$$

$$dS/dt = -k_1 ES + k_{-1}C \ , \quad dP/dt = k_2 C \ . \tag{2c,d}$$

Typical initial conditions are

$$E(0) = E_0, \quad S(0) = S_0, \quad C(0) = 0, \quad P(0) = 0 \ . \tag{3}$$

Since it follows that

$$E(t) + C(t) = E_0 \ , \tag{4}$$

the essence of the problem reduces to the solution of

$$dS/dt = -k_1(E_0 - C)S + k_{-1}C \ , \tag{5a}$$

$$dC/dt = k_1(E_0 - C)S - (k_{-1} + k_2)C \ , \tag{5b}$$

$$S(0) = S_0, \quad C(0) = 0 \ . \tag{5c,d}$$

Virtually all biochemistry texts continue with an invocation of a simplifying assumption, usually rather weakly supported, that is typically said to hold "if the concentration S of substrate is high enough." Under such circumstances it is argued that after a brief transient

$$dC/dt \sim 0 \ . \tag{6}$$

Then (5b) and (2d) yield

$$C = \frac{E_0 S}{K_m + S} \ , \quad \frac{dP}{dt} = \frac{k_2 E_0 S}{K_m + S} \ , \tag{7a,b}$$

where

$$K_m \equiv (k_{-1} + k_2)/k_1 \ . \tag{8}$$

Given (7b), measurements of the *reaction velocity* dP/dt can yield estimates of the kinetic parameters k_2 and K_m. Often the graph of dP/dt vs. S is not of the form predicted by (7b), indicating that more complex reaction mechnisms are involved. There is a large literature on these matters; Fersht (6) offers a good introduction.

The applied mathematician will at once sense that a singular perturbation phenomenon lies behind (6), which is now generally termed a *quasi-steady state assumption* (QSSA). Indeed, Heineken, Tsuchiya and Aris (7) made this point explicit twenty years ago. "Let us make our equations dimensionless," they wrote, introducing the variables

$$y \equiv S/S_0 \;,\;\; z \equiv C/E_0 \;,\;\; h \equiv k_1 E_0 t \;. \tag{9}$$

In terms of these variables, the equations (5a) and (5b) become

$$dy/dh = y + (y + \sigma^{-1} - \lambda)z \;,\;\; \epsilon_h \frac{dz}{dh} = y - (y + \sigma^{-1})z \;, \tag{10}$$

where the dimensionless parameters are given by

$$\epsilon_h = E_0/S_0 \;,\;\; \sigma = S_0/K_m \;,\;\; \lambda = k_2/k_1 S_0 \;. \tag{11}$$

Heineken et al. (7) then applied singular perturbation theory to (10), assuming that $0 < \epsilon_h \ll 1$. They cited a theorem of Vasil'eva to justify their calculations.

Given present day widespread familiarity with singular perturbation theory, the key step here is the introduction of the variables (9). But why *these*? Perhaps there are better variables. In any case the reader can rightfully demand motivation for a key substitution. Such motivation can be provided by the concept of *scaling* as discussed in detail by Lin and Segel (14, Section 6.3). Scaling is an art, and only recently has it become clear to me how to scale the variables of (5)[17].

The fast time scale t_C is found by solving (5b) under the assumption that the drop of S from its initial value, S_0, is negligible. One obtains

$$C(t) = \bar{C}[1 - \exp(-kt)] \;, \tag{12}$$

where

$$\bar{C} = E_0 S_0/(K_m + S_0) \;,\;\; k = k_1(S_0 + K_m) \;. \tag{13}$$

Thus

$$t_C \doteq k^{-1} \;. \tag{14}$$

The slow time scale, t_S, is found from a useful characterization of such a scale[14],

$$t_S \sim (S_{max} - S_{min})/|dS/dt|_{max} \ . \tag{15}$$

This gives

$$t_S = (K_m + S_0)k_2 E_0 \ . \tag{16}$$

Introduction of the scaled variables

$$s = S/S_0 \ , \quad c = C/\bar{C} \ , \quad T = t/t_S, \tag{17}$$

yields as an alternative to (10)

$$\frac{ds}{dT} = (\eta + 1)(\sigma + 1)\left[-s + \frac{\sigma}{\sigma + 1}cs + \frac{\eta(\eta + 1)^{-1}}{\sigma + 1}c\right] \tag{18a}$$

$$\epsilon \frac{dc}{dT} = (\eta + 1)(\sigma + 1)\left[s - \frac{\sigma}{\sigma + 1}cs - \frac{1}{\sigma + 1}c\right], \tag{18b}$$

where

$$\epsilon = \frac{E_0}{K_m + S_0} \ , \quad \sigma = \frac{S_0}{K_m} \ , \quad \eta = \frac{k_{-1}}{k_2} \ . \tag{19}$$

The new formulation gives the stronger result that the QSSA is expected to be valid when $E_0 \ll K_m + S_0$, not just $E_0 \ll S_0$. This extension is biochemically significant. Moreover, the new approximation is an improvement on the old in the matter of uniform validity with respect to the other two dimensionless parameters. For example, inspection of (10) and its counterpart in the initial layer leads one to doubt that the QSSA will be valid uniformly for $0 \leq \sigma$ as $\epsilon_h \to 0$. No such doubts are manifest upon inspection of (18). Indeed, M. Slemrod (unpublished) has proved that the QSSA is valid when $\epsilon \to 0$ uniformly for $0 \leq \sigma \leq \sigma_1 < \infty$. The reader is referred to Ref. (17) for considerable further discussion.

3. MODELING DYNAMICALLY CHANGING NETWORKS

3.1 Morphogenetic Pattern Formation

A major theme in modern theoretical biology is the attempt to shed light on *morphogenesis*, the creation of shape and form in a developing organism. One line of research stems from A. Turing's (19) demonstration that equations of chemical reaction and diffusion can give rise to spatially inhomogeneous "pre-patterns" that might trigger genes to yield the basis of observed structural regularities. Meinhardt (15) provides an excellent demonstration of the scope of this approach. An important conclusion is that spatial nonuniformity generally requires that activating influences (causing positive feedback) are relatively short-range compared to inhibitory influences. For reaction-diffusion equations, this requirement implies that the diffusion coefficient for an activator chemical is smaller than that for an inhibitor chemical.

A somewhat different approach to morphogenesis stresses that pattern formation might arise from the interplay of the complex inter- and intracellular forces that characterize living organisms[8].

Some years ago, the Hebrew University microbiologist Ilan Chet brought my attention to an interesting example of morphogenesis in fungi. Normally, a fungal colony growing radially in a laboratory petrie dish appears to be a homogeneously expanding disk. But under specially controlled conditions, concentric rings of heavy growth are formed. Here is a simple instance of pattern formation whose elucidation might be the first step in understanding how fungal colonies can generate the relatively complex three-dimensional patterns that characterize mushrooms.

To construct a model that might shed some light on Chet's observations, one must inquire into the nature of fungal colony. The colony turns out to be composed of a branching network of tiny "pipes" called *hyphae*. Growth takes place at hyphal tips. These may arise by budding and branching from hyphal walls and may disappear because of tip-tip or tip-wall collisions. Nutrient is absorbed from the ambient medium and transported through the hyphae toward the tips, to provide needed growth materials.

One natural approach to modeling a fungal colony is to ignore all internal structure and hence to treat the colony as an expanding disk[1]; alternatively, the expanding hyphal network can be simulated in detail on a computer[2]. An intermediate view seemed advantageous to me and my then graduate student, Leah Edelstein.

The essence of our approach was this. As one dependent variable, in a continuum model, we took $\rho(x, y, t)$, the hyphal length per unit area at point (x, y) and time t. As a second dependent variable, we selected $n(x, y, t)$, the number of tips per unit area. That is, we regarded the fungus as a continuum mixture of hyphae and an interspersed "fluid" of tips. The fungal density was taken to be governed by

$$\partial \rho / \partial t = nv - d\rho \ . \tag{20}$$

Here, v is the average elongation rate at a tip, while d is the probability per unit time that an element of hypha disintegrates or dies. A reasonable equation for tip density is

$$\partial n/\partial t = \nabla \cdot (nv) + \alpha\rho - \beta\rho n \quad . \tag{21}$$

The divergence term in (21) is standard in balance equations, α is the probability per unit time per unit hyphal length that a tip buds, and β is the proportionality factor in a term accounting for tip disappearance owing to tip-hypha collision.

In the simplest instance, v, d, α, and β can be taken to be constants. Then (20) and (21) can yield nonlinear travelling-wave solutions. Special methods for labelling tips can be employed to compare predicted and observed tip distribution profiles as a function of radius and, thereby, to narrow the range of possible models[3].

To approach the original motivating problem of fungal rings, account must be taken of nutrient absorption from the surrounding medium and its transport within the fungal network. The governing equations become considerably more complex – and more speculative. Rings can indeed be obtained, but apparently under fairly restricted circumstances. The interested reader is referred to Edelstein and Segel (4) for a full account.

It is appropriate here to make some general remarks on the contribution of the mathematical modeling to the biological understanding of fungal colony growth. (i) The simple model of (20) and (21) led to an appreciation of the importance of tip distribution and thereby to the new experiments mentioned above where the tips were fluorescently labelled and counted. (ii) The more complex models underlined the necessity for further experimental understanding of such processes as nutrient absorption and transport, which were demonstrated to play major roles in colony behavior. Only after such experiments are carried out will it become possible to check the feasibility of the proposed mechanism for ring formation. (iii) An important contribution of the modeling is to show in unambiguous fashion the interaction between local physiological behavior and global colony growth and morphogenesis.

From a general applied mathematical point of view, postulation of a continuum "sea" of tips forms what might be termed a *mecroscopic* approach that lies midway in complexity between the macroscopic option of ignoring the tips and the alternative of accounting for each of them in microscopic detail.

3.2 Cytoplasmic Mechanics

We have recently found the mecroscopic viewpoint of help in efforts, just beginning, to link underlying polymerization events to the dynamic overall behavior of cells and tissues. Let us characterize the problem area in the briefest possible terms. The living material that constitutes cells derives mechanical and rheological properties from the ever-shifting fibrous structures of the *cytoskeleton* (i.e., "cell skeleton"). Typical fibers are polar, with different polymerization and depolymerization rates at the two "tip" and "barb" ends. Among various special proteins that affect the network are some that, when activated by suitable "messenger" molecules such as calcium, lead to the severing of fibers or to their joining. From the information that is rapidly accumulating concerning polymerization biochemistry, one wishes to deduce the constitutive equations that characterize the macroscopic viscoelastic behavior of cells and cell assemblages, such as those at the heart of the morphogenesis theory of Jacobson *et al.* (8) mentioned above.

Our approach to cytoskeleton dynamics will be by an extension of our modeling of the fungal network. Now a "sea" of barbs as well as a sea of tips must be considered. It seems that we will have to divide fibers into four classes: free, linked at the barbed end, linked at the tip end, and linked at both ends. The number of the latter play an especially important role in determining, for example, whether the cytoplasm is a highly cross-linked "gel" or a liquid-like "sol". Although no solid achievements can as yet be reported in this new line of investigation, it seems likely that cytoplasmic mechanics is a natural arena for mecroscopic modeling.

4. STRUCTURE IN ASPECT

4.1 The Evolution of Prey Aspect Under Apostatic Predation

Another somewhat novel modeling approach resulted from a study[13], with S. Levin, of a certain type of predator-prey interaction that serves as a prototype problem for the extent of biological diversity, a major ecological question. It turns out that certain predators concentrate for a time on locating prey of a particular *aspect*, e.g., a particular color or spot pattern. When prey of this aspect become relatively scarce, the predators are likely to switch their searching to prey of a different aspect. This behavior is called *apostatic*.

Let $v(z,t)$ be the population density of victims (prey) of aspect z, at time t, and let $e(z,t)$ be the density of exploiters (predators) that are concentrating on such victims. The aspect variable, z, will generally be a multidimensional vector, but in a first model it is natural to restrict attention to one-dimensional aspect space.

Suppose exploiters enter the system at a constant rate, c, and switch from concentrating on a particular aspect at a rate that decreases linearly with an increase in the corresponding victim population. Then (for constants ν and μ)

$$\partial e/\partial t = c - e(\nu - \mu v) \ . \tag{22a}$$

Victims will be supposed to prefer mates of an aspect similar to their own. More precisely, a female victim of aspect z will be assumed to mate with a male of aspect z' with a probability distribution that is a gaussian function (α) of $z-z'$. The progeny of such a mating will be assumed to be gaussianly distributed (ϕ) about the mean parental type $(z + z')/2$. If exploiters eat victims according to "mass action" dynamics with proportionality factor k, then the victim equation is

$$\frac{\partial v}{\partial t}(z,t) = -kv(z,t)e(z,t) + r \iint v(\eta,t)v(\xi,t)\frac{\alpha(\eta-\xi)}{W(\eta,t)}\phi(\frac{\kappa+\xi}{2} - z)d\eta d\xi \tag{22b}$$

$$W(\eta,t) = \int \alpha(\eta-\xi)v(\xi,t)d\xi \ . \tag{23}$$

Here r is the reproduction rate, and it has been assumed that males and females are always equally numerous.

The system (22) possesses a steady state that is uniform in the aspect variable, z. If z is presumed to run over the infinite interval, than a linear stability analysis can easily be carried out with the aid of Fourier transforms[13]. Numerical calculations show instances wherein an unstable uniform solution evolves into a highly nonuniform steady state characterized by sharp peaks of exploiter and victim density[12]. In these instances, typified by heavy predation pressure, an originally highly diverse population breaks up into almost distinct aspect types. As with reaction-diffusion equations, relatively short range activation is a requirement—here expressed in terms of the widths of gaussian kernels.

From a general applied mathematical point of view, the analysis of prey aspect diversity illustrates the fact that familiar stability considerations can find application not only in studying the generation of spatial pattern but also in predicting patterns in a somewhat abstract but still highly applicable aspect space.

4.2 Stability and Instability in Immunological Shape Space

Recently, together with A. Perelson, I have begun to apply ideas of structure in aspect space to the construction and analysis of a representation of the immune system. The following very highly simplified introduction to immunology will provide the required background.

Our bodies protect us from invading *antigen* molecules (typically these invaders are parts of bacteria or viruses) by interposing an array of defending *antibody* molecules. Antibody molecules may float in the blood or may function as *receptors* (a kind of trigger) on the *lymphocyte* cells that form a central component of the immune defenses. If invading antigens bind to enough antibodies with suitable complementary or near-complementary shape, then a multi-faceted *immune response* is activated to repel the invaders. One characteristic of this response is the proliferation of lymphocytes with receptors that are closely complementary to the invading antigen. Various classes of these cells can kill pathogenic intruders or can modify the killing ability of other cells.

In a celebrated series of papers, N. Jerne (9,10,11) pointed out that the proliferation of selected antibodies will itself challenge the immune system and may lead to a sequence of reverberating responses and counter responses with important consequences for immune regulation. Molecular *shape* would seem to be the essential independent variable in a model of the Jernian *network theory*. Shape space is infinite dimensional, but it might well be adequately described by a space of fairly low dimension, N^{16}. In a first gross caricature, let us take $N = 1$. Continuing the construction of a hyper-simple immune network, let us model a selective detector system in which no distinction is made between antigen and antibody, and in which no attention is paid to whether antibody is free in the bloodstream or acts as receptor for lymphocytes.

Let $b(x,t)$ denote the concentration of detector cells associated with shape x, at time t (see Fig. 1). Let mutation and other factors supply such cells at constant rate s, and let d represent their death rate. In addition, let $F(a,i)$ be the probability that a cell is switched on to proliferate, as a function of activating influences a and inhibitory (suppressive) influences i:

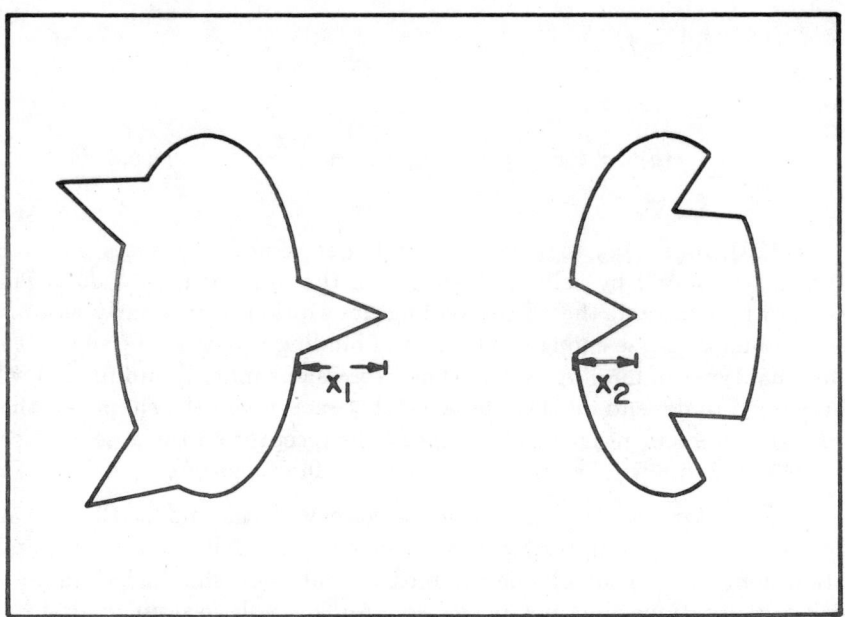

Fig. 1. A hypersimple immune system; "Cells" with wedge-shaped "receptors". If $x_2 = -x_1$ the receptors on the right-hand cell are perfectly complementary in shape to those on the left hand cell.

$$\partial F/\partial a > 0 \quad , \quad \partial F/\partial i < 0 \; . \tag{24}$$

We assume that the activating (inhibiting) influence is characterized by the amount of antigen binding to activation- (inhibition-) inducing receptors. Nonspecific activating or inhibiting factors are allowed to influence the birthrate of proliferating cells via the average population size

$$\bar{b}(t) \equiv \lim_{L \to \infty} \frac{1}{2L} \int_{-L}^{L} b(x,t) dx \; .$$

With all this, the governing equation for detector concentration takes the form

$$\frac{\partial b(x,t)}{\partial t} = s - d\, b(x,t)$$

$$+ r(\bar{b})b(x,t)F\left(\int m(x,y)b(y,t)dy\, ,\, \int n(x,y)b(y,t)dy\right) \quad .$$
(25)

In (25), $m(x,y)dy$ represents the number of activator sites on an x-cell that are bound by cells with shapes in the interval $(y, y+dy)$. This description of m, and the corresponding description of n, tacitly assumes that binding is sparse enough so that total binding is merely the sum of the individual types of binding events. The expressions $m(x,y)$ and $n(x,y)$ will be assumed to depend on the "distance" between y and the shape $-x$ that is exactly complementary to the shape of the receptor on the x-cells. More precisely, m and n will be taken to be gaussian functions of $y-(-x) = y+x$.

One matter that we plan to investigate with the aid of (25) [or its more realistic and comprehensive successors] is the following. A generalization concerning controllable systems is that they should be stable to small perturbations, but not overly so. An unstable system tends to go "out of control" when subjected to the mildest perturbation, but too stable a system suffers from the opposite fault of being only very weakly and slowly responsive to its controls. The immune system, too, presumably should be "just" stable. To generate such a system, one thinks naturally of constructing a slightly unstable system and then somewhat modifying the conditions. One path to destabilization, well known in the biological sciences and already encountered twice in this article, is short-range inhibition and long-range activation. In the present context "range" refers to shape space. A qualitative prediction at once emerges: the activating influences in the immune system require more specific chemical binding (hence are characterized by a shorter range in shape space) than inhibitory influences.

Our investigations of (25) and its generalizations have just begun, but already it appears clear that this type of model of interactions in shape space provides an interesting approach to important issues in theoretical immunology[18]. In particular, such models can suggest generalizations that either are capable of guiding experiments and/or can be sharpened to testable predictions with the aid of more detailed simulations such as those of Farmer, Packard, and Perelson (5).

References

1. Childress, S. and Keller, J.B. "Lichen growth." *J. Theoret. Biol* **82**, 157–65 (1980).
2. Cohen, D. "Computer simulation of biological pattern generation." *Nature* **216**, 246–248 (1967).
3. Edelstein, L., Hadar, Y., Chet, I., Henis, Y. and Segel, L.A. "A model for fungal colony growth applied to Sclerotium rolfsii." *J. gen. Microbiol.* **129**, 1873–1881 (1983).
4. Edelstein, L. and Segel, L.A. "Growth and metabolism in mycelial fungi." *Journal of theoretical Biology* **104**, 187–210 (1983).
5. Farmer, J.D., Packard, N.H. and Perelson, A.S. "The immune system, adaptation, and machine learning." *Physica* **22D**, 187–204 (1986).
6. Fersht, A. "Enzyme structure and mechanism". (2nd ed) W.H. Freeman, New York (1985).
7. Heineken, F.G., Tsuchiya, H.M. and Aris, R. "On the mathematical status of the pseudo-steady state hypothesis of biochemical kinetics". *Math. Biosci.* **1**, 95–113 (1967).
8. Jacobson, A., Odell, G., Oster, G. and Cheng, L. "Neurulation and the cortical tractor model for epithelial folding." *J. Emb. Exptl. Morph.* **96**, 19–49 (1986).
9. Jerne, N.K. (1973). "The immune system." *Sci. Am.* **229**, 52–60 (1973).
10. Jerne, N.K. "Towards a network theory of the immune system." *Ann. Immunol.* (Inst. Pasteur) **125C**, 373–389 (1974).
11. Jerne, N.K. "The immune system: A web of V-domains." *Harvey Lect.* **70**, 93–110 (1976).
12. Keshet, Y. and Segel, L.A. "Pattern formation in aspect." In: *Modeling of Patterns in Space and Time.* W. Jäger and J.D. Murray, editors. Springer-Verlag, Berlin, pp. 188–197 (1984).
13. Levin, S.A. and Segel, L.A. "Models of the influence of predation on aspect diversity in prey populations." *J. Math. Biol.* **14**, 542–548 (1982).
14. Lin, C.C. and Segel, L.A. "Mathematics applied to deterministic problems in the natural sciences." Macmillan, New York (1974).
15. Meinhardt, H. "Models of biological pattern formation." Academic Press, London (1982).
16. Perelson, A.S. and Oster, G.F. "Theoretical studies of clonal selection: minimal antibody repertoire size and reliability of self–non-self discrimination."

J. theor. Biol." **81**, 645–670. (1979).

17. Segel, L.A. "On the accuracy of the quasi-steady state assumption of enzyme kinetics." Unpublished (1987).

18. Segel, L.A. and Perelson, A.S. "Computations in shape space: an approach to network theory." In: Santa Fe Institute Studies in the Sciences of Complexity. Vol. 2. *Theoretical Immunology*. A.S. Perelson, ed. Addison-Wesley, Reading, MA. In press.

19. Turing, A.M. "The chemical basis for morphogenesis." *Philos. Trans. R. Soc. Lond. [B]*. **237**, 37–72 (1952).

Looking at a Computational Physics Environment

Robert H. Berman
Lincoln Laboratory
Lexington, MA 02173

Abstract

Now that MACSYMA and other symbolic computing programs are becoming commonly available, it is of special interest to applied users to compare their performance. In this paper, I describe the results of several benchmark tests that characterize symbolic computing and numeric performance on several machines including supercomputers and Lisp machines. The viewpoint I adopt for these tests is that of an applied user, not a hardware specialist.

I also comment on two important capabilities that modern computer algebra systems have: (1) to perform symbolic or analytic calculations systematically that would otherwise be too complex or time consuming to perform by hand; and (2), the ability to transform large expressions into FORTRAN for efficient stable numerical evaluation. I will comment on these capabilities with examples from a a nonlinear turbulence problem and a computer program distributed among a FORTRAN machine and an artificial intelligence workstation used to prepare code for an N degree of freedom Kalman filter.

1 INTRODUCTION

It is a pleasure to congratulate Professor C.C. Lin on the occasion of his retirement. As a student of his, I learned an important lesson about mathematics – the formal manipulation of mathematical symbols, while technically amusing, is often only a small part of understanding the way things work. It is also necessary to ask important questions and interpret the answers. Over the last 10 years as I became professionally involved with very complicated analytical models in plasma turbulence and Kalman filtering, I ran into technical hardships arising from straightforward, but very lengthy, multidimensional analytical calculations. And so, I began to use symbolic computing machines to perform these analytical calculations automatically and routinely, including the task of generating many thousands of lines of FORTRAN code to test ideas numerically.

2 DISCUSSION AND RESULTS

It is the purpose of this paper to summarize several issues regarding the performance of symbolic computing environments. One such environment, consisting of a symbolic computing machine integrated with an efficient numerical machine (array processor), was described at the 2nd MACSYMA Users conference [4] Several attempts to implement this configuration, consisting of a Lisp machine

(CADR), VAX 780 or KL-10 at M.I.T., network technology, and a supercomputer (CRAY) at the National Magnetic Fusion Energy Computing Center, have been in existence for some time. With the wide availability of VAX-class machines, and the efforts to provide supercomputer services to university clients, similar computational environments will become increasingly common. Indeed, other configurations involving primarily VAX or Lisp machines and supercomputers are now (middle of 1987) more generally available at numerous sites including commercial ones.

In addition to performance in actually computing, there is also an increase in capability in *formulating* problems in a symbolic computing environment.

While it is conventional to measure the performance of numerical hardware by single numbers (*e.g.* cycle time), it is clear that a single statistic can be inadequate because of differences in hardware (pipelining, storage interleaving). Furthermore, parallelism, dynamic address translation, register-to-register instructions, memory, and instruction lookahead all affect performance. In addition to the speed of the central processing unit, the execution efficiency of the symbolic computation, the execution efficiency of the supporting system, and the algorithm all are factors [12]. In spite of the danger of having too many variables to control, the average run time in solving a problem is useful to applied users to meter their costs. In several cases, the result is surprising in that the speed of the computer as measured in solving realistic problems is different than the theoretical hardware speed or the speed based on operation counts (See also Ref. [10]).

I will summarize three tests relevant to symbolic calculations now commonly available on superminicomputers like the VAX 780 (REDUCE, SMP, MACSYMA), specialized machines like the Symbolics 3600, and supercomputers like the CRAY 1 (These tests are described in detail in [2] and [3].). The first two tests calibrate the raw numerical performance of each machine. The first test is a table lookup, a calculation which involves capabilities of indirect addressing and memory speeds. The second test is a fast Fourier transform, which emphasizes floating point numerical performance. The third test is a symbolic calculation of certain polynomials involving exact arithmetic, differentiation and recursion.

The results of the symbolic calculation illustrate several novel and desirable features of a symbolic system, including: (1), error checking with word processor or by graphical methods; and (2), some optimization for stable numerical evaluation and (3) generation of run-time efficient FORTRAN code for intensive numerical evaluation at later times or for other machines. In particular, a distinction arises between optimization for stable numerical evaluation and for run-time speed. This is especially true in two cases. First, for machines like the KL-10, VAX or Symbolics 3600, numerical evaluation of a symbolic calculation can be very time consuming even with translated or compiled code. Second, when the result of the symbolic calculation is a FORTRAN production that can run for hours of CRAY time (*e.g.* Refs. [5], [7]), efficient code can be as important as correct code.

The principal results of the these tests can be summarized as follows:

1. Supercomputers, such as the CRAY, seem to be limited by memory access speeds in their performance for symbolic computing and do not perform significantly faster than superminicomputers, such as the VAX 780. On the other hand, supercomputer performance for floating point computing can be as much as 1000 times faster for primarily numerical performance. Thus, symbolic manipulation programs may need substantial software improvements to take advantage of the supercomputer hardware.

2. Specialized symbolic computers, such as the Symbolics 3600 Lisp Machine, emphasize single user interactivity, large address space, and an integrated environment for graphics, symbolic computing and networking. By linking such symbolic computing machines to large numerical supercomputers with existing network technology, a very impressive assault on a wide variety scientific problems with computers is now feasible. Lack of interactivity is a severe, if not fatal, handicap to this end.

3. Low cost virtual memory symbolic machines functioning as front-end workstations to large numerical machines provide an effective expenditure of resources to improve scientific productivity. This is feasible because large numerical (*e.g.* FORTRAN) codes can be more automatically generated and conveniently manipulated in an interactive symbolic machine while network technology makes intermachine communication practical.

4. In considering the automatic generation of code, there is an important distinction between optimizing expressions for stable numerical evaluation and for run time efficiency. This is particularly important when there are nonintuitive, nonlinear transformations of code that can result in more FORTRAN statements and operations, but with substantial improvements in performance on a CRAY-class machine. Operation count is not necessarily an adequate figure of merit for comparing algorithms in numerical analysis, especially when nonlinear improvements in execution speed are possible in the hardware.

5. Cost-performance estimates for symbolic computing suggest that substantial savings can be gained by automating the development and optimization of large physics production codes when compared with the cost in man-years for development or for production runs.

A supercomputer can service the needs of a few thousand scientists and engineers, while a symbolic computer typically services a few users. This difference arises from the resources and needs of the users. In a supercomputer environment, the total number of floating point operations is most important, while symbolic computing tends to emphasize rapid turnaround and interactivity. Thus, a supercomputer environment puts a premium on delivering CPU cycles to a large number of users running large or long calculations, while a symbolic computing environment emphasizes bandwidth.

It is also interesting to note that the capital costs of a CRAY installation are approximately \$20M while a VAX 780 is \$0.5M and a Symbolics 3600 is \$0.1M. Their speeds on floating point problems (FFT) are 1 CRAY = 1000× VAX 780 and 1 CRAY = 1000× Symbolics 3600 while their speeds for symbolic problems are 1 CRAY = 20× VAX 780 or 6× Symbolics 3600. Thus, the cost of the specialized vector hardware and fast memory design on the CRAY is justified by this type of average arithmetic performance improvement of some 10 to 1000 over the VAX or Symbolics 3600. Further the cost of Lisp machines will be incrementally declining rapidly during the next year or two, probably by at least a factor of two in the next year alone (in several steps), which cannot be said of VAXs with the same performance of a 780 (although some new VAXs, like the 785, are more powerful, they cost more). Then, of course, this estimate does not take into account all the support features (*e.g.* bit-map displays, laser printers) with the Symbolics 3600.

A more effective expenditure of money to improve scientific productivity may be in low cost large address space virtual memory symbolic machines functioning as front-end workstations to numerical machines. Modern network technology makes this link feasible. This trend is becoming more apparent (mid 1987) as large address space machines (personal computers with large disks, memory, fast chips like the Intel 80386) are readily available.

A second reason why a symbolic computing − supercomputer link is important is because it provides the capability of automatically generating numerically intensive codes. While this capability has existed for some 10 years with MACSYMA, it has recently been applied in plasma physics to the development of codes of some thousands of lines for magnetohydrodynamic problems [17], [8].

For large physics production codes, one can estimate the cost of development as 1 man-year of a computational physicist (100 k\$) while the run-time costs of such a code might be 100 CPU hours (100 hrs. × 2k\$/hr = 200 k\$). Numerical optimization techniques have been successfully applied to such large codes (*e.g.* a factor of 2 improvement occurred in a three dimensional magnetohydrodynamic code in toroidal geometry, see Ref [7]). Although a factor of 2 may be conservative, it represents a savings of the cost of 50 CPU hours. Therefore, these techniques can pay for themselves successfully the longer a code is in production. Other earlier efforts [1] were later adopted for one-dimensional codes. These techniques also made possible certain nonlinear turbulence calculations that illustrated for the first time an important nonlinear instability mechanism operating in linearly stable plasma [6].

In problems where capability is an issue, but capacity is not, the use of an automatic program generator is very important. For problems using an N degree of freedom Kalman filter, there is an important barrier to formulating efficient code and checking the resulting algorithms when there are expressions involving potentially many hundreds or thousands of terms. Here a computer algebra code generator that systematically keeps track of these terms can perform some numerical optimizations and enables one to write thousands of lines of FORTRAN in man-hours that would otherwise take man-months. Furthermore, the code is immediately transportable to the mainframe environment for use with production code.

It is perhaps the flexibility and integrated environment that a symbolic computing facility like MACSYMA provides that makes it an important avenue for overcoming the FORTRAN barrier [16] to the computerization of science. Earlier efforts at symbolic generation of large production codes (symbolic Algol as described in Ref. [15]; OLYMPUS from the Computer Physics Communications Library, Belfast. See also the example of OLYMPUS code cited in Ref. [11]) have not achieved wide popularity in part because of the rigidity of their structural stylistic conventions, and because of the difficulties in controlling their execution time efficiencies. This later difficulty is especially true in multiprocessor environments [14], [13]. By using a more personally convenient symbolic computing environment, these objections might be easily overcome.

3 ACKNOWLEDGMENTS

It is a pleasure to thank John Aspinall, Francois Brunel, George Carrette, Richard Fateman, Ellen Golden, Jeff Greif, Charles Karney, Leo Harten, Tony Hearn, John Kulp, Jean-Noel Leboeuf, Ralph Lewis, and Jim O'Dell for assistance in running these benchmarks or for useful discussions. Some of the benchmarks reported in this paper were initially described at a Workshop on Symbolic Computing held at Los Alamos in 1982.

This work is supported in part by the National Science Foundation, the Department of Energy and the Office of Naval Research. Certain discussions at the Institute for Fusion Studies were supported by N.S.F. grant ATM-82-14730.

References

[1] R.H. Berman, *Buffer*, **3**, 12 (1979).

[2] R.H. Berman, "A Computational Physics Environment," 3rd *Macsyma* Users' Conference, Schenectady (1984).

[3] R.H. Berman, "Symbolic Computing in Plasma Physics," 10th Conference on the Numerical Simulation of Plasma, San Diego (1983).

[4] R.H. Berman and J.L. Kulp, "A Computational Physics Environment," 2nd *Macsyma* Users' Conference, Washington (1979).

[5] R.H. Berman and J.N. Leboeuf, *Annual Sherwood Theory Meeting*, Lake Tahoe (1984).

[6] R.H. Berman, D.J. Tetreault, and T.H. Dupree, *Phys. Fluids*, **26**, 2437 (1983).

[7] F. Brunel, J.N. Leboeuf, T. Tajima, J.M. Dawson, M. Makino, and T. Kamimura, *J. Comp. Phys.*, **43**, 268 (1980).

[8] G. Cook, Ph. D. thesis, UCRL-53324, U. California (1982).

[9] A.C. Hearn, *SIGSAM Bull.*, **24**, 14 (1972).

[10] R.W. Hockney, 1978, in *Fast Poisson Solvers and Applications*, ed. U. Schumann, (Advance Publications: London), 78 - 97.

[11] R.W. Hockney and J.W. Eastwood, 1981, *Computer Simulation Using Particles*, (McGraw Hill: New York).

[12] R.D. Jenks and J.H. Griesmer, *SIGSAM Bull.*, **24**, 3 (1972).

[13] D.J. Kuck and M. Wolfe, *Physics Today*, **37**, 67 (1984).

[14] J. R. McGraw, *Physics Today*, **37**, 66 (1984).

[15] M. Petravic. G. Kuo-Petravic, and K.V. Roberts, *J. Comp. Phys.*, **10**, 503 (1972).

[16] K.G. Wilson, *Proc. I.E.E.E.*, **72**, 6 (1984).

[17] M.C. Wirth, Ph. D. thesis, UCRL-52996, U. California (1980).

SOUND WAVES IN FLUIDS; MATHEMATICAL MODELS AND PHYSICAL REALITY

Garret Birkhoff

Mathematical models of sound waves in fluids are of five different kinds:

#1. Purely *geometrical* models based on variational principles of Fermat and Huygens.
#2. Homogeneous and inhomogeneous *elastic fluid* models based on Euler's partial differential equations.
#3. *Thermodynamic* models in which heat is created by compression and diffused by conduction.
#4. *Viscous fluid* models in which kinetic energy is dissipated by viscosity.
#5. *Molecular* models which try to derive macroscopic laws by statistical considerations from the analysis of particle interactions.

This paper is concerned with the *consistency* of such models with each other and with physical reality.

Sections 1-6 are concerned with models of types #1 and #2, which are used to explain diffraction, refraction and reflection; they *conserve* acoustic energy. Sections 7 and 8 will take up models of types #3 and #4, which *dissipate* acoustic energy. Sections 9-11 will treat particle and digital models. In general, this paper summarizes and extends results presented in greater detail in Refs. 1-3.

A. ENERGY-CONSERVING CONTINUUM MODELS

1. *"The" wave equation.* By far the most successful model of sound waves is provided by Euler's purely mechanical concept of a *homogeneous elastic fluid*, in which the pressure $p = p(\rho)$ is determined by the density ρ. Variations in pressure is such a fluid are transmitted as "small oscillations" in the sense of Lagrange, who also derived plausibly from Euler's equations, in the last section of his famous *Mécanique Analytique*,[1] the constant-coefficient wave equation

$$\phi_{tt} = c^2 \nabla^2 \phi, \quad c^2 = dp/d\rho = p'(\rho), \tag{1.1}$$

for the velocity potential ϕ.

---

[1] First ed., 1788; see vol. ii, pp. 323-40 of the fourth ed. There, in Art. 6, Lagrange comments specifically on the assumption that $p = p(\rho)$.

According to this model, sound waves in any homogeneous gas or liquid should progress at a speed c that is independent of the wave length λ (and associated frequency $f = c/\lambda$). This prediction of *no dispersion* is physically realistic in the audible range. Moreover, in water, formula (1.1) and measurements of the isothermal bulk compressibility $dp/pd\rho$ can be used to predict the observed sound speed to within $\pm 0.5\%$, at any given temperature and pressure.

In air, however, the same procedure underpredicts c by over 15%. This discrepancy was explained by Laplace around 1810, using the then new concepts of a *thermoelastic* fluid heated by adiabatic compression.

2. *Geometrical acoustics.* Being associated with inhomogeneities, *refraction* cannot be predicted by Euler's original mathematical model of a homogeneous elastic fluid. To explain refraction, it is simplest to *assume* (with Fermat and Huygens) that the sound speed $c = c(\mathbf{x})$ varies with the position $\mathbf{x}$ in some known way, for example because of temperature variations.[2] To a static fluid with such a variable c, one can then apply either of two purely geometrical variational principles: (i) Fermat's principle of *least time*, or (ii) Huygen's principle that successive nearby *wave fronts* are separated by a distance $c(\mathbf{x})dt$. Originally proposed to explain optical refraction, these principles explain acoustic refraction in the same way.

The same principles can also be used to explain the refraction of sound in moving fluids, with wind (current) velocity $\mathbf{U}(\mathbf{x})$. The analysis is simplest in *stratified* fluids, in which $c = c(y)$ and the (horizontal) wind speed $U(y)$ depend only on the altitude (or depth in water).

Hamilton-Jacobi theory. For 150 years, the 'corpuscular' concept (of a photon or phonon) underlying Fermat's principle was thought to be antithetical to Huygens' 'undulatory' concept of a wave front. However, Hamilton-Jacobi theory reveals them as simply dual aspects of the same mathematical model. In a static fluid, the formula $ds = c(\mathbf{x})dt$ defines from any assumed smoothly varying $c(\mathbf{x})$ a *Riemannian manifold*. In this manifold, the acoustic paths are *geodesics*, while the wave fronts of the sound waves emanating from a point source P are the Riemannian 'spheres' with center P. The resulting 'ray theory' of acoustics is easily visualized, because such spheres necessarily cut the geodesics ('acoustic rays') emanating from P orthogonally.

The analogous explanation of refraction by wind requires using the much less familiar concept of a *Finsler manifold*. One must imagine the infinitesimal spheres with 'center' $\mathbf{x}$ as displaced through $\mathbf{U}(\mathbf{x})dt$, where $\mathbf{U}$ is the local vector wind velocity. Geodesics and spheres with center P are still defined in the resulting Finsler manifold, but the 'shortest' (i.e., quickest)

---

[2] In air, c is nearly proportional to $\sqrt{T}$; see Sec. 3.

path from Q to P is not in general the reverse of the shortest path from P to Q; moreover the relation of orthogonality is not symmetric. Hence ordinary geometric intuition is not applicable.

Although light waves are always bent down in the atmosphere, because matter slows down light, this is not true of sound. Thus sound waves usually bend up, because the temperature tends to decrease with altitude. Winds also refract sound waves upwards 'upwind', but they refract sound downwards 'downwind', making it more audible.[3]

3. *Thermodynamic models.* Euler's original concept of a (homogeneous) elastic fluid ignored two related physical facts: (a) the pressure $p = p(\rho, T)$ in all real fluids depends not only on the density but also on the (absolute) temperature T, and (b) compression heats matter. In water, heating by compression has a very small effect on the speed of sound, but in air, it increases c by about 18%.

Specifically, over normal atmospheric temperature ranges, air and its major constituents (O_2 and N_2) are nearly *ideal gases*. By definition, ideal gases satisfy the special equation of state $pV = RT$, where R is a universal constant, and they have constant specific heats C_V and $C_P = \gamma C_V$. In any ideal gas, clearly isothermal compression will make $c^2 = (dp/d\rho)_T = RT = p/\rho$, whereas adiabatic compression can be shown to make $p = k\rho^\gamma$, so that $dp/d\rho = \gamma p \rho$. (In air, $\gamma \simeq 1.4$, which increases c by about 18%.)

Both of these extreme cases, associated with infinite and zero conductivity respectively, give what may be called *thermoelastic* fluids satisfying Eulerian equations of state. Correspondingly, motions of both are *reversible* in time.[4] The transition from one extreme case to ther other (so-called *dispersion*), which takes place in the far ultrasonic range, is inconsistent with Euler's model; see Secs. 7-11.

4. *Diffraction.* In Secs. 4-6, I will demonstrate the consistency of geometrical acoustics with the concept of an elastic (or thermoelastic) fluid. This is already a very deep and difficult question.

In famous papers of 1882-83, Kirchhoff showed that the wave equation (1.1) is consistent with Huygen's Principle in the following sense. The disturbance produced outside a closed surface Γ in an infinite homogeneous elastic fluid by sources inside Γ is the same as would be produced by a fictitious distribution of sources on Γ of strength $-\partial \phi/4\pi \partial n$ per unit area, together with normally directed dipoles of strength $\phi/4\pi$. From this result, Kirchhoff deduced a "general theorem ... that geometrical optics is a

---

[3] Effects due to winds and currents will be neglected in the rest of this paper.

[4] The *Carnot cycle*, basic for considerations of thermodynamic efficiency, brings out the significance of this fact.

limiting form of physical optics ... as the wave length ... tends to zero."[5]

He then applied similar ideas to Fresnel's theory of the diffraction of light after passing through apertures in otherwise opaque barriers. "In 1816, Fresnel showed that the varied phenomena of diffraction are readily explained ... by the interference of light waves taken in connection with Huygens' principle."[6]

To relate Fresnel's explanation to Euler's model, Kirchhoff concentrated his attention on *simply harmonic* (alias 'monochromatic') waves. In Euler's model these have velocity potentials of the form

$$\phi(\mathbf{x}; t) = \Phi(\mathbf{x}) \cos \omega t + \Psi(\mathbf{x}) \sin \omega t . \tag{4.1}$$

Such velocity potentials are solutions of (1.1) if and only if Φ and Ψ satisfy the Helmholtz equation

$$\nabla^2 \Phi + k^2 \Phi = \nabla^2 \Psi + k^2 \Psi = 0, \quad k = \omega_c . \tag{4.2}$$

Loosely speaking, Kirchhoff proposed treating each point of the aperture as a source of hemispherical waves having the frequency, intensity, and phase of the incident wave.

Unfortunately, the treatment of diffraction by the so-called Fresnel-Kirchhoff equation is not exact. Neither Fresnel nor Kirchhoff considered optical *polarization* in depth; moreover, as Poincaré showed in 1892 (*Théorie Mathématique de la lumière 2*, pp. 187-8), Kirchhoff's boundary conditions are inconsistent. Nevertheless, "The standard diffraction calculations of classical optics are all based on the Kirchhoff assumption. It should be obvious that the recipe can have only very approximate validity."[7]

Actually, Sommerfeld solved in 1895 (Math. Annalen 47, 317-74) the only diffraction problem for which an exact solution of the wave equation (4.2) is known: diffraction past the half-plane $y = 0, x > 0$. Although definite integrals expressing the velocity potential are known, the (elliptic) particle orbits which they determine seem never to have been computed.

5. *Particle orbits.* Whether or not it has a velocity potential, a velocity field is called "simply harmonic" (or 'monochromatic') when it is of the form

$$\mathbf{u}(\mathbf{x}; t) = \mathbf{U}(\mathbf{x}) \cos \omega t + \mathbf{V}(\mathbf{x}) \sin \omega t . \tag{5.1}$$

---

[5] B.B. Baker and E.T. Copson, *Huygens' Principle*, Oxford, 1949, esp. Chap. I, §§4.2 and 5.4 and Chap. II, §2.2.

[6] A.L. Kimball's *College Physics* (Henry Holt, 1923), p. 662.

[7] John D. Jackson, *Classical Electrodynamics* (Wiley, 1962), p. 282.

This clearly implies that all particle orbits

$$\mathbf{x}(t) = \frac{1}{\omega}[\mathbf{U}(\mathbf{x})\sin\omega t - \mathbf{V}(\mathbf{x})\cos\omega t] \tag{5.1'}$$

are *ellipses*.

An interesting example concerns the *diffraction* of normally incident plane sound waves by a half-plane. This special case of Sommerfeld's exact solution (see Sec. 4) has been lucidly discussed by Lamb, in Sec. 308 of his *Hydrodynamics*. Over most of the domain, particle orbits are straight line segments parallel to the direction of propagation. However, in the parabolic zone $y > (kx^2/2\omega)$ and its image under reflection $-(\omega^2/2k)$ in the extended half-plane, where "interference fringes" are expected in optics, they are truly elliptic. Using modern computers, it should be straightforward to compute and plot sample orbits using modern computers and Lamb's formulas, and it would be interesting to compare the predicted orbits with those observed in laboratory experiments.

The analogous particle orbits computed by Kirchhoff's approximate method for monochromatic sound waves transmitted through a circular aperture should be of even greater scientific interest; see the end of Sec. 7.

"*Simple*" *waves*. As proposed in [1, p. 644], a sound wave in an otherwise static fluid may be called "simple" when its particle orbits $\delta\mathbf{x}(t)$ are all *rectilinear*, so that[8]

$$\delta\mathbf{x}(t) = \mathbf{A}(\mathbf{x})\cos[\omega t - \tau(\mathbf{x})] \,. \tag{5.2}$$

Since $\delta\mathbf{x}_t = \mathbf{u}$, (5.2) implies (5.1) with

$$\mathbf{\Omega} = \omega\cos\tau(\mathbf{x})\mathbf{A}(\mathbf{x}), \quad \mathbf{V} = \omega\sin\tau(\mathbf{x})\mathbf{A}(\mathbf{x}) \,, \tag{5.2'}$$

so that $\mathbf{U}(\mathbf{x})\|\mathbf{V}(\mathbf{x})\|\mathbf{A}(\mathbf{x})$. Since $\mathbf{u}_t$ must therefore also be parallel to $\mathbf{A}(\mathbf{x})$, it follows that the *isobaric surfaces* of constant pressure in simple waves must also be orthogonal to the acoustic paths at all times, so that energy must also be transmitted along acoustic paths.

In a *homogeneous* elastic fluid (Euler's model), such simple waves must have velocity potentials of the form (4.1), where Φ and Ψ satisfy the Helmholtz equation (4.2).

By definition, all simply harmonic *standing waves* in any elastic fluid satisfy (5.2) with constant $\tau(\mathbf{x}) = \tau_0$; hence they are "simple". If the fluid is homogeneous, so that $\mathbf{u} = \nabla\phi$, a celebrated theorem of Poincaré asserts

---

[8] This condition is unrelated to the concept of a "simple" one-dimensional wave defined in Courant-Friedrichs, *Supersonic Flow and Shock Waves*.

much more. Namely, in any bounded enclosure Γ with rigid boundary Γ, the standing wave solutions $\phi = \Phi(\mathbf{x})\cos(\omega t - \tau_0)$ of (4.1)-(4.2) satisfying $\partial\Phi/\partial n = 0$ on Γ provide a complete normal orthogonal basis of standing waves. These are, of course, just the *normal modes* of 'small oscillation' of the enclosed fluid, and Hilbert space methods presumably enable one to generalize Poincaré's theorem to inhomogeneous elastic fluids.

In an infinite homogeneous elastic fluid, all cylindrical and all spherical waves, being radial, are "simple". Moreover, the simply harmonic *far field* generated by any bounded set of oscillators oscillating at the same frequency f is asymptotically radial, hence "simple". But of greatest interest is the extent to which analogous results also hold in *inhomogeneous* elastic fluids, the model which I will discuss next.

6. *Chaplygin fluids.* The mathematical concept of an inhomogeneous elastic fluid, of variable elasticity $dp/\rho dp$ and hence sound speed c, should logically be defined in terms of Lagrangian coordinates. Any static such fluid, in equilibrium at constant pressure p_0, is equivalent to an associated *Chaplygin fluid* whose (tangential) equation of state $p = P(\mathbf{a}) - I^2(\mathbf{a})/\rho$ has everywhere the local equilibrium density $\rho_0(\mathbf{a})$, speed of sound $c(\mathbf{a})$, and hence local impedance $I(\mathbf{a})$ of the original fluid. This will be the case if $P(\mathbf{a}) = p_0 + \rho_0(\mathbf{a})c^2(\mathbf{a})$.

Plane waves. In one space dimension (i.e., for plane waves), Lagrangian coordinates work well, because one can let the material coordinate a signify cumulative mass. This choice of $\mathbf{a} = a$ gives the *linear* equation of motion

$$x_{tt} = [I^2(a)x_a]_a, \quad x_a = 1/\rho . \qquad (6.1)$$

If $I(a) = I_0$ is a constant in (6.1), progressive waves with $x = f(a - I_0 t)$ can pass through the fluid in one direction, even if $\rho_0(a)$ and $c(a) = I_0/\rho_0(a)$ are variable.

Unfortunately, in more than one space dimension, it seems impossible to treat displacements exactly. One can at best define $\mathbf{a}$ as the initial equilibrium position of a particle, so that $\mathbf{x}(\mathbf{a}, 0) \equiv \mathbf{a}$, and hope that the net displacement of fluid particles is negligible.

Furthermore, one cannot assume that a velocity potential exists. This unfortunate fact, which seems to be overlooked in the literature [3, Sec. 9], complicates attempts to correlate the 'ray theory' of refraction (Fermat-Huygens) with the concept of an inhomogeneous elastic fluid.

Refraction of "simple" waves. However, one can artificially confine particle orbits in any elastic fluid to quickest paths ('rays') emanating from any source., by imaginary frictionless constraints. If s denotes distance from the source along any ray, and $S(s)$ denotes the infinitesimal cross-sectional area of any "acoustic tube" bounded by such rays, then (letting accents signify differentiation with respect to s of functions of s alone), the equations of continuity and motion are

$$u_t = -p_s/\rho \quad \text{and} \quad \rho_t = -\rho[S(s)u]_s/S(s) \ . \tag{6.2}$$

Hence, since $dp/d\rho = c^2(s)$,

$$\delta p_t = -\rho c^2[u_s + (S'/S)u] \ . \tag{6.3}$$

Moreover simply harmonic waves, satisfying by definition

$$\delta p = P(s)\cos\omega t + Q(s)\sin\omega t$$
$$u = U(s)\cos\omega t + V(s)\sin\omega t \tag{6.4}$$

must therefore satisfy

$$P'' + [(\rho S)'/(\rho S)]P' + (\omega^2/c^2)P = 0 \ , \tag{6.5}$$

and similarly for Q. The derivation of (6.5) from Eqs. (6.2)-(6.4) is straightforward.

Asymptotically, for a given total pressure variation $\delta p_{\max} - \delta p_{\min} = 2\delta p_{\max}$, as the wave length λ tends to zero, the pressure gradient will have a component $\delta p_{\max}/\lambda$ parallel to the acoustic tube, and $\delta p_{\max}$ perpendicular to the tube. Hence the artificial constraint suppressing accelerations perpendicular to the tube are asymptotically negligible: the "simple" wave assumption is consistent, because the *surfaces of constant phase are asymptotically isobars.*

The preceding discussion also shows, without assuming the existence of a velocity potential, that the concept of an *eikonal function* is asymptotically valid for sound waves in an (otherwise static) elastic fluid, in the short wave limit.[9] Moreover, the partial differential equations corresponding to (6.2) and (6.3) can be expanded in an asymptotic series in λ.

More precisely, the 'simple wave' concept gives a self-consistent zeroth order approximation, from which it should be possible to determine the amplitudes and directions of transverse oscillations by perturbation methods to a first order approximation.

7. *Viscosity.* Although the Navier-Stokes equations provide the most widely accepted mathematical model for the physics of fluids, their contribution to our understanding of acoustics is really very limited. In particular, although Stokes' derivation of these equations clearly defines a

---

[9] As was stated earlier, however, it does assume that fluid particles have a negligible net displacement.

hypothetical *bulk viscosity* μ_B, he and most later authors have neglected it (set it equal to zero).

In contrast, the *shear viscosity* μ, which alone influences the motion of "incompressible" fluids, has been precisely measured in many fluids, and much is known about its effects, especially in the asymptotic *boundary layer* approximation.

Viscosity and diffraction. The effects of viscosity on the diffraction of sound seem not to have been explored, either experimentally or theoretically. An interesting example to study is provided by Sommerfeld's model solution for plane waves impinging normally on a rigid half-plane (see Secs. 4-5). Having determined the elliptic particle orbits associated with this model in the neighborhood of the edge, as suggested in Sec. 5, it should be easy to determine the oscillating boundary layer formed along the half-plane. The question of flow separation from the edge is more difficult, however, and it may be hard to find a fluid, a frequency, and a sound intensity for which the effect of viscosity on diffraction is appreciable!

Like heat conduction, shear viscosity provides a plausible *qualitative* explanation for variations in the speed of sound with the wave length $\lambda = 2\pi c/\omega$. However, until 1920, it was mainly used in *combination* with heat conduction as a boundary layer *correction*, to explain why measurements of the speed of sound in Kundt tubes deviated from the adiabatic fluid model of §3. Nothing was known before then about the transition from the observed adiabatic sound speed $c = \sqrt{7p/5\rho}$ in air to a hypothetical "isothermal" sound speed $\sqrt{p/\rho}$. Moreover, to rationalize observations, the Navier-Stokes viscoelastic model was also combined with the kinetic theory of gases in ways to be described in §9.

8. *Kirchhoff-Langevin model.* Indeed, it was only after 1950 that a thorough analysis was made of the propagation of plane sound waves of infinitesimal amplitude in a general *viscous thermoelastic fluid*, endowed with constant shear viscosity μ, bulk viscosity μ_B, and thermal conductivity K.

For this analysis, the scientific world is greatly indebted to Clifford Truesdell, who concluded that no choice of these empirical constants fitted the *absorption* and *dispersion* effects which by that time had been measured in laboratories.

To explain the observed rates of absorption and dispersion, one must appeal to *molecular physics*, in ways that I shall now try to explain.

9. *Kinetic theory of gases.* Although hypothetical particle models of fluids were invoked by earlier authors, including Daniel Bernoulli and Navier, the first impressive (and still the most successful) particle model of fluids is the *kinetic theory of gases*, as developed by Maxwell and Boltzmann. This not only rationalized the equation of state of an ideal gas, including the

formula $p = k\rho^\gamma$ for adiabatic compression, it also gave a rationale for the approximate value $\gamma = 1.4$ in diatomic gases like air. Furthermore, it explained why the (shear) viscosity $\mu = \mu(T)$ of a gas at a given temperature is nearly independent of the pressure, and why the Prandtl number (the ratio $\mu C_V/K$ of viscous to thermal diffusivity) is about 0.74 in diatomic gases.

Indeed throughout the nineteenth century, the kinetic theory of gases was generally regarded as reinforcing and refining the continuum models of Euler-Laplace and Navier-Stokes, rather than as explaining different phenomena. Only diffusion itself was recognized as a phenomenon not foreseen by the differential equations of these continuum models.

Especially relevant for acoustics is the prediction that the rate of absorption of sound energy per wave length, $\gamma\alpha$, is (cf. [3, (3.1")])

$$\lambda\alpha = Kf/p, \quad \text{where} \quad K = 8\pi^2 m\mu(T)/3\gamma . \tag{9.1}$$

p being the pressure and m the molecular mass.

The success of the classical kinetic theory of dilute gases is especially impressive when one recalls the simplicity of its assumptions. It assumed the *motion* of gas molecules to be governed by Newtonian force laws of repulsion between point-masses, such as $F = a/r^5$ (Maxwell) or $F = a/r^\alpha - b/r^\beta$ with $\beta > \alpha$ (Lennard-Jones).[10] The thermal energy of the preceding "dynamical" model supplemented its *translational* energy of $mv^2/2$ by hypothetical *vibrational* and *rotational* energies, each degree of freedom of each species of molecule having the same molecular energy (Maxwell's equipartition principle). In the range of audible sound (25 Hz-2500 Hz) and atmospheric temperatures (roughly 225°K-325°K), theory and observation seemed reasonably consistent.

However, as the range of experiments broadened, basic flaws in classical kinetic theory because recognized. First, measurements of *radiation* inspired Planck's *quantum* hypothesis. This also helped to rationalize observed variations in γ for diatomic gases at low temperatures. Most relevant for acoustics was the discovery, around 1925, of the fact that the absorption rate of ultrasonic waves in CO_2 might be 80 times that predicted by kinetic theory (formula (9.1)). To rationalize this discrepancy, the concept of a *relaxation time* required to transfer rotational to translational energy was invoked.

---

[10]S. Chapman and D.C. Cowling, "The Mathematical Theory of Non-Uniform Gases", second edition, Cambridge University Press, 1958. See especially p. 36, Chap. X, and the "Notes added in 1951". These last suggest the transition to modern ideas about molecular physics discussed in Sec. 10 below.

10. *Modern molecular models.* For example, consider still air at 20°C and atmospheric pressure having a relative humidity of 20%.[11] Relaxation times of 6 and .08 milliseconds for rotational energy transfer in N_2 and O_2, respectively, are believed to absorb 10-100 times as much acoustic energy in the audible range as (shear) viscosity and heat conduction. In the ultrasonic range above 100 kHz, however, classical theory is believed to predict the observed rate of absorption.

The concept of an empirical relaxation time is generally invoked today to explain substantial deviations from general formulas like (9.1). I do not know to what extent molecular physics has been successful in *predicting* such relaxation times and their effect on acoustic absorption, in gases or liquids.

However, it is clear that modern molecular models of gases differ dramatically from those of classical kinetic theory. In modern physics, each molecule is pictured as a collection of atomic nuclei and negatively charged electrons, with "states" and energy levels given by the Schrödinger equation. Moreover polyatomic gases cease to be "perfect" at higher temperatures, because dissociation and ionization increase the number of their "degrees of freedom", which changes the meaning of Maxwell's equipartition principle.

It is also clear that the mechanisms of sound absorption in liquids can be much more varied than in gases (see [3, Appendix A]). The noble concept of a universal theory of sound absorption, applicable to all gases and all liquids at all wave lengths, seems to be no longer tenable.

11. *Digital models.* Finally, I turn to digital models of sound waves in fluids, discretized in time as well as in space. Many such digital models have been developed. Each of the *analytical* models discussed earlier has many discretizations.[12] In principle, the active new field of *computational acoustics* is concerned with their *accuracy*, and with the *efficiency* of numerical methods for solving the resulting very large systems of (perhaps linear) algebraic equations.

My concern is not with the practical *utility* of such digital models, which is best judged by the users themselves, but with their *scientific* significance. This is usually very hard for outsiders to judge, because the most useful computer programs often incorporate empirical corrections.

From a scientific standpoint, the simpler and more naive a digital model

---

[11] For the data referred to here, see Allan D. Pierce, *Acoustics*, McGraw-Hill, 1981, p. 560. The sensitivity of α to humidity was discovered in the 1930s.

[12] For discretizations of (1.1), see Sec. 10 of my article in SIAM Rev.25 (1983), 1-34; for a discretization of (4.2), see the article by A. Bayliss, C. I. Goldstein, and E. Turkel in Ref. 1, pp. 655-65.

is, the better. Thus computations which show that *numerical dispersion* can lead to spurious *reflection* in finite element models of sound waves are very interesting from a scientific standpoint (see [1, pp. 733-46]). Likewise, computations of particle orbits associated with the elastic fluid model should make this more meaningful. Moreover comparisons of computed orbits with laboratory observations, as suggested in Sec. 7, would help to determine the influence of viscosity on acoustic diffraction.

Far more difficult, but correspondingly rewarding, would be the construction of computer models of colliding O_2, N_2 and H_2O molecules from well-established basic physical principles, and to deduce from such models the rates of sound and ultrasonic wave absorption in air mentioned in Sec. 10.

References
1. Martin H. Schultz and Ding Lee (eds.), *Computational Ocean Acoustics*, Pergamon Press, 1985.
2. G. Birkhoff, "Sound waves in fluids", to appear in R. Vichnevetsky (ed.), *Proceedings of a Workshop on Numerical Fluid Dynamics*, North-Holland, 1987.
3. G. Birkhoff, "Consistency of models for sound waves in fluids" in Ding Lee, R. L. Sternberg, and M.H. Schultz (eds.), *Computational Acoustics*, 2 vols., North-Holland, 1988.

A CLASS OF EXACT SOLUTIONS IN VISCOUS INCOMPRESSIBLE MAGNETO-HYDRODYNAMICS

A. D. D. Craik

Department of Applied Mathematics,
University of St Andrews
St Andrews, Fife, KY16 9SS, Scotland, U. K.

1. INTRODUCTION

Recently, Craik & Criminale [1] described a procedure for finding new classes of solutions of the Navier-Stokes equations. The solutions are conveniently decomposed into a 'basic flow' and a 'disturbance', but the latter need not be small compared with the former. The solutions are unbounded in space and basic flows have uniform rates of strain. The disturbance is of transverse plane-wave form, but with a wavenumber that changes in magnitude and orientation with time.

Craik & Criminale (see also Bayly [2]) gave a number of explicit solutions corresponding to various basic flows. It is here shown that there exist corresponding classes of exact solutions in magneto-hydrodynamics, that have not previously been reported.

2. BASIC FORMULATION

The equations of incompressible magneto-hydrodynamics may be expressed in symmetric form as (see for example Cowling [3])

$$(\partial/\partial t + \mathbf{Q} \cdot \nabla)\mathbf{P} = -\nabla \pi + \nu_1 \nabla^2 \mathbf{P} + \nu_2 \nabla^2 \mathbf{Q} ,$$
$$(\partial/\partial t + \mathbf{P} \cdot \nabla)\mathbf{Q} = -\nabla \pi + \nu_1 \nabla^2 \mathbf{Q} + \nu_2 \nabla^2 \mathbf{P} ,$$
$$\nabla \cdot \mathbf{P} = \nabla \cdot \mathbf{Q} = 0, \quad \pi \equiv p/\rho + (2\mu\rho)^{-1}(\mathbf{B} \cdot \mathbf{B})$$
$$\mathbf{P}, \mathbf{Q} \equiv u \pm (\mu\rho)^{-1/2}\mathbf{B}, \quad \nu_{1,2} \equiv (\nu \pm \eta)/2 .$$

(1, a, b, c, d)

The magnetic diffusivity is related to the magnetic permeability μ and electrical conductivity σ by $\eta = (\mu\sigma)^{-1}$.

Just as for the Navier-Stokes equations, we seek 'basic' states of the form

$$\mathbf{P}^\circ = \mathbb{P}\mathbf{x} + \mathbf{P}^\circ, \quad \mathbf{Q}^\circ = \mathbb{Q}\mathbf{x} + \mathbf{Q}^\circ ,$$
$$\pi = \pi^\circ, \quad \mathbb{P} = \{p_{ij}\}, \quad \mathbb{Q} = \{q_{ij}\} \quad (3 \times 3) .$$

(2)

The matrices $\mathbb{P}$, $\mathbb{Q}$ and vectors $\mathbf{P}^\circ$, $\mathbf{Q}^\circ$ are independent of the space coordinates $\mathbf{x}$, but may depend on time t. The basic state must satisfy equations (1).

Defining
$$\tilde{\pi} \equiv \pi^0 + (P^0_{i,t} + Q^\circ_k p_{ik})x_i$$

yields
$$\frac{d\mathbb{P}}{dt} + \mathbb{P}\mathbb{Q} = -\mathbb{M}, \quad \frac{d\mathbb{Q}}{dt} + \mathbb{Q}\mathbb{P} = -\mathbb{M} \tag{3}$$

where $\mathbb{M}$ is symmetric, with elements $\partial^2 \tilde{\pi}/\partial x_i \partial x_j$.

For steady basic flows it is seen that
$$\mathbb{P}\mathbb{Q} = \mathbb{Q}\mathbb{P} = -\mathbb{M}(\text{symmetric}),$$
$$\text{trace}(\mathbb{P}) = \text{trace}(\mathbb{Q}) = 0. \tag{4a,b}$$

Let the basic state be disturbed, to
$$[\mathbf{P}, \mathbf{Q}, \pi] = [\mathbf{P}^\circ, \mathbf{Q}^\circ, \pi^\circ] + \text{Re}\left\{[\hat{\mathbf{p}}(t), \hat{\mathbf{q}}(t), \hat{\pi}(t)]e^{i(\boldsymbol{\alpha}\cdot\mathbf{x}+\delta)}\right\} \tag{5}$$

with $\boldsymbol{\alpha} = \boldsymbol{\alpha}(t), \delta = \delta(t)$. Then
$$\frac{d\hat{\mathbf{p}}}{dt} + i\left[\frac{d\boldsymbol{\alpha}}{dt}\cdot\mathbf{x} + \frac{d\delta}{dt} + \mathbf{Q}^\circ\cdot\boldsymbol{\alpha}\right]\hat{\mathbf{p}} + (\hat{\mathbf{q}}\cdot\nabla)\mathbf{P}^\circ = -i\boldsymbol{\alpha}\hat{\pi} - (\boldsymbol{\alpha}\cdot\boldsymbol{\alpha})(\nu_1\hat{\mathbf{p}} + \nu_2\hat{\mathbf{q}}) \tag{6}$$

from (1a), and a similar equation with p's and q's interchanged from (1b). Also, (1c) yield $\boldsymbol{\alpha}\cdot\hat{\mathbf{p}} = \boldsymbol{\alpha}\cdot\hat{\mathbf{q}} = 0$. It readily follows that
$$d\boldsymbol{\alpha}/dt = -\mathbb{Q}^{tr}\boldsymbol{\alpha} = -\mathbb{P}^{tr}\boldsymbol{\alpha}, \quad \hat{\pi} = \frac{i}{\boldsymbol{\alpha}\cdot\boldsymbol{\alpha}}\left[\boldsymbol{\alpha}^{tr}(\mathbb{Q}\hat{\mathbf{p}} + \mathbb{P}\hat{\mathbf{q}})\right] \tag{7a,b}$$

and $\hat{\mathbf{p}}(t), \hat{\mathbf{q}}(t)$ satisfy first-order ordinary differential equations with coefficients depending on $\boldsymbol{\alpha}(t)$. Here 'tr' denotes transpose.

For consistency, $(\mathbb{P}^{tr} - \mathbb{Q}^{tr})\boldsymbol{\alpha} = 0$: it follows that any $\mathbf{x}$-dependent part of $\mathbf{B}^\circ$ must remain perpendicular to the wavenumber vector $\boldsymbol{\alpha}(t)$.

3. ADMISSIBLE BASIC STATES

(a) *Constant magnetic fields.* When the field $\mathbf{B}^\circ$ is constant, $\mathbb{P} = \mathbb{Q}$. There are then the same admissible basic velocity fields as were found by Craik & Criminale [1] when no magnetic field is present. There are two families: two-dimensional flows with uniform strain rates and constant vorticity, and three-dimensional irrotational flows with uniform strain rates. An arbitrary initial wavenumber vector $\boldsymbol{\alpha}(0)$ then evolves according to (7a), just as with no field present; but the disturbance amplitude functions and pressure are affected by any constant field.

(b) *Variable magnetic fields.* Now, take $\boldsymbol{\alpha}(t) = [\alpha_1(t), \alpha_2(t), 0]$ and let the x-dependent part of the magnetic field $\mathbf{B}^\circ$ be wholly the x_3-direction, with magnitude $\bar{a}x_1 + \bar{b}x_2$). The admissible states are limited to two classes, if B° is independent of time t.

Class (i) has velocity and magnetic field depending linearly on just a single variable $(\bar{a}x_1 + \bar{b}x_2)$. The velocity is therefore a plane Couette flow. Class (ii) has admissible basic velocities (apart from additive constants) that are axisymmetric irrotational stagnation-point flows. Choosing the coordinate axes to lie along principal directions of rate-of-strain yields the basic state in its simplest form, namely

$$\mathbf{U}^\circ = a\Big[x_1, -\frac{1}{2}x_2, -\frac{1}{2}x_3\Big], \quad \mathbf{B}^\circ = \bar{b}[0, 0, x_2] \ .$$

There are other admissible states that depend on time t. These include a steady two-dimensional flow with elliptical streamlines (c.f. Bayly [2]), for which the basic spatially-varying magnetic field is time-periodic.

4. DISCUSSION

For any admissible state, the wavenumber of the disturbance may be found explicitly from (7a). Then the amplitude functions $\hat{p}$ and $\hat{q}$, which determine the velocity and field perturbations, may be calculated from their linear differential equations. The pressure disturbance follows from (7b). Numerical solution is normally necessary, but analytic solutions can be found for particular cases.

When each streamline of the basic flow extends to infinity, the perturbation wavenumber normally grows exponentially and dissipative processes are then ultimately dominant. But initially-growing disturbances exist for all basic flows. When streamlines are closed, the perturbation is either periodic in time or display continuous growth due to Floquet instability, in the non-dissipative limit $\nu_1 = \nu_2 = 0$. Despite dissipation, the Floquet instability will persist for certain wavenumbers, as recently shown for the Navier-Stokes equations by Landman & Saffman [4].

References
1. Craik, A. D. D. & Criminale, W. O., Evolution of wavelike disturbances in shear flows: a class of exact solutions of the Navier-Stokes equations. *Proc. Roy. Soc. Lond.* **A406** (1986), 13.
2. Bayly, B. J., Three-dimensional instability of elliptical flow. *Phys. Rev. Lett.* **57** (1986), 2160.
3. Cowling, T. G., *Magnetohydrodynamics*. Adam Hilger, 1976.
4. Landman, M. J. & Saffman, P. G. *Phys. Fluids.* (to appear 1987).

STRESS SINGULARITIES AT A
RIM OF CIRCULAR CYLINDERS[*]

Yihan Lin
Institute of Applied Mathematics
University of British Columbia
Vancouver, B.C. V6T 1Y4
Canada

Frederic Y. M. Wan
Department of Applied Mathematics
University of Washington, FS-20
Seattle, WA 98195
U. S. A.

1. INTRODUCTION

Stress singularities at a corner of an elastic sheet where boundary conditions undergo abrupt changes have been studied in numerous investigations since the publication of [5]. Corner stress singularities may also exist at the upper and lower rims of a flat plate. For applications to certain plate problems (see [1] for example), we consider in this paper the stress singularity problem for a semi-infinite homogeneous circular cylinder occupying the region $\{Z < 0, X^2 + Y^2 < R_0^2\}$ in axisymmetric deformation. In the neighborhood of the rim, the end face $Z = 0$ of the cylinder is traction free while the cylindrical surface $X^2 + Y^2 = R_0^2$ is constrained from displacements. The equation for determining the severity of the stress singularity will be obtained and numerical results needed in applications [4] will be generated for a range of material parameter values for both isotropic and transversely isotropic cylinders.

2. AXISYMMETRIC LINEAR ELASTOSTATICS OF SEMI-INFINITE CIRCULAR CYLINDERS

Consider a homogeneous elastic circular cylinder occupying the region $\{Z < 0, R^2 = X^2 + Y^2 < R_0^2\}$ and free of distributed external loads in

---

[*]The research of the first author is supported by an Izaak Walton Killam memorial predoctoral fellowship at UBC. The research of the second author is supported in part by an NSF grant No. DMS-8606198.

the cylinder interior. The elasticity of the cylinder is characterized by the following orthotropic stress strain relations [3].

$$\varepsilon_{rr} = a_{11}\sigma_{rr} + a_{12}\sigma_{\theta\theta} + a_{13}\sigma_{zz}, \quad \varepsilon_{\theta\theta} = a_{21}\sigma_{rr} + a_{22}\sigma_{\theta\theta} + a_{23}\sigma_{zz},$$
$$\varepsilon_{zz} = a_{31}\sigma_{rr} + a_{32}\sigma_{\theta\theta} + a_{33}\sigma_{zz}, \quad \varepsilon_{rz} = \varepsilon_{zr} = a_{44}\sigma_{rz} \tag{2-1}$$

relevant to axisymmetric deformations. For simplicity, we limit our consideration to the special case of transverse isotropy with $a_{ij} = a_{ji}$, $a_{11} = a_{22}$ and

$$a_{11} = \frac{1}{E}, a_{12} = -\frac{\nu}{E}, a_{13} = a_{23} = -\frac{\nu'}{E'}, a_{33} = \frac{1}{E'}, a_{44} = \frac{1}{G'}. \tag{2-2}$$

For infinitesimally small axisymmetric deformations the strain components in (2-1) are defined in terms of the radial and axial displacement components u_r and u_z by

$$\varepsilon_{rr} = u_{r,R}, \quad \varepsilon_{\theta\theta} = \frac{u_r}{R}, \quad \varepsilon_{zz} = u_{z,Z}, \quad \varepsilon_{rz} = u_{r,Z} + u_{z,R}. \tag{2-3}$$

with $(\ ), t = \partial(\ )/\partial t$.

Static equilibrium of the cylinder is satisfied by the following stress function representation:

$$\sigma_{rr} = -\frac{\partial}{\partial Z}\left[\frac{\partial^2}{\partial R^2} + \frac{\alpha_2}{R}\frac{\partial}{\partial R} + \alpha_1\frac{\partial^2}{\partial Z^2}\right]\psi,$$
$$\sigma_{zz} = \frac{\partial}{\partial Z}\left[\alpha_3\Delta_R + \alpha_4\frac{\partial^2}{\partial Z^2}\right]\psi,$$
$$\sigma_{\theta\theta} = -\frac{\partial}{\partial Z}\left[\alpha_2\frac{\partial^2}{\partial R^2} + \frac{1}{R}\frac{\partial}{\partial R} + \alpha_1\frac{\partial^2}{\partial Z^2}\right]\psi,$$
$$\sigma_{rz} = \sigma_{zr} = \frac{\partial}{\partial R}\left[\Delta_R + \alpha_1\frac{\partial^2}{\partial Z^2}\right]\psi \tag{2-4}$$

where $\Delta_R = (\ ),_{RR} + R^{-1}(\ ),_R$ and

$$\beta\alpha_1 = a_{13}(a_{11} - a_{12}), \quad \beta\alpha_2 = a_{13}(a_{13} + a_{44}) - a_{12}a_{33},$$
$$\beta\alpha_3 = a_{13}(a_{11} - a_{12}) + a_{11}a_{44}, \quad \beta\alpha_4 = a_{11}^2 - a_{12}^2 \tag{2-5}$$

with $\beta = a_{11}a_{33} - a_{13}^2 > 0$ and $\alpha_4 > 0$ by the positive definiteness of the strain energy. The stress function ψ is required by the compatibility conditions of the strain measures to satisfy

$$\left[\Delta_R + \frac{1}{s_1^2}\frac{\partial}{\partial Z^2}\right]\left[\Delta_R + \frac{1}{s_2^2}\frac{\partial^2}{\partial Z^2}\right]\psi = 0 \tag{2-6}$$

where
$$\{s_1^2, s_2^2\} = \frac{1}{2\alpha_4}\{(\alpha_1 + \alpha_3) \pm \sqrt{(\alpha_1 + \alpha_3)^2 - 4\alpha_4}\} . \qquad (2\text{-}7)$$

For an isotropic medium, we have $2\alpha_4 = \alpha_1 + \alpha_3 = 2$ and therewith $s_1 = s_2 = 1$ so that equation (2-6) reduces correctly to a biharmonic equation with axisymmetry. With $\alpha_1 = \alpha_2 = -\nu/(1-\nu)$ and $\alpha_3 = (2-\nu)/(1-\nu)$, the representation (2-4) also reduces correctly to the corresponding representation for the isotropic case (cf. (5-2) in [1]).

We are interested here in a semi-infinite circular cylinder which is free of traction on the face $Z = 0 (0 < R_1 < R < R_0)$ and is constrained against any displacement along the cylindrical surface $R = R_0 (Z_0 < Z < 0)$. In that case we have

$$Z = 0 : \begin{cases} \frac{\partial}{\partial Z}\left[\alpha_3 \Delta_R + \alpha_4 \frac{\partial^2}{\partial Z^2}\right]\psi = 0, & (R_1 < R < R_0) . \\ \frac{\partial}{\partial R}\left[\Delta_R + \alpha_1 \frac{\partial^2}{\partial Z^2}\right]\psi = 0, \end{cases} \qquad (2\text{-}8)$$

$$R = R_0 : u_r = 0, \quad u_z = 0 \quad (Z_0 < Z < 0) . \qquad (2\text{-}9)$$

The displacement conditions (2-9) require

$$R = R_0 : \varepsilon_{zz} = u_{z,Z} = 0, \quad \varepsilon_{rz,Z} - \varepsilon_{zz,R} = u_{r,ZZ} = 0 \quad (Z_0 < Z < 0) \qquad (2\text{-}10)$$

which we will use in the subsequent development.

The cylinder is loaded elsewhere but we will not specify the precise nature of the external load except that it should be axisymmetric. It is known (see [2] for example) that the abrupt change from the traction-free conditions (on the end surface) to the no displacement conditions (on the cylindrical edge) across the geometrically non-smooth rim may give rise to a stress singularity along the rim $\{R = R_0, Z = 0\}$. We are interested in the nature of the singularity there. The determination of the severity of the singularity along the rim does not require a precise specification of the load.

3. LOCAL POLAR COORDINATES AND STRESS SINGULARITIES

For the study of stress singularities, we are only concerned with the solution behavior in a narrow region near $R = R_0$ and $Z = 0$, so that $R_0 - R \leq l$ and $Z \leq -l$ with $\varepsilon \equiv l/R_0 \ll 1$. Let $x = (R_0 - R)/l$ and $y = -Z/l$. We have then

$$\Delta_R = l^{-2}\left[\frac{\partial^2}{\partial x^2} - \frac{\varepsilon}{1-\varepsilon x}\frac{\partial}{\partial x}\right] = l^{-2}\left[\frac{\partial^2}{\partial x^2} + 0(\varepsilon)\right] \qquad (3\text{-}1)$$

so that the leading term of the asymptotic expansion for ψ as $\varepsilon \to 0$ satisfies

$$\Psi_{,xx} + s_1^{-2}\Psi_{,yy} = 0 , \quad \psi_{,xx} + s_2^{-2}\psi_{,yy} = \Psi . \qquad (3\text{-}2a,b)$$

The corresponding leading term approximations for the stress components are given by

$$\begin{aligned}
\sigma_{rr} &= \frac{1}{l^3}\frac{\partial}{\partial y}\left[\frac{\partial^2}{\partial x^2} + \alpha_1\frac{\partial^2}{\partial y^2}\right]\psi , \\
\sigma_{\theta\theta} &= \frac{1}{l^3}\frac{\partial}{\partial y}\left[\alpha_2\frac{\partial^2}{\partial x^2} + \alpha_1\frac{\partial^2}{\partial y^2}\right]\psi \\
\sigma_{zz} &= -\frac{1}{l^3}\frac{\partial}{\partial y}\left[\alpha_3\frac{\partial^2}{\partial x^2} + \alpha_4\frac{\partial^2}{\partial y^2}\right]\psi, \\
\sigma_{rz} &= \sigma_{zr} = -\frac{1}{l^3}\frac{\partial}{\partial x}\left[\frac{\partial^2}{\partial x^2} + \alpha_1\frac{\partial^2}{\partial y^2}\right]\psi .
\end{aligned} \qquad (3\text{-}3)$$

In (3-2) and (3-3), we have used the same symbols to denote the corresponding leading term asymptotic expressions. We will continue to do so in the subsequent development as only leading term approximations will be involved.

For (3-2a), we set $x = \rho\cos\phi$ and $s_1 y = \rho\sin\phi$ and transform the equation for Ψ into Laplace's equation in the polar coordinates (ρ,ϕ) with solutions of the form

$$\Psi = \rho^\lambda\{\cos(\lambda\phi), \sin(\lambda\phi)\} \qquad (3\text{-}4)$$

for any constant λ. For (3-2b), we set $x/s_2 = r\cos\theta, y = r\sin\theta$ and transform the equation for ψ into Poisson's equation in the polar coordinates (r,θ):

$$\psi_{,rr} + r^{-1}\psi_{,r} + r^{-2}\psi_{,\theta\theta} = s_2^2\Psi . \qquad (3\text{-}5)$$

With (r,θ) and (ρ,ϕ) related by

$$\rho = r\sqrt{s_1^2\sin^2\theta + s_2^2\cos^2\theta} \equiv rA_\theta , \quad \phi = \tan^{-1}\left(\frac{s_1}{s_2}\tan\theta\right) \equiv \omega_\theta , \qquad (3\text{-}6)$$

we have

$$\rho^\lambda\{\cos(\lambda\phi), \sin(\lambda\phi)\} = r^\lambda A_\theta^\lambda\{\cos(\lambda\omega_\theta), \sin(\lambda\omega_\theta)\} . \qquad (3\text{-}7)$$

A particular solution ψ_c of (3-2b) for $\Psi = \rho^\lambda\cos(\lambda\phi)$ may be taken in the form $r^{\lambda+2}f_\lambda(\theta)$ with f_λ determined by

$$f_\lambda'' + (\lambda+2)^2 f_\lambda = A_\theta^\lambda\cos(\lambda\omega_\theta) , \quad (\ )' \equiv \frac{d(\ )}{d\theta} . \qquad (3\text{-}8)$$

It follows that

$$\psi_c = s_2^2 r^{\lambda+2} f_\lambda(\theta), \quad f_\lambda(\theta) = \int_0^\theta A_t^\lambda \cos(\lambda\omega_t)\sin((\lambda+2)(\theta-t))dt . \quad (3\text{-}9)$$

Similarly, we have as a particular solution ψ_s for (3-2b) with $\Psi = \rho^\lambda \sin(\lambda\phi)$

$$\psi_s = s_2^2 r^{\lambda+2} g_\lambda(\theta), \quad g_\lambda(\theta) = \int_0^\theta A_t^\lambda \sin(\lambda\omega_t)\sin((\lambda+2)(\theta-t))dt . \quad (3\text{-}10)$$

The complementary solution ψ_h of (3-2b) is of the form $r^\mu\{\cos(\mu\theta), \sin(\mu\theta)\}$. While the general solution for ψ is a superposition of these three types of solutions $\{\psi_h, \psi_c, \psi_s\}$ for all λ and μ, it suffices for our analysis of rim-corner singularities to consider one typical term in this general solution in the form:

$$\psi(r,\theta) = r^{\lambda+2}\{c_1 \cos((\lambda+2)\theta) + c_2 \sin((\lambda+2)\theta) + c_3 f_\lambda(\theta) + c_4 g_\lambda(\theta)\} \quad (3\text{-}11)$$

where λ and c_i are constants to be determined by the boundary conditions. For an isotropic medium, we have $s_1 = s_2 = A_\theta = 1$ and $\omega_\theta = \theta$ so that f_λ and g_λ reduce to linear combinations of $\cos(\lambda\theta), \sin(\lambda\theta), \cos((\lambda+2)\theta)$ and $\sin((\lambda+2)\theta)$.

For the boundary conditions, we note the relations

$$\frac{\partial}{\partial x} = \frac{1}{s_2}\left(\cos\theta \frac{\partial}{\partial r} - \frac{\sin\theta}{r}\frac{\partial}{\partial\theta}\right), \quad \frac{\partial}{\partial y} = \sin\theta\frac{\partial}{\partial r} + \frac{\cos\theta}{r}\frac{\partial}{\partial\theta} \quad (3\text{-}12)$$

which may be applied repeatedly to get expressions for second and third order partial derivatives for ψ. These expressions are needed for the application of the boundary conditions but will not be listed here. With (3-3) and (3-11), the traction-free conditions (2-8) become

$$\beta_1(\lambda+1)c_1 + s_2^\lambda c_3 = 0, \quad \beta_2(\lambda+1)c_2 + s_1 s_2^{\lambda-1} c_4 = 0 \quad (3\text{-}13, 14)$$

where

$$\beta_1 = \frac{1}{\alpha_1 s_2^2} - 1, \quad \beta_2 = \frac{\alpha_3}{\alpha_4 s_2^2} - 1. \quad (3\text{-}15)$$

The boundary conditions (2-10) on the strain measures can also be expressed in terms of ψ by way of the stress strain relation for ε_{zz} and the stress function representation. With (3-11), these conditions become

$$c_1 \beta_3(\lambda+1)\cos\left((\lambda+2)\frac{\pi}{2}\right) + c_2 \beta_3(\lambda+1)\sin\left(+(\lambda+2)\frac{\pi}{2}\right)$$
$$+ c_3 \left\{\beta_3(\lambda+1)f_\lambda\left(\frac{\pi}{2}\right) + s_1^\lambda \cos\left(\lambda\frac{\pi}{2}\right)\right\}$$
$$+ c_4 \left\{\beta_3(\lambda+1)g_\lambda\left(\frac{\pi}{2}\right) + s_1^\lambda \sin\left(\lambda\frac{\pi}{2}\right)\right\} = 0 \quad (3\text{-}16)$$

$$c_1 \sin\left((\lambda+2)\frac{\pi}{2}\right) - c_2 \cos\left((\lambda+2)\frac{\pi}{2}\right) - c_3 \hat{f}_\lambda\left(\frac{\pi}{2}\right) - c_4 \hat{g}_\lambda\left(\frac{\pi}{2}\right) = 0 \quad (3\text{-}17)$$

where

$$\beta_3 = \frac{2\alpha_1 a_{13} - \alpha_4 a_{33}}{(\alpha_2+1)a_{13} - \alpha_3 a_{33}} s_2^2 - 1, \quad (3\text{-}18)$$

$$\hat{f}_\lambda(\theta) = \int_0^\theta A_t^\lambda \cos(\lambda \omega_t) \cos((\lambda+2)(\theta-t))dt,$$

$$\hat{g}_\lambda(\theta) = \int_0^\theta A_t^\lambda \sin(\lambda \omega_t) \cos((\lambda+2)(\theta-t))dt \quad (3\text{-}19)$$

For the linear system (3-13), (3-14), (3-16), and (3-17) to have a nontrivial solution, the determinant of the coefficient matrix must vanish giving an equation for the remaining unknown λ in the form $F(\lambda) = 0$ where

$$\begin{aligned}
F(\lambda) = & \frac{1}{2}(\lambda+1)^2 \int_0^{\frac{\pi}{2}} \int_0^{\frac{\pi}{2}} A_\psi^\lambda A_\phi^\lambda \sin(\lambda[\omega_\psi - \omega_\phi]) \sin([\lambda+2][\phi-\psi]) d\phi d\psi \\
& + \frac{1}{\beta_1} s_2^\lambda (\lambda+1) \int_0^{\frac{\pi}{2}} A_\psi^\lambda \sin(\lambda \omega_\psi) \cos([\lambda+2]\psi) d\psi \\
& - \frac{1}{\beta_2} s_1 s_2^{\lambda-1} (\lambda+1) \int_0^{\frac{\pi}{2}} A_\psi^\lambda \cos(\lambda \omega_\psi) \sin([\lambda+2]\psi) d\psi \\
& + \frac{1}{\beta_3} s_1^\lambda (\lambda+1) \int_0^{\frac{\pi}{2}} A_\psi^\lambda \sin\left(\lambda\left[\frac{\pi}{2} - \omega_\psi\right]\right) \cos\left([\lambda+2]\left[\frac{\pi}{2} - \psi\right]\right) d\psi \\
& + \frac{1}{\beta_2 \beta_3} s_1^{\lambda+1} s_2^{\lambda-1} \cos\left(\lambda \frac{\pi}{2}\right) \cos\left([\lambda+2]\frac{\pi}{2}\right) \\
& + \frac{1}{\beta_1 \beta_3} s_1^\lambda s_2^\lambda \sin\left(\lambda \frac{\pi}{2}\right) \sin\left([\lambda+2]\frac{\pi}{2}\right) - \frac{1}{\beta_1 \beta_2} s_1 s_2^{2\lambda-1}
\end{aligned}$$
$$(3\text{-}20)$$

For an isotropic material, the expression for $F(\lambda)$ simplifies to

$$F(\lambda) = \cos^2\left(\lambda \frac{\pi}{2}\right) - \frac{4\nu(1-\nu)}{3-4\nu} + \frac{(1-\lambda^2)}{3-4\nu}. \quad (3\text{-}21)$$

Observe that $F(0) = 4(1-\nu)^2/(3-4\nu) > 0$ and $F(1) = -4\nu(1-\nu)/(3-4\nu) < 0$; they imply the existence of a real root in the range $0 < \lambda < 1$ and thus a stress singularity for the problem. Upon setting $\lambda = a + ib$, we may write $F(\lambda) \equiv R(a,b) + iI(a,b)$ where R and I are the real and imaginary part of $F(\lambda)$, respectively, with $I < 0$ for $b > 0$ and $I > 0$ for $b < 0$. It follows from the Argument Principle in complex function theory (applied over a rectangular contour Γ consisting of the line segments $R_e(\lambda) = 0$, $R_e(\lambda) = 1$ and $I_m(\lambda) = \pm H$) that $F(\lambda)$ has no other root inside the strip $0 < R_e(\lambda) < 1$ of the complex λ-plane. We are not interested in roots of $F(\lambda) = 0$ with a real part not greater than zero or not less than unity. In the first case, the singularity is too severe and gives rise to unbounded strain energy which is not acceptable. In the second, the corresponding stress components are bounded everywhere including the rim of the cylinder end and hence are not singular.

4. NUMERICAL RESULTS

For the more general orthotropic cylinder, the nonlinear equation (3-20) has been solved numerically for real roots* between 0 and 1 by a standard library subroutine using a combination of bisection and Newton's method. After one root λ_0 has been found, the interval $(0, \lambda_0 - \varepsilon)$ is searched to make sure there is no smaller root. The program is checked by substituting the root found for the isotropic case into (3-21) verifying $F(\lambda) = 0$ is satisfied to a high degree of accuracy (10^{-5}).

For the isotropic case, we have the values in Table I for the exponent of the stress singularity $\lambda - 1$. Note that λ depends only on ν and the singularity becomes increasingly more severe with increasing ν. Since R_0 does not enter into equation (3-21), it can be shown that the small region near the rim is in fact in a plane strain state as far as the stress singularity is concerned.

TABLE I: Stress Singularity Exponent $(\lambda - 1)$ for Isotropic Materials

ν	0	0.1	0.2	0.3	0.4	0.5
$\lambda - 1$	0	-0.1330	-0.2189	-0.2888	-0.3501	-0.4054

For orthotropic materials, the exponent $\lambda - 1$ is first computed for Douglas firs with $E' = 1.56 \times 10^6$ lbs/in^2, $E = 0.05E'$, $G' = 0.078E'$, $\nu = 0.287$ and $\nu' = 0.449$. It can be shown that $\lambda - 1$ is unchanged if the values of the elastic parameters in the transverse and in-plane direction are interchanged [4]. The exponent is then computed for several other combinations of E/E' and G'/E' (with ν and ν' kept fixed at their values for Douglas firs) to indicate the effects of the parameter changes on $\lambda - 1$. For the cases given in Table II, we have verified numerically that the plate is also in a plane strain state near the rim as in the isotropic case. Note that the singularity for $\{E/E' = 1, G'/E' = 1/3\}$ in Table II $(\lambda - 1 = -0.3510)$ is weaker than the corresponding isotropic cylinder with the same G'/E' ratio $(\lambda - 1 = -0.4054)$.

---

*Complex roots with a real part in the interval (0,1) will not be given here as the corresponding highly oscillatory stress behavior in the neighborhood of the rim appears to be unrealistic.

TABLE II: Stress Singularity Exponent $(\lambda - 1)$ for Orthotropic Cylinders

G'/E \ E/E'	0.005	0.05	0.5	1.0	1.5
0.0078	-0.0851	-0.1637	-0.2434	-0.2692	-0.2860
0.078	Complex s_i	-0.1544	-0.2813	-0.3203	-0.3443
1/3	Complex s_i	Complex s_i	Complex s_i	-0.3510	-0.3840

If $u_r = 0$ at $R = R_0$ is replaced by $\sigma_{rr} = 0$, it has been shown by a similar analysis [4] that we have $F(\lambda) = \sin(\lambda \pi)$ for an isotropic cylinder in that case. Hence, there is no corner stress singularity for an isotropic plate with the particular type of mixed edge conditions. No stress singularity (with a real exponent λ) could be found numerically for the corresponding transversely isotropic cylinder [4].

References
1. R. D. Gregory and F. Y. M. Wan, "On Plate Theories and Saint Venant's Principle," *Int. J. Solids Structures* 21, 1985, 1005-1024.
2. V. A. Kondrat'ev and O. A. Oleinik, "Boundary Value Problems for Partial Differential Equations in Non-Smooth Domains," *Russian Math. Surveys* 38, 1983, 1-86.
3. S. G. Leknitskii, *Theory of Elasticity of an Anisotropic Body*, Holden-Day, Inc., San Francisco, 1963.
4. Y. H. Lin, "A Mathematical Theory of Elastic Orthotropic Plates in Plane Strain and Axisymmetric Deformations," Ph.D. dissertation, Institute of Applied Mathematics, University of British Columbia, Vancouver, 1987.
5. M. L. Williams, "Stress Singularities Resulting from Various Boundary Conditions in Angular Corners of Plates in Extension," *J. Appl. Mech.* 52, 1952, 526-528.

II. STABILITY & TURBULENCE

ON THE THEORY OF TURBULENCE FOR INCOMPRESSIBLE FLUIDS

Pei-Yuan Zhou(Chou) and Shi-Yi Chen

Department of Mechanics, Peking University, Beijing, CHINA

The theory of turbulence, based upon the Reynolds equations of mean motion and the dynamical equations of the velocity correlations of successive orders, derived from the equations of the turbulent velocity fluctuatuion, by using the condition of pseudo-similarity and the hypotheses on the viscous dissipation terms in the correlation equations, is developed by a method of successive approximation. As examples in the first order approximation, we have solved the turbulent flows through a channel, in a plane wake and in jets by using the equations of mean motion and of the double correlation, while the terms in the triple correlation have been neglected. The agreements between the calculated values and the experiments are satisfactory.

In the present paper the channel flow for the first approximation has been recalculated, yielding satisfactory agreement with experiment. The equations of the triple and quadruple correlations in addition to those used in the first order approximation are solved for the plane turbulent wake in the second order approximation by the method of substitution, starting from the solution of the first order approximation. Agreements between theory and existing experiments for the triple velocity correlation are also satisfactory. The theory has yielded the components of the quadruple correlation which can be tested by experiment and can also be used to obtain correlations of still higher orders.

INTRODUCTION

The theory of turbulence for incompressible fluids, based upon the Navier-Stokes equations of motion, was first put forward by Reynolds in 1985. His important contribution consists of pointing out that a fully developed turbulent flow is composed of two parts, the mean motion and the turbulent fluctuation, and the derivation of the equations of mean motion by taking the average of the Navier-Stokes equations of motion. Due to the non-linearity of these latter equations, there appears the apparent stress, known as the Reynolds stress, in the equations of mean motion. On account of the Reynolds stress being unknown, the dynamical equations of mean motion thus derived are not closed. The subsequent mixture length theories of the momentum or vorticity transport proposed by Prandtl, Taylor and Von Kàrmàn in the early part of this century had the objective to relate the turbulent velocity fluctuations to the derivatives of the mean motion in order to make the set of equations of mean motion closed, while the equations of turbulent fluctuation which are the differences of the Navier-Stokes equations of motion and the Reynolds equations of mean motion were iqnored since Reynolds' time.

The theory of turbulence, based upon the solutions of the dynamical equations of the velocity correlations of the double and triple orders derived from the equations of turbulent velocity fluctuation, and the equations of mean motion was first put forward in 1940[1] and then further developed in 1945[2]. In the 1945 paper it was also pointed out that the direct approach to the solution of the general turbulence problem is to solve simultaneously the equations of mean motion and of turbulent fluctuation together with their corresponding equations of continuity. But this is a set of non-linear integro-partial differential equations and to seek their solutions for general shear turbulent motions of fluids is very difficult.

However, for simpler problems like the homogeneous isotropic turbulence and the homogeneous shear turbulent flow, simultaneous solutions of the equations of mean motion and of turbulent fluctuation can be carried out. For homogeneous isotropic turbulence we introduce the condition of pseudo-similarity into the solution in order to choose the right kind of vortex to be the turbulence element. From the solutions of the turbulent velocity fluctuation we can then build the velocity correlations of any order to be compared with experimental measurements. For homogeneous turbulent shear flows special types of flow also satisfy this conditions of pseudo-similarity.

For general shear turbulent flows, two methods to solve the problem have been developed without resorting to solve the equations of mean motion and turbulent fluctuation directly. The first method follows the line initiated in the above mentioned 1940 paper[1]. Due to the non-linearity of the equations of turbulent fluctuation, the dynamical equations of the correlations thus built are not closed, similar to the equations of mean motion, containing the unknown Reynolds stress which is a turbulent velocity correlation of the second order mentioned before. To overcome this difficulty in this method is to assume a relation between the velocity correlation of a given order with that of a lower one. Further hypotheses upon the pressure gradient and velocity fluctuation correlations and the terms in viscous decay have to be assumed in order to obtain definite solutions of the given turbulence problem. This line of approach was subsequently developed by Rotta (1951)[3] and a number of investigators in the sixties and seventies, especially after the computer was invented[4].

The second method of approach is to consider the derivation of the equations of mean motion, the building of the turbulent velocity correlations of successive orders and seeking the solutions of their corresponding dynamical equations as a method of successive approximation. This point of view was first brought out in 1945[2] and further developed in a recent paper[5]. Here we must point out that the condition of pseudo-similarity for the homogeneous isotropic turbulence has to be generalized for the general shear turbulent flows. The turbulent pressure gradient and velocity correlations and the dissipation terms in the dynamical equations of the double correlation have been treated before[2], while those for the equations of the triple and quadruple correlations can be treated in a similar way.

This second method of solution theoretically has the advantage over the first for being able to obtain velocity correlations of higher orders, if we carry out the process of successive approximation further, thereby yielding solutions to general turbulent shear flows which satisfy the condition of pseudo-similarity.

As the first order approximation we have used the equations of mean motion and the equations of the double velocity correlation to solve the problems of the flow in a channel, the plane wake[5], the plane and axial jets[6], while terms involving triple velocity correlations have been neglected. Agreeements between the theoretical values of the mean velocities and double velocity correlations calculated with experiments are satisfactory.

In the present paper the channel flow has been recalculated in the first order approximation by one of us, Shi-Yi Chen. Agreement between theory and experiment had been improved than the previous calculation[5]. We also give the solution of the plane wake problem to the next order of approximation by using the dynamical equations for the velocity correlations of the triple and quadruple orders, while neglecting the fifth order correlation. Agreements between the calculations and the existing experimental data for the components of the triple velocity correlation are satisfactory. Since the present theory has given the values of those of the quadruple correlations which are not yet known, experiments can be carried out to test their validity.

1. EQUATIONS OF MOTION

We first put down the equations of motion and start with the Navier-Stokes equations and the equation of continuity for incompressible fluids:

$$\frac{\partial u_i}{\partial t} + u^j u_{i,j} = -\frac{1}{\rho} p_{,i} + \nu \nabla^2 u_i,$$
$$u^j_j = 0.$$
(1.1)

in which u_i is the velocity vector, p, the pressure, and ν, the kinematic coefficient of viscosity. Since we are using the rectangular system of coordinates x^i, the contravariant vector u^i is the same as its covariant form u_i and the comma sign followed by the coordinate x^j under u_i like $u_{i,j}$ denotes the covariant partial differentiation of u_i with respect to the coordinate x^j.

Following Reynolds, the vector u_i and pressure p in a fully developed turbulent flow can be separated into the mean motion and the turbulent fluctuation:

$$u_i = U_i + w_i, \qquad p = \bar{p} + \omega$$
(1.2)

in which U_i and $\bar{p}$ are the mean values of the velocity u_i and pressure p respectively, while w_i and ω are their turbulent fluctuations of which the mean values are both zero:

$$\bar{w}_i = 0, \qquad \bar{\omega} = 0.$$

By introducing (1.2) into (1.1) and taking the average, we obtain the Reynolds equations and the equation of continuity for the mean motion,

$$\frac{\partial U_i}{\partial t} + U^j U_{i,j} = -\frac{1}{\rho} \bar{p}_{,i} + \frac{1}{\rho} \tau^j_{i,j} + \nu \nabla^2 U_i,$$
$$U^j_j = 0,$$
(1.3)

with Reynolds stress,

$$\tau^j_i = -\rho \overline{w_i w^j}.$$

By subtracting the Reynolds equations and the equations of continuity from the Navier-Stokes equations (1.1), we obtain the dynamical equations and the equation of continuity for the turbulent velocity fluctuation:

$$\frac{\partial w_i}{\partial t} + U^j w_{i,j} + w^j w_{i,j} + w^j U_{i,j}$$
$$= -\frac{1}{\rho} \omega_{,i} - \frac{1}{\rho} \tau^j_{i,j} + \nu \nabla^2 w_i,$$
$$w^j_j = 0.$$
(1.4)

From the equations of turbulent velocity fluctuation we can derive the dynamical equations of the double, triple and quadruple velocity correlations as follows:

$$\frac{\partial \overline{w_i w_k}}{\partial t} + U_{i,j}\overline{w^j w_k} + U_{k,j}\overline{w^j w_i} + U^j(\overline{w_i w_k})_{,j} + \overline{(w^j w_i w_k)}_{,j}$$
$$= -\frac{1}{\rho}(\overline{\omega_{,i} w_k} + \overline{\omega_{,k} w_i}) + \nu \nabla^2 \overline{w_i w_k} - 2\nu g^{mn} \overline{w_{i,m} w_{k,n}}, \quad (1.5)$$

$$\frac{\partial \overline{w_i w_k w_l}}{\partial t} + U_{i,j}\overline{w^j w_k w_l} + U_{k,j}\overline{w^j w_l w_i} + U_{l,j}\overline{w^j w_i w_k}$$
$$+ U^j(\overline{w_i w_k w_l})_{,j} + \overline{(w^j w_i w_k w_l)}_{,j}$$
$$= -\frac{1}{\rho}(\overline{\omega_{,i} w_k w_l} + \overline{\omega_{,k} w_l w_i}) + \overline{\omega_{,l} w_i w_k}) \quad (1.6)$$
$$+ \overline{(w^j w_i)}_{,j}\overline{w_k w_l} + \overline{(w^j w_k)}_{,j}\overline{w_l w_i} + \overline{(w^j w_l)}_{,j}\overline{w_i w_k}$$
$$+ \nu g^{mn}(\overline{w_i w_k w_l})_{,mn}$$
$$- 2\nu g^{mn}(\overline{w_{i,m} w_{k,n} w_l} + \overline{w_{k,m} w_{l,n} w_i} + \overline{w_{l,m} w_{i,n} w_k}),$$

$$\frac{\partial \overline{w_i w_k w_l w_p}}{\partial t} + U_{i,j}\overline{w^j w_k w_l w_p} + U_{k,j}\overline{w^j w_l w_p w_i} + U_{l,j}\overline{w^j w_p w_i w_k}$$
$$+ U_{p,j}\overline{w^j w_i w_k w_l} + U^j(\overline{w_i w_k w_l w_p})_{,j} + \overline{(w^j w_i w_k w_l w_p)}_{,j}$$
$$= -\frac{1}{\rho}(\overline{\omega_{,i} w_k w_l w_p} + \overline{\omega_{,k} w_l w_p w_i} + \overline{\omega_{,l} w_p w_i w_k} + \overline{\omega_{,p} w_i w_k w_l})$$
$$+ \overline{(w^j w_i)}_{,j}\overline{w_k w_l w_p} + \overline{(w^j w_k)}_{,j}\overline{w_l w_p w_i} + \overline{(w^j w_l)}_{,j}\overline{w_p w_i w_k}$$
$$+ \overline{(w^j w_p)}_{,j}\overline{w_i w_k w_l} + \nu g^{mn}(\overline{w_i w_k w_l w_p})_{,mn}$$
$$- 2\nu g^{mn}(\overline{w_{i,m} w_{k,n} w_l w_p} + \overline{w_{k,m} w_{l,n} w_i w_p} + \overline{w_{l,m} w_{p,n} w_i w_k}$$
$$+ \overline{w_{i,m} w_{l,n} w_k w_p} + \overline{w_{k,m} w_{p,n} w_i w_l} + \overline{w_{i,m} w_{p,n} w_k w_l}). \quad (1.7)$$

In the above differential equations $\omega_{,i}$ is the solution of the Poisson equation obtained before[2] and g_{ik} is the metric tensor. In the rectangular system of coordinates $g_{ik} = 1$, for $i = k$, and $g_{ik} = 0$, for $i \neq k$.

For the wall-bound and free turbulent flows like the channel, wakes and jets, we choose the solution of the equation of velocity turbulent fluctuation (1.4) of the type[5]:

$$w_i = q\phi_i(\frac{x}{\lambda}, t) \quad (1.8)$$

with

$$q^2 = \overline{w_j w^j} = q^2(x, t), \qquad \lambda = \lambda(x, t).$$

Here q is the magnitude of the velocity fluctuation and λ is the generalized Taylor's microscale of turbulence.

The velocity correlations of the second, third, and quadruple orders between two distinct points P and P' from (1.8) can be written as:

$$\begin{aligned}
\overline{w'^n w_i} &= \overline{w^n w_i} + q^2 \phi_i^n(\eta, x, t), \\
\overline{w'^m w'^n w_i} &= \overline{w^m w^n w_i} + q^3 \psi_i^{mn}(\eta, x, t), \\
\overline{w'^n w_i w_k} &= \overline{w^n w_i w_k} + q^3 \phi_{ik}^n(\eta, x, t), \\
\overline{w'^m w'^n w_i w_k} &= \overline{w^m w^n w_i w_k} + q^4 \psi_{ik}^{mn}(\eta, x, t), \\
\overline{w'^n w_i w_k w_l} &= \overline{w^n w_i w_k w_l} + q^4 \phi_{ikl}^n(\eta, x, t).
\end{aligned} \quad (1.9)$$

In above equations we have

$$\xi^i = x'^i - x^i, \qquad \eta^i = \frac{\xi^i}{\lambda}. \quad (1.10)$$

The functions ϕ_i^n, ψ_i^{mn}, ϕ_{ik}^n, ψ_{ik}^{mn} and ϕ_{ikl}^n all vary rapidly with ξ^i, the coordinate difference between the two points P and P', and vary slowly with x^i. Hence partial derivatives of the functions defined in (1.9) with respect to ξ^i are much greater than those with respect to x^i.

The turbulent pressure gradient and velocity fluctuation correlations, and the dissipation terms in the dynamical equations of correlations in (1.5), (1.6) and (1.7) are written below. Their calculations are explained in the Appendix.

For the pressure gradient and velocity fluctuation correlation we have:

$$\frac{1}{\rho}(\overline{\omega_{,i} w_k} + \overline{\omega_{,k} w_i}) = q^2 a'^n_{mik} U^m_{,n} + q^2 \lambda a'^{nr}_{mik} U^m_{,nr} + \frac{q^3}{\lambda} b'_{ik} \quad (1.11)$$

in which

$$\begin{aligned}
a'^n_{mik} &= \frac{1}{2\pi} \iiint [\phi^n_{i,mk} + \phi^n_{k,mi}] \frac{1}{r} dV', \\
a'^{nr}_{mik} &= \frac{1}{2\pi\lambda} \iiint \{\phi^n_{i,m} \delta^r_k + \phi^n_{k,m} \delta^r_i + \xi^r [\phi^n_{i,mk} + \phi^n_{k,mi}]\} \frac{1}{r} dV', \\
b'_{ik} &= \frac{\lambda}{4\pi} \iiint [\psi^{mn}_{i,mnk} + \psi^{mn}_{k,mni}] \frac{1}{r} dV';
\end{aligned} \quad (1.12)$$

$$\frac{1}{\rho}(\overline{\omega_{,i} w_k w_l} + \overline{\omega_{,k} w_l w_i} + \overline{\omega_{,l} w_i w_k}) = q^3 b'^n_{mikl} U^m_{,n} \\
+ q^3 \lambda b'^{nr}_{mikl} U^m_{,nr} + \frac{q^4}{\lambda} c'_{ikl}, \quad (1.13)$$

in which

$$b'^n_{mikl} = \frac{1}{2\pi} \iiint [\phi^n_{ik,ml} + \phi^n_{kl,mi} + \phi^n_{li,mk}] \frac{1}{r} dV',$$

$$b'^{nr}_{mikl} = \frac{1}{2\pi\lambda} \iiint \{\phi^n_{ik,m}\delta^r_l + \phi^n_{kl,m}\delta^r_i + \phi^n_{li,m}\delta^r_k$$

$$+ \xi^r [\phi^n_{ik,ml} + \phi^n_{kl,mi} + \phi^n_{li,mk}]\} \frac{1}{r} dV', \qquad (1.14)$$

$$c'_{ikl} = \frac{\lambda}{4\pi} \iiint [\psi^{mn}_{kl,mni} + \psi^{mn}_{li,mnk} + \psi^{mn}_{ik,mnl}] \frac{1}{r} dV';$$

$$\frac{1}{\rho}(\overline{\omega_{,i}w_k w_l w_p} + \overline{\omega_{,k}w_l w_p w_i} + \overline{\omega_{,l}w_p w_i w_k} + \overline{\omega_{,p}w_i w_k w_l})$$

$$= q^4 c'^n_{miklp} U^m_{,n} + q^4 \lambda c'^{nr}_{miklp} U^m_{,nr} + \frac{q^5}{\lambda} d'_{iklp}, \qquad (1.15)$$

in which

$$c'^n_{miklp} = \frac{1}{2\pi} \iiint [\phi^n_{ikl,mp} + \phi^n_{klp,mi} + \phi^n_{lpi,mk} + \phi^n_{pik,ml}] \frac{1}{r} dV',$$

$$c'^{nr}_{miklp} = \frac{1}{2\pi\lambda} \iiint \{\phi^n_{ikl,m}\delta^r_p + \phi^n_{klp,m}\delta^r_i + \phi^n_{lpi,m}\delta^r_k + \phi^n_{pik,m}\delta^r_l$$

$$+ \xi^r [\phi^n_{ikl,mp} + \phi^n_{klp,mi} + \phi^n_{lpi,mk} + \phi^n_{pik,ml}]\} \frac{1}{r} dV', \qquad (1.16)$$

$$d'_{iklp} = 0.$$

All the partial differentiations under the above integrals are taken with respect to the coordinate ξ^i and all the tensors a', b', c', d' with various ranks defined integrals (1.12) - (1.16) are approximately constants proved before[2].

The viscous dissipation terms in (1.5), (1.6) and (1.7) are given by (cp. Appendix)

$$2\nu g^{mn} \overline{w_{i,m} w_{k,n}} = -\frac{2\nu}{3\lambda^2}(k-5)q^2 \delta_{ik} + \frac{2\nu k}{\lambda^2}\overline{w_i w_k}, \qquad (1.17)$$

$$2\nu g^{mn}(\overline{w_{i,m} w_{k,n} w_l} + \overline{w_{k,m} w_{l,n} w_i} + \overline{w_{l,m} w_{i,n} w_k})$$

$$= \frac{2\nu}{\lambda^2}\{c_1[\overline{w_i w_j w^j}\delta_{kl} + \overline{w_k w_j w^j}\delta_{li} + \overline{w_l w_j w^j}\delta_{ik}] + c_2 \overline{w_i w_k w_l}\}, \qquad (1.18)$$

$$2\nu g^{mn}(\overline{w_{i,m} w_{k,n} w_l w_p} + \overline{w_{k,m} w_{l,n} w_i w_p} + \overline{w_{l,m} w_{p,n} w_i w_k}$$

$$+ \overline{w_{i,m} w_{l,n} w_k w_p} + \overline{w_{k,m} w_{p,n} w_i w_l} + \overline{w_{i,m} w_{p,n} w_k w_l})$$

$$= \frac{2\nu}{\lambda^2}\{d_1[\overline{w_i w_k w_m w^m}\delta_{lp} + \overline{w_k w_l w_m w^m}\delta_{ip} + \overline{w_l w_p w_m w^m}\delta_{ik}$$

$$+ \overline{w_i w_l w_m w^m}\delta_{kp} + \overline{w_k w_p w_m w^m}\delta_{il} + \overline{w_i w_p w_m w^m}\delta_{kl}]$$

$$+ d_2 \overline{w_i w_k w_l w_p}\}. \qquad (1.19)$$

2. THEORY OF HOMOGENEOUS ISOTROPIC TURBULENCE AND THE CONDITION OF PSEUDO-SIMILARITY

In a wind-tunnel the mean velocity of the air-stream behind the grid is constant. For an observer moving with this mean velocity, the equations of motion for the turbulent fluctuation (1.4) are the same as the Navier-Stokes equations of motion. The theory of homogeneous isotropic turbulence is based upon the vortex solution of the Navier-Stokes equations which forms the element of turbulence. This vortex is axially symmetrical and is randomly distributed as regards to its position and orientation of its axis of symmetry in the turbulent fluid. Then the turbulent velocity correlation of any order at one point or between two distinct points can be calculated by averaging the products of the velocity components at two points within the vortex over its position and the orientation of its axis in the fluid.

But the Navier-Stokes equations are non-linear partial differential equations of the second order and have different kinds of solutions. The theory was first applied to the case in the final period of decay in which the non-linear terms in the Navier-Stokes equations can be neglected[7], and the similarity condition was assumed to select the right kind of vortex to represent the turbulence element. The already well-known double velocity correlation could be thus explained. Furthermore, the theory predicted the triple velocity correlation which was published in 1965[8] and verified experimentally by Bennett and Corrsin ten years later[9].

For very high Reynolds number flows in the initial period of decay near the grid in the wind-tunnel, there are solutions of the Navier-Stokes equations. By using another condition of similarity a different vortex solution was obtained, yielding the law of decay of turbulence and the spread of Taylor's scale of micro-turbulence λ in agreement with experiments. The double and triple velocity correlations thus calculated have also qualitative agreements with measurements[10].

The condition of pseudo-similarity was proposed to correlate the two kinds of similarity in the initial and final periods of decay for isotropic homogeneous turbulence[11,12]. A large number of theoretical calculations of the decay of turbulent energy, the spread of Taylor's scale of micro-turbulence λ, the double and triple velocity correlations from the initial to the final period of decay, together with the three-dimensional and one-dimensional energy spectrum functions and the energy trasfer spectrum function etc. all agree with experimental measurements very well.

This condition of pseudo-similarity for homogeneous isotropic turbulence can be written in terms of λ and R_λ, the turbulence Reynolds number, as follows[5]:

$$\frac{\lambda}{\nu}\frac{d\lambda}{dt} = \frac{1}{R_0}R_\lambda^2 + 2, \qquad (2.1)$$

with

$$R_\lambda = \frac{q\lambda}{\nu}, \qquad (2.2)$$

in which, R_0, a Reynolds number, is a constant. In this relation when R_λ is large, namely, in the initial period of decay, the nember 2 in (2.1) can be ignored. Then we have λ^2 and q^2 satisfying,

$$\lambda^2 = 10\nu t, \qquad q^2 \sim \frac{1}{t}. \qquad (2.3)$$

In the final period of decay, R_λ in (2.1) can be neglected and we have

$$\lambda^2 = 4\nu t, \qquad q^2 \sim \frac{1}{t^{5/2}}. \qquad (2.4)$$

The experimental proof of the condition (2.1) from the initial to the final period of decay has been carried out in the turbulence wind-tunnel of Peking University (to be published).

For general shear turbulent flows the condition of pseudo-similarity (2.1) can be put in the following form,

$$\frac{\lambda}{\nu}\left(\frac{\partial \lambda}{\partial t} + U^j \lambda_{,j}\right) = \frac{1}{R_0}\{R_\lambda^2 - \left[R_1 + \frac{\kappa_1}{\nu}\lambda^2 \sqrt{g^{kl}U_{,k}^j U_{j,l}}\right.$$
$$\left. + \frac{\kappa_2}{\nu}\lambda^3 \sqrt{g^{ij}g^{kl}g^{mn}\Omega_{ik,m}\Omega_{jl,n}}\right]^2 + 2R_0\}, \qquad (2.5)$$

where $\Omega_{ik} = U_{i,k} - U_{k,i}$, and κ_1, κ_2, R_0 and R_1 are constants, and q and R_λ are defined by (1.8) and (2.2) respectively.

In the first order approximation for the channel flow, the wake and jets, the equations of mean motion (1.3), the double velocity correlation (1.5) and the condition of pseudo-similarity (2.5) have been used, while terms in the triple correlation have been ignored[5].

3. THE TURBULENT CHANNEL FLOW

The turbulent channel flow was calculated in the first order approximation before[5]. Here is the result of Shi-Yi Chen's recalculation by using a computer. The theoretical difference between the two calculations is that in the former calculation the viscous term in the equation of mean motion (1.3) was neglected, while at present it is retained. The terms in viscosity, the viscous decay and the triple correlation in the dynamical equations of double correlation (1.5) are also neglected as before. The agreements of the calculated values of the mean velocity and the mean values of the squares of the three turbulent velocity fluctuations with Laufer's measurements[13] are all well (cf. Fig. 1 - 5) for the Reynolds number $R_e = 12,300$ flow. since there are no measurements of the scale of micro-turbulence λ for this flow in Laufer's experiment we only give ites theoretical value in Fig. 6.

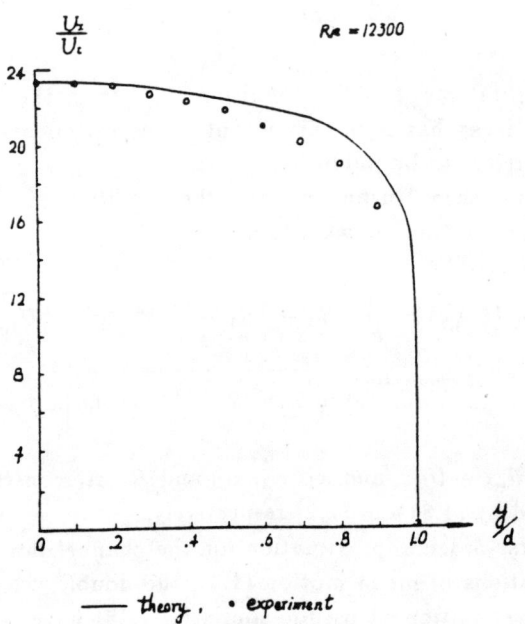

—— theory , • experiment

Fig. 1

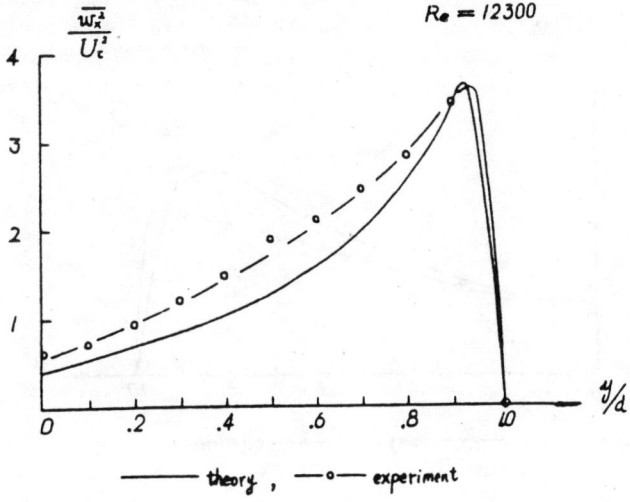

Fig. 2

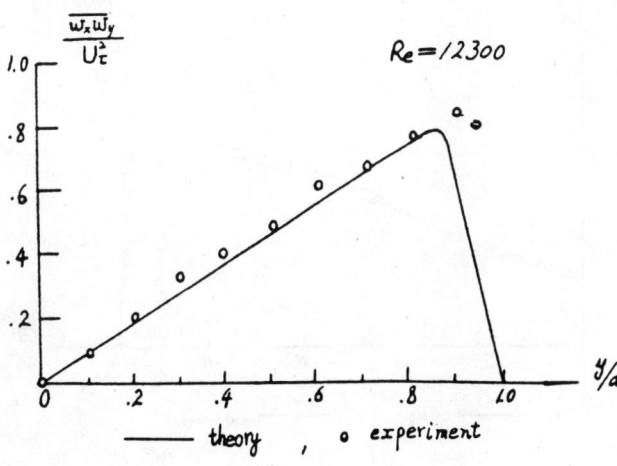

Fig. 3

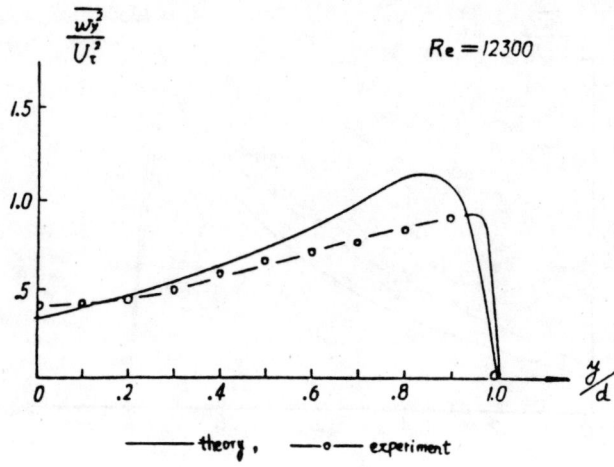

Fig. 4

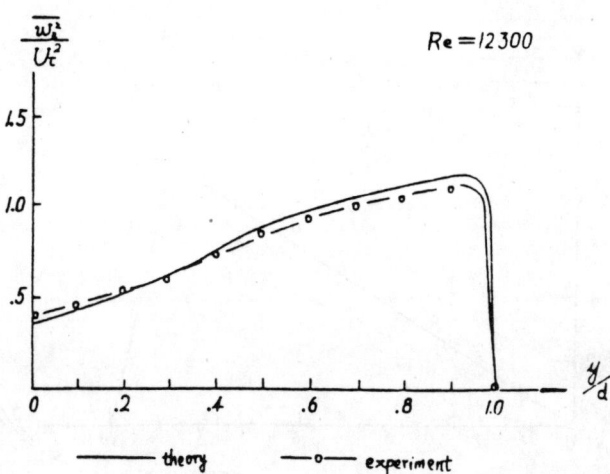

Fig. 5

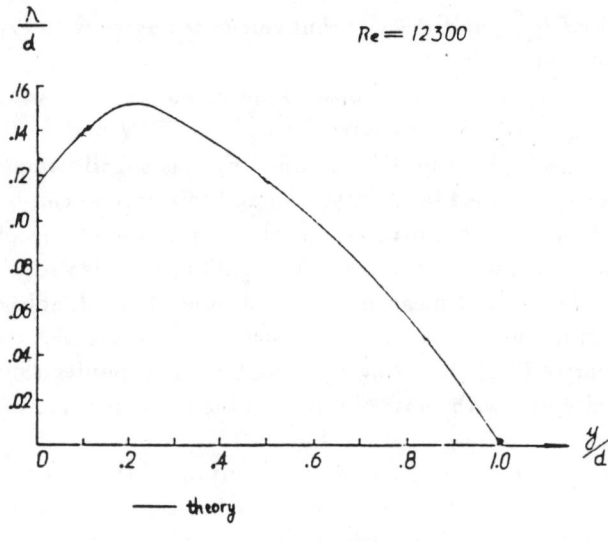

Fig. 6

4. THE PLANE TURBULENT WAKE

Now we consider the problem of the plane turbulent wake in which there are already measurements of the triple velocity correlation. For the second order approximation, besides the equations of mean motion and the double velocity correlation, and the condition of pseudo-similarity, we have in addition to use the dynamical equations of the triple and quadruple orders (1.6) and (1.7). The rigorous method of solution is to solve simultaneously for the mean velocity U_i, the double, triple and quadruple velocity correlations and the turbulence scale λ from all the equations, (1.3), (1.5), (1.6), (1.7), and (2.5) with (1.11) - (1.19), while neglecting the terms in the quintuple correlation. But this method of solution is complicated.

A simple method of approach is to find the solution by successive substitution. This method was used in the first order approximation of the solution for the plane wake[14]. To start with in the first step we find the approximate expression of the Reynolds shear τ_{xy} to be proportional to the gradient of the mean velocity $\partial U/\partial y$, a result which was obtained before[15]. This gives us $\overline{w_1 w_2}^{(0)}$. Then the mean velocity denoted by $U_i^{(0)}$ is solved and the other components of the double velocity correlation, $\overline{w_i w_k}^{(0)}$, and the scale of turbulence $\lambda^{(0)}$ can be obtained from the equations of the double velocity correlation and the condition of pseudo-similarity, the terms in the triple correlation being ignored. The theoretical values of the above functions $U_i^{(0)}$, and $\overline{w_i w_k}^{(0)}$ thus calculated agree with experimental measurements very well.

The next step in this first order approximation is to use the above values of $\overline{w_i w_k}^{(0)}$ and $\lambda^{(0)}$ to solve for $U_i^{(1)}$, $\overline{w_i w_k}^{(1)}$ and $\lambda^{(1)}$. The new functions $U_i^{(1)}$, $\overline{w_i w_k}^{(1)}$ and $\lambda^{(1)}$ obtained by this substitution differ very littel from those obtained in the first step and this process can be repeated.

In this first order approximation for the solution of the plane wake, Shi-Yi Chen, who carried out this investigation, has also solved simultaneously the equations of mean motion (1.3) and of the double correlation (1.5), neglecting the terms in triple correlation, under the condition of pseudo-similarity (2.5), by using a computer. The results thus obtained also agree quite well with those obtained in the above method of successive substitution.

The rigorous method of finding the solution of the plane wake in the second order approximation, as mentioned before, is to solve simultaneously the equations of mean motion, of the double, triple and quadruple correlations together with the condition of pseudo-similarity. To avoid this, we use again the method of successive substitution. Since in the first order approximation we have already obtained values of the mean velocity and double correlation which agree with experiment quite well, we can use them as known values and put them into the equations of $\overline{w_i w_k w_l}$ and $\overline{w_i w_k w_l w_p}$ and look for their solutions.

But to solve the equations for the triple and quadruple correlations together is still complicated. Hence we can separate their solutions by assuming an often used relation between the quadruple correlation with the double correlation in homogeneous isotropic turbulence based upon the quasi-normal distribution condition, which is given by:

$$\overline{w_i w_k w_l w_p} = \overline{w_i w_k}\,\overline{w_l w_p} + \overline{w_i w_l}\,\overline{w_p w_k} + \overline{w_i w_p}\,\overline{w_k w_l} \qquad (4.1)$$

and solve the equations of triple correlation first. All of the six non-vanishing components of the triple correlation for the plane wake thus computed agree with experiment very well (cp. Figs. 7 - 12)[16]. In these figures, for comparison, we have also put down the theoretical curves according to the closure condition put forth by Launder, Reece, and Rodi[17] for the triple correlation,

$$\overline{w_i w_j w_k} = -C \frac{q^2}{\epsilon}(\overline{w_i w_{j,l} w^l w_k} + \overline{w_j w_{k,l} w^l w_i} + \overline{w_k w_{i,l} w^l w_j}), \qquad (4.2)$$

with $\qquad \epsilon = -2\nu g^{mn} g^{rs} \overline{w_{r,m} w_{s,n}}, \qquad C = 0.055,$

in which ϵ is derived from (1.17).

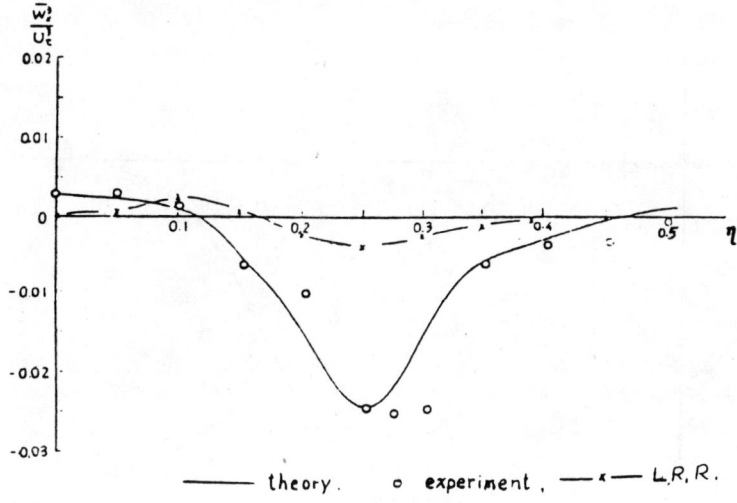

Fig. 7

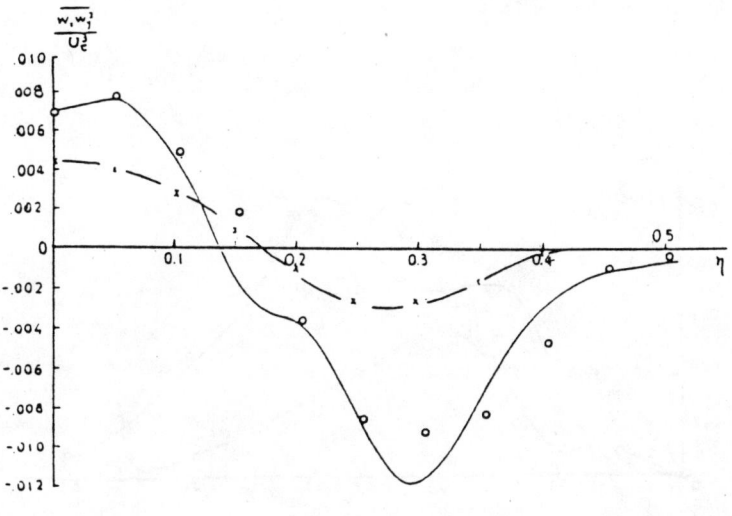

Fig. 8

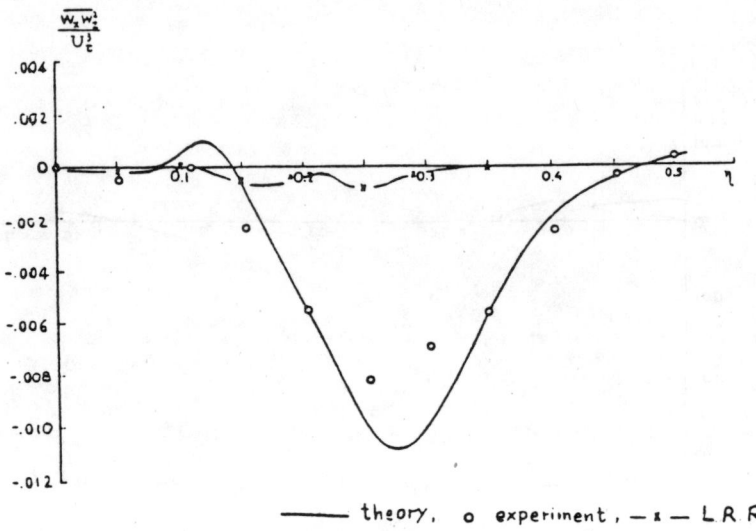

——— theory, ○ experiment, — × — L.R.R

Fig. 9

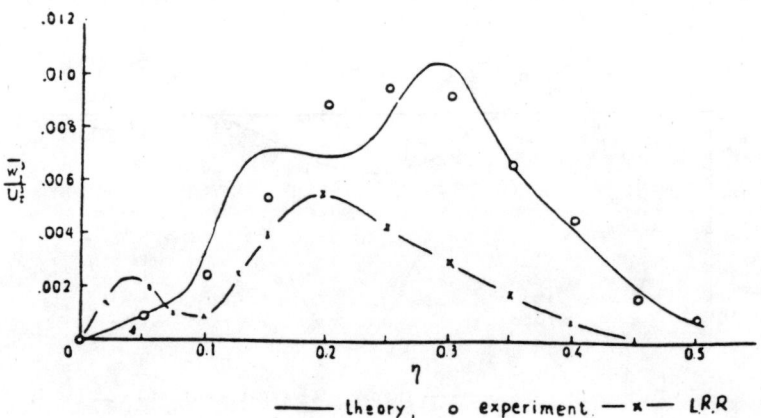

——— theory, ○ experiment — × — L.R.R

Fig. 10

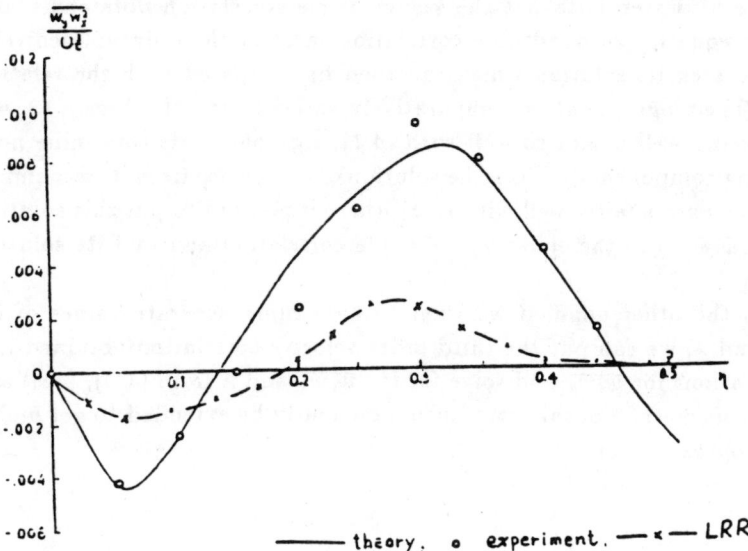

Fig. 11

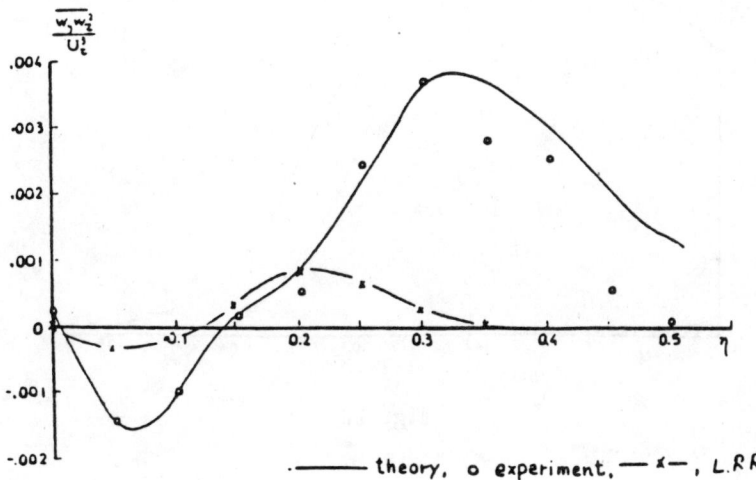

Fig. 12

The next step is to put the known triple correlation obtained above into the equation of quadruple correlation, neglect the quintuple correlation and seek its solution which can then be compared with the relation (4.1). Their agreement is comparatively satisfactory (cf. Figs. 13, 14, 15, agreeing well or not so well with (4.1), e.g., out of its total nine non-vanishing components). Since the solution of $\overline{w_i w_k w_l w_p}$ from its dynamical equations agrees fairly well with (3.1), there is no need to put this solution of $\overline{w_i w_k w_l w_p}$ into the equations of triple correlation and find its solution again.

On the other hand, if we want to have more accurate values of U_i, $\overline{w_i w_k}$ and λ, we can put the third order velocity correlation obtained into the equations for $\overline{w_i w_k}$ and solve for U_i, $\overline{w_i w_k}$ and λ from (1.3), (1.5) and (2.5). This method of substitution can obviously be extended to get higher order approximations.

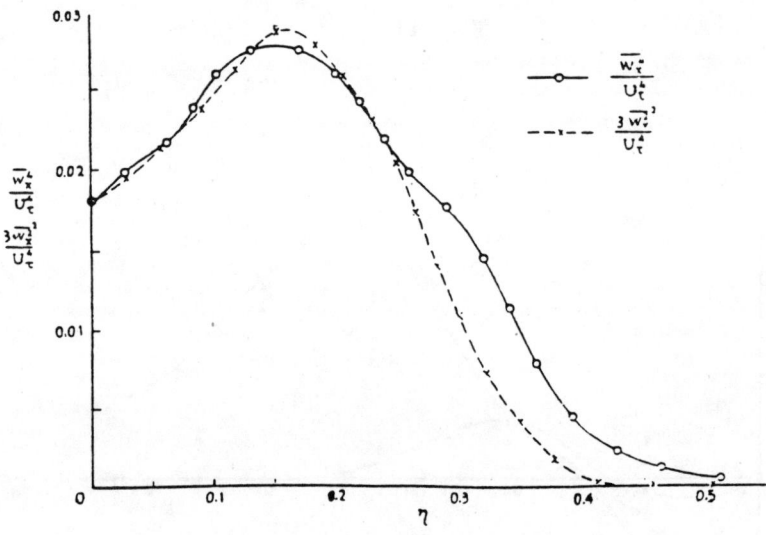

Fig. 13

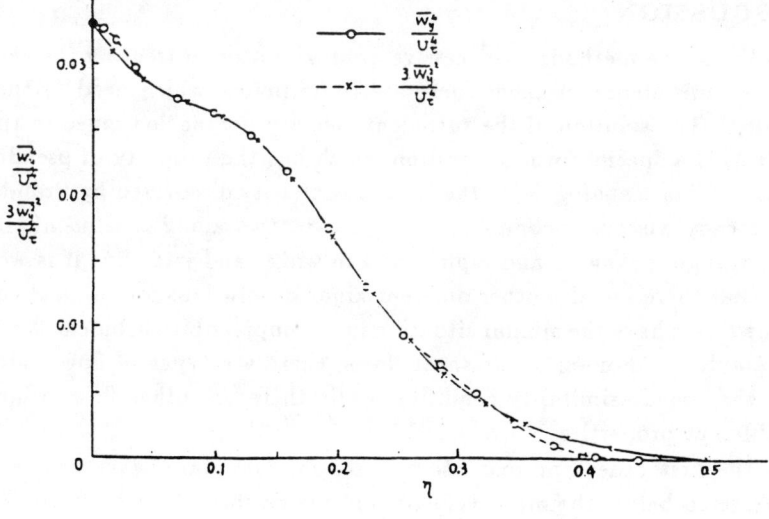

Fig. 14

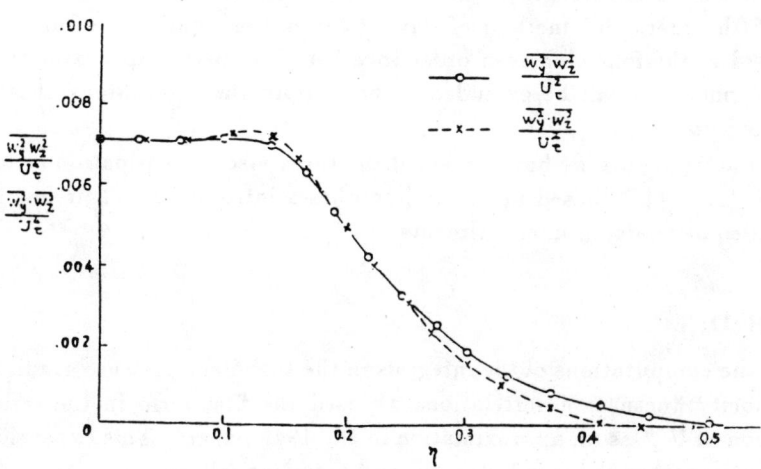

Fig. 15

5. DISCUSSION

In the above methods of successive approximation in treating the general shear turbulence problem there are some points which need further discussion. The solution of the turbulent velocity fluctuation given in the form (1.8) is a special form of solution which has the property of pseudo-similarity. This is analogous to the kind of similarity discovered by Prandtl in the steady viscous incompressible laminar flows along a semi-infinite plane, through a channel and a pipe, and in wakes and jets. But it is well known that there are also other different kinds of solutions for laminar viscous flows. We have the similar situation in incompressible turbulent flows. For example, for homogeneous shear flows, there are types of flow which satisfy the pseudo-similarity condition while there are other flows which have different properties[18].

In the first order approximation in solving the plane wake problem, we considered before the mean velocity and the double velocity correlation together, neglected the terms in the triple correlation and found their solutions from their equations under the condition of pseudo-similarity. Likewise, in the second order approximation we now consider the correlations of the triple and quadruple orders together, add their dynamical equations to the first set and find their solutions, while neglecting correlation terms of the fifth order. this method of solving correlation equations of an odd order and of the following even order together gives better approximation of the former and can be extended to find approximate solutions of still higher orders.

In the Appendix we have obtained the three viscous dissipation terms in (1.5), (1.6), (1.7) based upon the hypotheses introduced. Their justification also depends upon experiments.

APPENDIX

In the computations of the integrals in the turbulent pressure gradient and velocity fluctuation correlations we used the first term in the series expansion of $U'^{m}_{,n}$ as an approximation in the 1945 paper[2]. This expansion has been improved to consider the second term[5] as follows:

$$U'^{m}_{,n} = U^{m}_{,n} + \xi^{r} U^{m}_{,nr}. \tag{1}$$

Then in (1.5) we have

$$[U'^{m}_{,n}(\overline{w'^{n} w_i})'_{,m}]'_{,k} = U^{m}_{,n}(\overline{w'^{n} w_i})'_{,mk} \\ + U^{m}_{,nr}[(\overline{w'^{n} w_i})'_{,m}\delta^{r}_{k} + \xi^{r}(\overline{w'^{n} w_i})'_{,mk}. \tag{2}$$

Similar in (1.6) and (1.7), we have respectively

$$[U_{,n}^{\prime m}(\overline{w'^n w_k w_l})'_{,m}]'_{,i} = U_{,n}^{m}(\overline{w'^n w_k w_l})'_{,mi} \\ + U_{,nr}^{m}[(\overline{w'^n w_k w_l})'_{,m}\delta_i^r + \xi^r(\overline{w'^n w_k w_l})'_{,mi}], \quad (3)$$

$$[U_{,n}^{\prime m}(\overline{w'^n w_k w_l w_p})'_{,m}]'_{,i} = U_{,n}^{m}(\overline{w'^n w_k w_l w_p})'_{,mi} \\ + U_{,nr}^{m}[(\overline{w'^n w_k w_l w_p})'_{,m}\delta_i^r + \xi^r(\overline{w'^n w_k w_l w_p})'_{,mi}]. \quad (4)$$

Putting the above relations into the integrals of (1.5), (1.6) and (1.7) respectively and considering the functions defined in (1.9), we then obtain (1.11), (1.12), (1.13), (1.14), (1.15) and (1.16). Here we note that the constant d'_{iklp} in (1.16) involves differentiations leading to macro-lengths in the denominator and quintuple correlations under its integral. Therefore d'_{iklp} can be set equal to zero.

We found before[2] the viscous dissipation terms in the equations of double correlation (1.5) by assuming that the double velocity correlation between two points P and P', $\overline{w_i w'_k}$, could be expanded in powers of ξ^j and the coefficients of $\xi_l \xi_m$ in the expansion should be linear functions of $\overline{w_n w_p}$, δ_{rs} and their products. The double correlation $\overline{w_i w'_k}$ should furthermore satisfy the equation of continuity,

$$(\overline{w_i w'^j})_{,j} = 0. \quad (5)$$

From the $\overline{w_i w'_k}$ the dissipation term in (1.17) was obtained by taking partial differentiation of $\overline{w_i w'_k}$ with respect to x^m and x'^n, setting x'^n equal to x^n and then contracting by g^{mn}.

The viscous dissipation terms in (1.18) and (1.19) can be obtained in the same way. For the terms in (1.18) we assume that in the expansion of $\overline{w_i w_k w'_l}$ in powers of ξ^j, the coefficients of $\xi_m \xi_n$ should be linear combinations of $\overline{w_p w_q w_r}$, δ_{st} and their products. Likewise $\overline{w_i w_k w'_l}$ should satisfy the equation of continuity,

$$(\overline{w_i w_k w'^l})_{,l} = 0. \quad (6)$$

By the method used to obtain (1.17), we can get (1.18).

Formula (1.19) can be obtain by the same method and similar assumption explained for getting the dissipation terms in the equations of the double and triple correlations in (1.17) and (1.18) respectively. This method can be extended to find similar viscous dissipation terms for still higher approximations.

References

1. Chou, P. Y., *Chin. Journ. of Phys.* **4**, 1-33 (1940).
2. Chou, P. Y., *Quart. of Appl. Math.* **3**, 38-54 (1945).
3. Rotta, J. C., *Zeit. für Phys.* **129**, 547-572 (1951); **131**, 51-77 (1951).
3. Launder,B. E., "Stress Transport Closure-into the Third Generation", *Turbulent Shear Flows*, 259-266 (1979). Springer-Verlag (Berlin) for the references there in.
5. Zhou (Chou), Pei-yuan, *Scientia Sinica (A)* **28**, 405-421 (1985).
6. Wu, Z., "The Solution of the Turbulent Plane and Axial Jets under the Condition of Pseudo-Similarity", M. S. Thesis, Dept. of Mechanics, Peking University, Beijing, China. 1986.
7. Chou, P. Y. & Tsai, S. T., *Acta Mechanica Sinics (Chin. Journ. of Mech.)* **1**, 3-14 (1957).
8. Huang, Y. N., *Acta Mechanica Sinica* **8**, 122-132 (1965).
9. Bennett, J. C. & Corrsin, S., *Phys. of Fluids* **21** 2129-2140 (1978).
10. Chou, P. Y., Shih, H. K. & Li, S. N., *Acta Scientiarum Naturalium Universitatis Pekinensis* **10**, 39-52 (1965).
11. Chou, P. Y. & Huang, Y. N., *Scientia Sinica* **18**, 199-222 (1975).
12. Huang, Y. N. & Zhou (Chou), P. Y., *Scientia Sinica* **24**, 1207-1230 (1981); also *Proc. Indian Acad. of Sci. (Eng. Sc.)* **4** 177-197 (1981).
13. Laufer, J., *NACA TN*, 2123 (1950).
14. Chen, S. Y., "The Solution of the Plane Turbulent Wake under the Condition of Pseudo-Similarity", M. S. Thesis, Dept. of Mechanics, Peking University, Beijing, China. 1984.
15. Chou, P. Y., *Scientia Sinica* **8**, 1095-1119 (1959).
16. Fabris, G., *Phys. of Fluids* **26**, 422-427 (1983).
17. Launder,B. E., Reece, G. J. & Rodi, W., *Jour. of Fluid Mech.* **68**, 537-566 (1975).
18. Huang, Y. N., *Papers on Theoretical Physics and Mechanics in Commemoration of Professor Zhou (Chou) Pei-yuan's Eightieth Birthday*, 137-159, 1982, Science Press (Beijing);
    Turbulence and Chaotic Phenomena in Fluids (edited by T. Tatsumi), 357-364, 1984, North-Holland (Amsterdam).

NONLINEAR EULER PARTIAL DIFFERENTIAL EQUATIONS: SINGULARITIES IN THEIR SOLUTION

J. T. Stuart

*Department of Mathematics,
Imperial College, London*

Dedication

It is a pleasure to dedicate this lecture to C. C. Lin, who has done so much to foster applied mathematics and associated areas of engineering science and the physical sciences, and whose contributions to the theory of hydrodynamic stability and transition to turbulence have been so significant. The present paper has its application in that same field, and I would like to pay tribute to the influence which C. C. has had on my studies in fluid motion and stability, especially during two separate years spent at M. I. T. during the last three decades.

The problem of the evolution of wave motion in a boundary layer or shear flow, particularly in relation to the development of turbulence, is intimately connected with the natural presence of three-dimensional vortex structures in the flow. These longitudinal vortices, whose axes are aligned parallel to the main stream, convect, stretch and tilt the basic vorticity and create thereby local shear layers. It is from such local regions that turbulent "spots" are observed to emerge.

Associated with the nonlinear form of these structures is the mathematical possibility that the flow field can develop a singularity in a finite time. This line of thought is pursued here and it is shown definitely that this can happen, at least for particular flow structures. The possible relationship to the eruption and bursting of longitudinal filaments from turbulent sublayers is also discussed.

1. INTRODUCTION

The problem of transition to turbulence in a boundary layer has been with us for nearly 100 years but significant theoretical progress was initiated by Prandtl and his school and by Heisenberg more than 60 years ago.[1] Experimental confirmation of the existence of the theoretically-predicted wave motions came first in the 1940s with the beautiful experiments of of Schubauer and Skramstad at the National Bureau of Standards in Washington.[1] C. C. Lin[2] was thus stimulated to study further the theoretical problem and to produce his influential and comprehensive account of the subject, with its many nuances and ramifications substantially clarified. This and other work was later summarized by Lin in his monograph.[3]

Somewhat later Schubauer, Klebanoff and their colleagues at the National Bureau of Standards returned to an experimental attack,[1] involving studies both of turbulent "spots" and of three-dimensional aspects more generally, of the natural or artificially-stimulated vortex motions superimposed on waves in a boundary layer, their work being reinforced by that of Kovasznay and his colleagues.[1] Most particularly, and relevant to this paper, we note the discovery that relatively-weak secondary flows, with a strength no more than 1% or 2% of the mainstream velocity, play a very prominent role in the development of embryonic turbulent spots. It was this knowledge that led Benney and Lin[4] to study a theoretical model for the development of the three-dimensional secondary flows in the form of longitudinal vortices, this work being taken further by Benney.[5]

A concomitant of the three-dimensional vortex structure is the fact of local shear layers being stimulated to appear in the outer part of the boundary layers, at spanwise locations (or "peaks") where the secondary flow is directed away from the wall. Kovasznay[1] suggested that vortex stretching plays a crucial part in the development of such shear layers, and this idea was pursue by the writer[6] through a study of the Euler equations for flows which are independent of the streamwise co-ordinate x, but depend on both y and z and on t. In this work, the velocity component u in the x direction is given by a time-dependent linear partial differential equation of the first order, provided the y and z components of velocity (v, w) are already known. These given v and w components represent in fact the secondary flow, or longitudinal-vortex flow discovered by Klebanoff. It was shown[6] that the vortex-stretching process, which is intrinsic even to this linear problem, does lead to the development of local intense shear layers, at the appropriate "peak" locations and on the very short time scales of the experimental observations.

The idea of studying an equation made linear by x-independence, together with specification of the secondary flow in the $y - z$ plane, has since been exploited by Pearson and Abernathy,[7] Moore[8] and Yang[9] in the somewhat different context of a single longitudinal vortex but with inclusion of viscosity. Moreover, inviscid studies relevant to the boundary-layer transition problem have been made by Ellingsen and Palm[10] and by Landahl.[11]

The object of the present account is to relax the concept[6] of linearity of the u component of velocity, by allowing the velocity field to depend in some way on x. The three velocity components are then coupled, and it is natural to enquire as to the differences which will emerge, as compared with that earlier work. In particular, it is noted here that the local shear layer, which has already been described, becomes more and more intense as t becomes large. The flow, however, is not singular, except in the limit $t \to \infty$. The point of our enquiry is to find out whether this feature is modified and whether a singularity can develop in some way in a finite time. A preliminary account of this investigation was given in a lecture in Kyoto in 1983,[12] while further developments were reported in a lecture in 1984 to the London Mathematical Society.[13] During this symposium, moreover, Professor M. Landahl drew the author's attention to independent and related work of Russell and himself[14] of 1984. We now pursue our investigations in the following sections.

2. BASIC EQUATIONS: EULERIAN AND LAGRANGIAN DESCRIPTION

Consider a semi-infinite domain bounded by $y = 0$ with a main velocity component (u) parallel to the x coordinate. We shall suppose that there are also velocity components (v, w) parallel to the y and z (spanwise) axes respectively. The pressure is denoted by p and t is the time. In this situation the relevant Euler differential equations are

$$\frac{\partial u_i}{\partial t} + u_j \frac{\partial u_i}{\partial x_j} = -\frac{\partial p}{\partial x_i}; \quad \frac{\partial u_i}{\partial x_i} = 0 ; \qquad (2\text{-}1)$$

here x_i represents (x, y, z) and u_i represents (u, v, w).

The above statement, with appropriate initial and boundary conditions, is of course written from an Eulerian point of view and is connected with observations of a field of flow which would be measured at a fixed point. The author's first work[12] with this problem followed this line. During some fascinating and most helpful discussions, however, Professor L. E. Fraenkel suggested to the author that the results which had been obtained, including a solution with a singularity occurring at a finite time, might be more easily obtained, and perhaps illuminated, by a discussion in terms of a Lagrangian description of fluid flow. His advocacy, therefore, guided the author to study the problem afresh from this point of view.

In the Lagrangian framework we follow the motion of a marked particle, which is characterised by three numbers (a_i) whose values denote the Cartesian coordinates of the particle at $= 0$. The current coordinates of the particle at time t are x_i. Then the equation of motion of the particle is[15]

$$\frac{\partial^2 x_j}{\partial t^2}\frac{\partial x_j}{\partial a_i} = -\frac{\partial p}{\partial a_i}, \qquad (2\text{-}2)$$

while the continuity condition is

$$\det\left[\frac{\partial x_j}{\partial a_i}\right] \equiv J \equiv \frac{\partial(x_1,x_2,x_3)}{\partial(a_1,a_2,a_3)} = 1. \qquad (2\text{-}3)$$

The velocity field (u_i) of (2-1) is given by

$$u_i(a_k,t) = \frac{\partial}{\partial t}x_i(a_k,t). \qquad (2\text{-}4)$$

Of crucial importance, as Edward Fraenkel pointed out to the author, is that the vorticity (ω_i) associated with the particle at time t is given in terms of the vorticity at $t=0$ by Cauchy's formula,[15] namely

$$\omega_i(a_k,t) = \omega_j(a_k,0)\frac{\partial x_i}{\partial a_j}, \qquad (2\text{-}5)$$

where the vorticity is given by the usual curl formula

$$\omega_i(a_k,t) = \varepsilon_{ijn}\frac{\partial u_n}{\partial x_j}. \qquad (2\text{-}6)$$

In the next two sections we proceed to pose and analyze a particular problem in both the Eulerian and Lagrangian frames.

3. A SPECIAL STRUCTRE: EULERIAN DESCRIPTION

Following the earlier work[6] we consider the flow neighbouring a plane of symmetry, $z=0$, taking w to be proportional to z and u,v independent of z. In terms of the z structure, this uncouples the local field near to $z=0$ from the far field, which might be periodic in z. In addition, and to be specific, we impose also a stagnation-point structure in x, so that u is proportional to x and v and w are independent of x. This uncouples the local field near to $x=0$ from the far field. It can be seen that these assumptions put the field of flow into the general nonlinear class described by C. C. Lin,[16] but with viscosity excluded in the present case.

Thus we have

$$\begin{aligned} u(x_i,t) &= axu(y,t), \\ v(x_i,t) &= v(y,t), \\ w(x_i,t) &= bzw(y,t), \\ p(x_i,t) &= -\frac{1}{2}a^2x^2 - \frac{1}{2}b^2z^2p_2 + p(y,t), \end{aligned} \qquad (3\text{-}1)$$

where a and b are constant rates of strain.

Already comment has been made on the importance of the vorticity, but we emphasize that the pressure has its own importance also. In particular, we are constrained by our assumptions on the velocity field to have the pressure quadratic in x and z, but with coefficients at most t-dependent.[16] In the present paper we shall later set the multiplicative constant p_2 equal to zero, so that there is no gradient of pressure across $z = 0$.

Setting (3-1) in (2-1) we have

$$u_t + vu_y + a(u^2 - 1) = 0, \tag{3-2}$$

$$w_t + vw_y + b(w^2 - p_2) = 0, \tag{3-3}$$

$$au + v_y + bw = 0, \tag{3-4}$$

$$v_t + vv_y + p_y = 0. \tag{3-5}$$

The initial and boundary conditions are

$$t = 0, u = u_0(y), w = w_0(y);$$

$$y = 0, v = 0;$$

$$y \to \infty, u \to 1, w \to 0. \tag{3-6}$$

Immediately we note that, if $a \equiv 0$, (3-3) to (3-5) can be solved separately for v, w and p, and then u is given by the linear equation (3-2). This calculation has been reported[6] and leads to substantial agreement with Klebanoff's experiments as to the time taken for the local shear layer to develop.

In the more general case of $a \neq 0$ the situation is more complex, but we can still note one simplifying factor, namely that (3-5) determines the pressure once (3-2) to (3-4) have been solved for u, v and w. We turn now to the question of achieving this. Our aim is as follows: if (3-2) and (3-3) were of the form

$$\frac{\partial \chi}{\partial t} + v \frac{\partial \chi}{\partial y} = 0, \tag{3-7}$$

we could immediately infer that $\chi = \chi(\phi)$, where $\phi(y, t)$ would be a characteristic of

$$\frac{dt}{1} = \frac{dy}{v(y, t)}. \tag{3-8}$$

Since we do not know $v(y,t)$, this may not yet seem helpful. However, it transpires that ϕ can be determined simultaneously with v from the continuity condition (3-4), after certain transformations have been made. But, first of all, we need to obtain two forms like (3-7) to replace (3-2) and (3-3).

To this end, we write

$$u(y,t) = \tanh[u_2(y,t) + at], \qquad (3\text{-}9)$$

$$w(y,t) = [w_2(y,t) + bt]^{-1}, \qquad (3\text{-}10)$$

and then (3-2) and (3-3), with $p_2 = 0$, become

$$u_{2t} + vu_{2y} = 0 \qquad (3\text{-}11)$$

$$w_{2t} + vw_{2y} = 0. \qquad (3\text{-}12)$$

(If p_2 were not equal to zero, (3-10) would be replaced by a transformation akin to (3-9).) Thus from (3-11) and (3-12)

$$u_2(y,t) = u_2(\phi) \qquad (3\text{-}13)$$

$$w_2(y,t) = w_2(\phi) \qquad (3\text{-}14)$$

where $\phi(y,t)$ is the characteristic function of (3-8), if indeed it can be found in due course.

The continuity equation (3-4) becomes

$$a\tanh(u_2(\phi) + at) + v_y + b[bt + w_2(\phi)]^{-1} = 0. \qquad (3\text{-}15)$$

Our next object is to solve this equation. Since $\phi(y,t) = \text{const}$ is a solution of (3-8) it can be inverted to give $y = Y(t, \phi)$, and then

$$v = \frac{\partial}{\partial t} Y(t, \phi). \qquad (3\text{-}16)$$

Thus, by differentiation

$$\frac{\partial v}{\partial y} = \left[\frac{\partial^2}{\partial \phi \partial t} Y(t, \phi)\right] \left[\frac{\partial}{\partial y} \phi(y,t)\right]$$

$$= \left[\frac{\partial^2 Y}{\partial \phi \partial t}(t, \phi)\right] \left[\frac{\partial Y}{\partial \phi}(\phi, t)\right]^{-1}, \qquad (3\text{-}17)$$

since t is kept fixed for the inversion between ϕ and Y; and therefore

$$\frac{\partial v}{\partial y} = \frac{\partial}{\partial t}\log\frac{\partial Y}{\partial \phi}(\phi, t). \tag{3-18}$$

We now can substitute (3-18) in (3-15), and integrate with respect to t keeping ϕ fixed to obtain

$$\frac{\partial Y}{\partial \phi} = \frac{G(\phi)\operatorname{sech}(u_2(\phi)+at)}{[bt+w_2(\phi)]}, \tag{3-19}$$

where $G(\phi)$ is an arbitrary function.

We note that ϕ can be interpreted as the particle coordinate, and so $\phi = y$ at $t = 0$. Moreover, since $v = 0$ at $y = 0$ by (3-6), a particle on $y = 0$ never lifts away, so $\phi = 0$ when $y = 0$. Also (3-6), (3-9) and (3-10) imply that

$$u_2(\phi) = \tanh^{-1} u_0(\phi) \text{ and } w_2(\phi) = [w_0(\phi)]^{-1}. \tag{3-20}$$

Thus (3-19) yields

$$Y(t,\phi) = \int_0^\phi \frac{d\phi}{[1+btw_0(\phi)][\cosh at + u_0(\phi)\sinh at]}. \tag{3-21}$$

Inversion of this formula gives the characteristic function ϕ in terms of $y(\equiv Y)$ and t, while v, u and w are given by (3-9), (3-10), (3-16) and (3-20) as

$$v = \frac{\partial Y}{\partial t}(t,\phi) \tag{3-22}$$

$$u = \frac{\tanh at + u_0(\phi)}{1 + u_0(\phi)\tanh at} \tag{3-23}$$

$$w = \frac{w_0(\phi)}{1 + btw_0(\phi)}. \tag{3-24}$$

The velocity field given by (3-21) - (3-24) satisfies (3-2) - (3-4) and (3-6). Before analyzing its meaning, we go on to describe the analysis by the Lagrangian scheme.

4. A SPECIAL STRUCTURE: LAGRANGIAN DESCRIPTION

In this scheme we need to consider (2-2) - (2-6) by, first of all, imposing a structure analogous to (3-1) as follows:

$$x_1 = a_1 \xi(a_2, t), \tag{4-1}$$

$$x_2 = Y(a_2, t), \tag{4-2}$$

$$x_3 = a_3 \varsigma(a_2, t). \tag{4-3}$$

By use of (2-4), followed by substitution of a_1 and a_3 from (4-1) and (4-3), we have further

$$u_1 = a_1 \xi_t(a_2, t) = x_1 \xi_t / \xi, \tag{4-4}$$

$$u_2 = Y_t(a_2, t), \tag{4-5}$$

$$u_3 = a_3 \varsigma_t(a_2, t) = x_3 \varsigma_t / \varsigma. \tag{4-6}$$

These last three formulae are seen to be of the form (3-1) with $x_1 \equiv x$, $u_1 \equiv u$, $\xi_t/\xi \equiv au(y,t)$ and so forth.

Clearly the initial conditions should include

$$\xi(a_2, 0) = \varsigma(a_2, 0) = 1, \quad Y(a_2, 0) = a_2, \tag{4-7}$$

since the particle numbers a_1, a_2, a_3 and the Cartesian coordinates must be coincident at $t = 0$. There are, however, other conditions, which relate to the initial velocity (and vorticity) fields (3-6) and to the pressure in the form (3-1). These conditions will be considered in detail later.

The continuity condition (2-3) yields

$$\xi \varsigma Y_\phi = 1, \tag{4-8}$$

where we use $\phi \equiv a_2$ as in Section 3. If ξ and ς can somehow be calculated, then Y follows from (4-8) by integration. In the meantime we concentrate on a calculation of the three vorticity components as given by (2-6), to find

$$\omega_1 = a_3 \xi(\varsigma \varsigma_{\phi t} - \varsigma_\phi \varsigma_t) \tag{4-9}$$

$$\omega_2 \equiv 0 \tag{4-10}$$

$$\omega_3 = a_1 \varsigma(-\xi \xi_{\phi t} + \xi_\phi \xi_t). \tag{4-11}$$

At $t = 0$ the vorticity field is given by (2-6), (3-6), (4-1) and (4-3) and is of the form

$$[ba_3 w_{0\phi}(\phi), 0, -aa_1 u_{0\phi}(\phi)] . \qquad (4\text{-}12)$$

An appeal to Cauchy's vorticity formula (2-5) then implies that

$$\xi \xi_{\phi t} - \xi_\phi \xi_t = a u_{0\phi}(\phi) , \qquad (4\text{-}13)$$

$$\zeta \zeta_{\phi t} - \zeta_\phi \zeta_t = b w_{0\phi}(\phi) . \qquad (4\text{-}14)$$

Our object now is to solve each of these two partial differential equations, in which the right-hand sides are known. There is an immediate difficulty in that the conditions (4-7) are to be imposed on a characteristic, $t = 0$, so that Cauchy's initial-value theorem does not apply. We shall need to keep this difficulty in mind.

We now consider the solution of (4-13), since (4-14) is quite similar. The most noticeable thing about (4-13) is the "Wronskian"-like character of the operator. This feature we now exploit in two different ways in order to find the general solution.

I. Let
$$\xi_\phi = a u_{0\phi} \eta(\phi, t) , \qquad (4\text{-}15)$$

so that (4-13) becomes
$$\xi \eta_t - \eta \xi_t = 1 , \qquad (4\text{-}16)$$

from which we can infer that $\xi(\phi, t)$ and $\eta(\phi, t)$ are linearly independent solutions of the equation

$$\xi_{tt} + \rho(t, \phi) \xi = 0 , \qquad (4\text{-}17)$$

where ρ is an arbitrary function. Substitution into the t differential of (4-13) next shows that $\rho_\phi = 0$, so that

$$\xi_{tt} + \rho(t) \xi = 0 . \qquad (4\text{-}18)$$

The solution of (4-18) for ξ is thus a linear combination of two solutions $\alpha(t)$ and $\beta(t)$, but with coefficients depending on ϕ.

II. Let
$$\xi_t = Q(\phi, t) , \qquad (4\text{-}19)$$

so that (4-13) becomes
$$\xi Q_\phi - Q \xi_\phi = a u_{0\phi}(\phi) \qquad (4\text{-}20)$$

from which we can infer that $\xi(\phi, t)$ and $Q(\phi, t)$ are linearly independent solutions of the equation

$$\xi_{\phi\phi} - \frac{u_{0\phi\phi}}{u_{0\phi}} \xi_\phi + \sigma(\phi, t) \xi = 0 , \qquad (4\text{-}21)$$

where σ is an arbitrary function. Substitution into the ϕ differential of (4-13) shows that $\sigma_t = 0$ so that

$$\xi_{\phi\phi} - \frac{u_{0\phi\phi}}{u_{0\phi}}\xi_\phi + \sigma(\phi)\xi = 0 \ . \tag{4-22}$$

The solution of (4-22) for ξ is thus a linear combination of two solutions $A(\phi)$ and $B(\phi)$, but with the coefficients depending on t.

In summary, therefore, the two arguments **I** and **II** show that the general solution of (4-13) is

$$\xi = A(\phi)\alpha(t) + B(\phi)\beta(t) \ , \tag{4-23}$$

where $\alpha(t)$ and $\beta(t)$ satisfy (4-18) and

$$\alpha\beta_t - \beta\alpha_t = 1 \ , \tag{4-24}$$

while $A(\phi)$ and $B(\phi)$ satisfy (4-22) and

$$AB_\phi - BA_\phi = au_{0\phi}(\phi) \ . \tag{4-25}$$

We have two arbitrary functions to choose, namely $\rho(t)$ and $\sigma(\phi)$.

We turn now to the initial-value problem. A calculation of the pressure from (2-2) shows that there is a contribution of the form $(1/2)x_1^2\rho(t)$ which is equal to the corresponding term in (3-1) iff $\rho(t) = -a^2$. Thus (4-18) yields

$$\alpha(t) = \cosh at, \quad \beta(t) = \frac{1}{a}\sinh at \ . \tag{4-26}$$

Next, we know that $\xi(\phi, 0) = 1$, so that (4-23) at $t = 0$, namely $\xi(\phi, 0) = A(\phi)$, implies that $A(\phi) \equiv 1$ is one solution of (4-22). Therefore $\sigma(\phi) \equiv 0$, and

$$A(\phi) = 1, \quad B(\phi) = au_0(\phi) \ . \tag{4-27}$$

Thus, we have determined ξ as

$$\xi(\phi, t) = \cosh at + u_0(\phi)\sinh at \ . \tag{4-28}$$

Similar arguments from (4-14) show that

$$\varsigma(\phi, t) = 1 + btw_0(\phi) \ . \tag{4-29}$$

If we now substitute for ξ in (4-4), for ς in (4-6) and integrate (4-8) subject to $Y = 0, \phi = 0$, we obtain formulae equivalent to (3-23) and (3-24) with (3-1) for $u_1 \equiv u(x_i, t)$ and $w_1 \equiv w(x_i, t)$, with $Y(\phi, t) \equiv y$ given by (3-21). Then $u_2 \equiv v(x_i, t)$ follows as (3-22).

Thus the equivalence of the Eulerian and Lagrangian descriptions becomes clear. For completeness we write down the formula for the pressure (for the case $p_2 \equiv 0$ of (3-1)), as derived from (2-2):

$$p = -\frac{1}{2}a^2 x_1^2 - \int^\phi Y_{tt}(t,\phi)[\xi(\phi,t)\varsigma(\phi,t)]^{-1} d\phi . \qquad (4\text{-}30)$$

A brief analysis, which need not be repeated here, shows that this can be derived also from (3-5) when appropriate transformations are used.

Having described the Eulerian and Lagrangian solutions for the special structure imposed by (3-1) and (4-1) – (4-3) we can now discuss the mathematical and physical implications. These will be pursued in the next section.

5. MATHEMATICAL AND PHYSICAL IMPLICATIONS

We return now to the solution for the velocity field given by (3-21) – (3-24). In order to assess the nature of the solution as it depends on y and t, we need to be specific about the initially-prescribed functions, $u_0(y)$ and $w_0(y)$, which become $u_0(\phi)$ and $w_0(\phi)$ in the solution for $t > 0$. The matter of whether or not the velocity field remains bounded depends on the signs of $u_0(y)$ and $w_0(y)$. If one or the other of these functions has a negative region then u or w could become singular for some positive t, even though both denominators (3-23, 3-24) are unity for $t = 0$. Simultaneously, however, we need to note that the relation (3-21) between Y and ϕ may involve a singularity in the integrand.

For the above reasons we need to be quite explicit, perhaps by reference to relevant experimental observation, about the nature of $u_0(y)$ and $w_0(y)$. Typical examples of appropriate initial profiles are

$$u_0(y) = 1 - e^{-\mu y} , \qquad (5\text{-}1)$$

$$w_0(y) = \lambda e^{-\mu y}(1 - \alpha e^{-\mu y} + (\alpha - 1)e^{-2\mu y}) , \qquad (5\text{-}2)$$

where μ and λ are positive parameters. From (5-1), we see that the profile is of standard boundary-layer type. For $w_0(y)$, on the other hand, we note that $w_0(0) = w_0(\infty) = 0$ while there is another zero at $\mu y = \log(\alpha - 1)$, which is positive if $\alpha > 2$. If for example, $\alpha = \frac{11}{3}$, this zero is at $\mu y = \log \frac{8}{3}$, while there is a minimum which is negative and of magnitude $(-3\lambda/16)$ at $\mu y = \log \frac{4}{3}$; the maximum is at $\mu y = \log 6$ and has positive magnitude $25\lambda/324$. The negative region of $w_0(\phi)$ is significant.

In (3-21), with (5-1), we can see that the second bracket in the denominator is positive definite. On the other hand the first bracket in

(3-21) is positive only if t is less than some critical value, t_s, which is defined by the negative minimum of (5-2); for the numerical example quoted above $t_s = 16/3b\lambda$, and the first bracket in (3-21) goes to zero there. Thus the integrand of (3-21) becomes singular at $t = t_s$, this singularity occurring in (3-21) at the minimum of $w_0(\phi)$, namely $\phi = \mu^{-1} \log \frac{4}{3}$. A standard calculation then shows that the value of y at which the singularity (y_s) occurs is given asymptotically by

$$y_s = (t_s - t)^{-1/2} T(t), \quad t \to t_s, \quad (5\text{-}3)$$

where $T(t)$ tends to a finite value as t tends to t_s from below (This result is valid for any profile, $w_0(y)$, which has a smooth parabolic minimum).

The behaviour of the velocity field can also be described in the same limit. From (3-22) we see that v is singular as

$$v = (t_s - t)^{-3/2} V(t, \phi), \quad (5\text{-}4)$$

where V is finite as $t \to t_s$ from below, provided $(\phi_0 - \phi)(t_s - t)^{-1/2}$ is held fixed in that limit. (Here ϕ_0 denotes the position of the minimum of $w_0(\phi)$.) On the other hand (3-23) shows that u remains bounded, although the relationship (3-21) between ϕ, y and t becomes increasingly distorted as $t \to t_s$, as (5-3) shows, thus substantially modifying the profile of u as a function of y. Finally (3-4) shows that w becomes singular as $t \to t_s$ from below, the singularity being of the form

$$w = (t_s - t)^{-1} W(t, \phi), \quad (5\text{-}5)$$

where W is finite as $t \to t_s$, and $(\phi_0 - \phi)(t_s - t)^{-1/2}$ is held fixed in the limit.

There is much speculation at the present time as to the possibility of solutions of the Euler equations, even when subject to "sensible" initial conditions, developing a singularity in a finite time. The present work shows quite explicitly just that possibility (subject, however, to the need to invert (3-21)). It may be objected that the present solution requires a semi-infinite domain ($y \geq 0$). Be that as it may, the present contribution is believed by the author at least to be of value for illustration of the possibilities intrinsic in initial-value problems for the Euler equations.

One futher mathematical matter needs to be mentioned, namely that of the stagnation-point structure assumed in the x coordinate. To some extent, but not completely, this can be remedied as has been described earlier.[12] If we have a flow with general x, y and t variations, namely

$$u = u(x, y, t), \quad v = v(x, y, t), \quad w = bzw(x, y, t)$$

$$p = -\frac{1}{2} b^2 z^2 p_2 + p(x, y, t), \quad (5\text{-}6)$$

but retaining the z symmetry, it is not difficult to show the following by the techniques of Section 3, where ς is the z component of vorticity:

$$\left[\frac{\partial}{\partial t} + u\frac{\partial}{\partial x} + v\frac{\partial}{\partial y}\right](w\varsigma) = 0 , \qquad (5\text{-}7)$$

$$\left[\frac{\partial}{\partial t} + u\frac{\partial}{\partial x} + v\frac{\partial}{\partial y}\right]\left[\frac{1}{w} - bt\right] = 0 . \qquad (5\text{-}8)$$

(This has been given here for the case $p_2 = 0$).

If, then, $\phi(x,y,t)$ and $\psi(x,y,t)$ are characteristics of

$$\frac{dx}{u} = \frac{dy}{v} = dt , \qquad (5\text{-}9)$$

it then follows that

$$w = \frac{w_0(\phi,\psi)}{1 + btw_0(\phi,\psi)} , \qquad (5\text{-}10)$$

$$\varsigma = \varsigma_0(\phi,\psi)[1 + btw_0(\phi,\psi)] , \qquad (5\text{-}11)$$

where $w_0(x,y)$ and $\varsigma_0(x,y)$ are initially-prescribed functions at $t = 0$. Similar results can also be obtained by the Lagrangian scheme of Section 4.

Of course, we do not yet know ϕ and ψ, since they depend through (5-9) on u and v, which are themselves unknown. The continuity equation has to be considered. In the simpler, stagnation-point, case described in the present paper, the problem could be and has been solved, namely in the determination simultaneously in Sections 3 and 4 of the one characteristic and the field of v. The present generalised case provides substantial problems, principally perhaps through the relationship between pressure and the velocity field in a Poisson equation. The author is still working on this aspect from both Eulerian and Lagrangian points of view.

Even so, it can be seen that the possibility of a singularity developing is still there in (5-10), provided $w_0(x,y)$ has a negative region for some values of y. But, unlike the situation given explicitly in Sections 4 and 5, we cannot be sure that a singularity will develop in the Euclidean space because we cannot yet be explicit about the relationship between ϕ and ψ on the one hand and x and y on the other.

Finally, there is the matter of the relationship of the structures described in this paper to problems of (i) transition in a boundary layer and (ii) the sub-layer of a turbulent boundary layer. It may, for example, be appealing to speculate on the relevance of the singular structures described here to eruptions of longitudinal vortex filaments from the sub-layer into the main part of a turbulent boundary layer. Careful considerations would be required, especially since the singular value of y occurs in the outer regions of the flow, perhaps near the edge of the boundary layer. For transition (i) there is the continuing problem of the development of embryo turbulent spots from longitudinal-vortex flows. The ideas of the present paper may be relevant there also.

The present work was presented in summarized form in 1983,[12] but at about the same time and independently Calogero presented[17] an exact solution of a wave equation which has a strong connection with the present structures, although the detailed mathematics and physical background are different. Moreover, reference has been made already to the approximate but relevant theory of Russell and Landahl[14] for the transition problem. Somewhat earlier work which may be noted is the paper of Hoskins and Bretherton[18] on frontogenesis in Meteorology, and more recently the paper of Stern and Paldor[19] on transition processes. One further relevant association is that the operator is (4-13) and (4-14) is known by Hirota's name, as Professor A. C. Newell pointed out to me.

These remains the matter of viscosity, which has been ignored in this paper hitherto. It is possible that, according to the Navier-Stokes equations, the singularity will be prevented from occurring. On the other hand, the seminal work of Van Dommelen and Shen[20] on singular structures near to separation is worthy of note. Their work is a combination of numerical studies and asymptotic analysis based on boundary layer ideas, but the connection of the present paper with their concepts is quite strong.

6. ACKNOWLEDGEMENTS

I am particularly indebted to Professor L. E. Fraenkel, who directed my attention to the Lagrangian approach of Section 4, after hearing of my studies by Eulerian means. Moreover I would like to thank Dr. S. J. Cowley for suggesting the relevance of Van Dommelen's thesis and discussing it with me. This work was partially supported by the N. S. F. Fluid Dynamics and Hydraulics Program, Grant MSM83-20307 and by the DARPA/URI Applied and Computational Mathematics Program at Brown University, with assistance also from a NATO Research Grant for Travel and from M. I. T.

References

1. Stuart, J. T., "Instability of Flows and their Transition to Turbulence", Z. Flugwiss. Weltraumforsch. 10, 379-392 (1986): and Stuart, J. T., "Instability, Three-dimensional Effects and Transition in Shear Flows" in

"Perspectives in Turbulence", eds. P. Bradshaw and H. U. Meier, 1-25, Springer, Berlin (1987).
2. Lin, C. C., "On the Stability of Two-Dimensional Parallel Flows", *Quart. Appl. Math.* **3**, 117-142, 218-234 and 277-301 (1945).
3. Lin, C. C., "The Theory of Hydrodynamic Stability", Camb. Univ. Press (1955).
4. Benney, D. J. and Lin, C. C., "On the Secondary Motion Induced by Oscillations in a Shear Flow", *Phys. Fluids*, **3**, 656-657 (1960).
5. Benney, D. J., "A Non-Linear Theory for Oscillations in a Parallel Flow", *J. Fluid Mech.*, **10**, 202-236 (1961).
6. Stuart, J. T., "The Production of Intense Shear Layers by Vortex Stretching and Convection", AGARD Report 514 (1965).
7. Pearson, C. F. K. and Abernathy, F. H., "Evolution of the Flow Field Associated with a Streamwise Diffusing Vortex", *J. Fluid Mech.* **146**, 271-283 (1984).
8. Moore, D. W., "The Interaction of a Diffusing Line Vortex and an Aligned Shear Flow", *Proc. Roy. Soc.* (A) **399**, 367-375 (1985).
9. Yang, Z., A Single Streamwise Vortical Structure and its Instability in Shear Flows", Thesis for Ph.D., Harvard University (1987).
10. Ellingsen, T. and Palm, E., "Stability of Linear Flow", *Phys. Fluids* **18**, 487-488 (1975).
11. Landahl, M. T., "A Note on an Algebraic Instability of Inviscid Parallel Shear Flows", *J. Fluid Mech.* **98**, 243-251 (1980).
12. Stuart, J. T., "Instability of Laminar Flows, Non-Linear Growth of Fluctuations and Transition to Turbulence" in "Turbulence and Chaotic Phenomena in Fluids" (IUTAM Symposium Kyoto 1983), ed. T. Tatsumi, 17-26, North-Holland (1984).
13. Stuart, J. T., "Three-Dimensional Inviscid Developments in the Transition Process", Senior Whitehead Lecture to London Mathematical Society (1984).
14. Russell J. M. and Landahl, M. T., "The Evolution of a Flat Eddy near a Wall in an Inviscid Shear Flow", *Phys. Fluids* **27**, 557-570 (1984).
15. Lamb, H., "Hydrodynamics", 6th ed., 12-14 and 204-206, Camb. Univ. Press (1932).
16. Stuart, J. T., "Unsteady Boundary Layers", in Laminar Layers, ed. L. Rosenhead, 356-357, Clarendon Press, Oxford (1963).
17. Calogero, F., "A Solvable Nonlinear Wave Equation", *Stud. Appl. Math.* **70**, 189-199 (1984).
18. Hoskins, B. J. and Bretherton, F. P., "Atmospheric Frontogenesis Models: Mathematical Formulation and Solution", *J. Atmos. Sci.* **29**, 11-37 (1972).
19. Stern, M. E. and Paldor, N., "Large-amplitude Long Waves in a Shear Flow", *Phys. Fluids* **26**, 906-919 (1983).
20. Van Dommelen, L. L. and Shen, S. F., "The Genesis of Separation" in *Num. Phys. Aspects Aero. Flows*. (Proc. Symp. 1981), ed. T. Cebeci, 293-311, Springer, New York (1982).

NONLINEAR STABILITY OF ROTATING PIPE FLOW

T.R. Akylas and N. Toplosky*
*Massachusetts Institute of Technology
Cambridge, Massachusetts 02139, U.S.A.*

A numerical investigation of finite-amplitude, non-axisymmetric disturbances, in the form of traveling spiral waves, is made in pipe flow with superimposed solid-body rotation. Supercritically bifurcating neutral spiral waves are calculated for various axial and azimuthal Reynolds numbers and wavenumbers. For fixed axial mean pressure gradient, the axial mean flow induced by these waves gives rise to a significant mass flux defect, in certain cases as large as 40–50% of the undisturbed mass flux; the possible relevance of this finding to the phenomenon of vortex breakdown is pointed out. In non-rotating pipe flow, no neutral disturbances in the assumed form of spiral waves are found for moderate Reynolds numbers.

1. INTRODUCTION

It is now recognized that the linear stability properties of fully developed laminar flow in a circular pipe, subjected to rigid rotation about its axis, are markedly different from those of non-rotating pipe flow. Although in the absence of rotation pipe flow is entirely stable to infinitesimal perturbations, a relatively minute amount of rigid rotation is capable of causing linear instability[1,2]. Also, in the rapid-rotation regime, Pedley[3] first showed that rigid rotation has a strong destabilizing effect, giving rise to linear instability at axial Reynolds numbers as low as 83.

Our interest in the stability of rotating pipe flow was originally aroused by the rather unusual behaviour of this flow in the slow-rotation limit and by its possible connection to the instability of non-rotating pipe flow, which at present remains a theoretically unresolved problem. Experimentally, non-rotating pipe flow undergoes transition at a finite Reynolds number greater than about 2100 depending on the particular experimental conditions,[4,5] while, as remarked earlier, linear stability theory gives no evidence of instability.

---

* Permanent address: Naval Underwater Systems Center, New London, Connecticut 06320, U.S.A.

Furthermore, despite several attempts, no relevant nonlinear instability mechanism has been identified, with the only exception of the asymptotic theory of Smith & Bodonyi[6]. Using a nonlinear-critical-layer formalism, valid at large Reynolds numbers, Smith & Bodonyi[6] were able to find finite-amplitude, neutral, non-axisymmetric modes in the form of spiral waves. It appears that this class of waves exists only for azimuthal wavenumber equal to 1 and their phase speed and axial wavenumber are amplitude-dependent; in particular, as the amplitude decreases, the Reynolds number being large, the axial wavenumber tends to zero and the phase speed approaches a constant value. Interestingly enough, in slowly rotating pipe flow, linear instability also shows up at high axial Reynolds number, for spiral waves with azimuthal wavenumber equal to 1 and axial wavenumber approaching zero as the axial Reynolds number tends to infinity.[1,2] This suggests that perhaps one could reach the class of finite-amplitude modes of Smith & Bodonyi[6] as the pipe rotation is reduced to zero, by continuation along the branch of finite-amplitude neutral states that bifurcates from linear instability of slowly rotating pipe flow.[7] In order to explore the behaviour of nonlinear disturbances as the rotation is reduced to zero, and thereby examine the soundness of this conjecture, a numerical investigation is required.

Numerical computation of finite-amplitude traveling waves has proven a fruitful approach to the nonlinear stability of certain shear flows. The majority of the existing work has been confined to computations of two-dimensional disturbances on plane shear flows such as plane Poiseuille flow,[8,9] plane Poiseuille-Couette flow[10,11] and, more recently, boundary-layer flows[11]. However, as explained earlier, the problem at hand calls for the computation of non-axisymmetric traveling waves in cylindrical geometry. For this purpose, we use an extension of the spectral method proposed by Leonard & Wray,[12] which expands the velocity components in terms of a complete set of divergence-free functions that automatically satisfy the no-slip conditions at the wall of the pipe.

As already indicated, the present study was originally motivated by the instability of rotating pipe flow in the slow-rotation limit. However, nonlinear spiral waves are also of some interest in the rapid-rotation regime: the linear spiral instability waves, found by Pedley[3] in rapidly rotating pipe flow, twist in the opposite direction to the swirl of the basic flow, which makes them possible candidates for explaining the generation of similar backwards spiral disturbances often observed in vortex breakdown.[13,14] This proposal, which was first made by Ludwieg,[15,16] received some criticism from Hall[13] and others: it is not clear how weak spiral disturbances, arising from a linear instability mechanism, can lead to the dramatic flow changes, in particular axial-flow stagnation, that accompany vortex breakdown. In this respect, the changes induced in the basic flow by the presence of finite-amplitude spiral waves could provide some useful information about the possible relevance of the instability mechanism proposed by Ludwieg to the phenomenon of vortex breakdown.

2. FORMULATION AND NUMERICAL SOLUTION

Consider fully developed laminar flow in an infinitely long, circular pipe of radius r_0, $0 \leq r \leq r_0$, $-\infty < z < \infty$, subjected to rigid rotation about its axis with constant angular velocity ω. The basic steady flow, which is driven by a constant axial pressure gradient, is the combination of a parabolic velocity profile, $W_0(1 - r^2/r_0^2)$, with centreline velocity W_0 in the axial direction, and solid-body rotation, ωr, in the azimuthal direction. Our interest centres on finite-amplitude perturbations to this basic flow in the form of periodic spiral waves traveling with constant phase speed. Taking the axial wavelength to be $2\pi L$, three relevant dimensionless parameters are the axial Reynolds number R, the azimuthal Reynolds number Ω, and the dimensionless axial wavenumber μ:

$$R = \frac{W_0 r_0}{\nu}, \qquad \Omega = \frac{\omega r_0^2}{\nu}, \qquad \mu = \frac{r_0}{L}, \qquad (1)$$

where ν is the fluid kinematic viscosity. Earlier work[1] has shown that, in terms of these parameters, the slow-rotation regime (in which we seek the least amount of rotation needed to linearly destabilize a primarily axial flow) is defined by the limit

$$\Omega = O(1), \qquad \tilde{R} = \mu R = O(1), \qquad \mu \to 0. \qquad (2)$$

On the other hand, the rapid-rotation regime (in which we seek the least amount of axial flow needed to linearly destabilize a predominantly azimuthal flow) is realized in the limit[3]

$$R = O(1), \quad \tilde{\Omega} = \mu\Omega = O(1), \quad \mu \to 0. \tag{3}$$

The appropriate scalings of the perturbation-velocity components and pressure differ in these two extremes; we choose to scale according to the slow-rotation regime and we define dimensionless (primed) variables according to

$$r = r_0 r', \quad z = Lz', \quad t = \frac{L}{W_0} t', \quad (u, v) = \mu W_0 (u', v'),$$
$$w = W_0 w', \quad p = \mu^2 \rho W_0^2 p', \tag{4}$$

where t is time, $\boldsymbol{u} = (u, v, w)$ are the radial, azimuthal and axial perturbation-velocity components in cylindrical coordinates (r, θ, z), p is the perturbation pressure, and ρ is the fluid density.

Dropping the primes, finite-amplitude, periodic spiral-wave solutions of the governing equations of continuity and momentum can be formally expanded in Fourier series:

$$\boldsymbol{u} = \sum_{k=-\infty}^{\infty} \boldsymbol{u}^k(r) e^{ik\phi}, \quad p = \sum_{k=-\infty}^{\infty} p^k(r) e^{ik\phi} \quad (0 \leq r \leq 1), \tag{5}$$

where

$$\phi = z + \ell\theta - ct; \tag{6}$$

here ℓ is the (integer) azimuthal wavenumber, c is the constant real phase speed and, in view of the scalings chosen in (4), the axial wavelength is normalized to 2π. Also, since $\boldsymbol{u}$ and p are real,

$$\boldsymbol{u}^{-k} = \boldsymbol{u}^{k*}, \quad p^{-k} = p^{k*}, \tag{7}$$

where $*$ denotes complex conjugate, so that the sums in (5) may be restricted to $k \geq 0$.

For given values of the parameters R, Ω, μ and the wavenumber ℓ (plus a phase normalization), the unknown Fourier coefficients $\boldsymbol{u}^k$, p^k and the phase speed c are determined by requiring that the series (5) satisfy the continuity and momentum equations, subject to regularity conditions at the centre ($r = 0$) and the no-slip condition at the wall of the pipe ($r = 1$); this task has to be dealt with numerically. After the solution has been obtained, it is convenient to have a measure of the size of the perturbation and, for this purpose, we choose the relative kinetic energy of the fluctuating harmonics ($|k| \geq 1$) in (5), averaged over one axial wavelength:

$$E = \sum_{k=1}^{\infty} E_k , \qquad (8)$$

where

$$E_k = \frac{2\pi}{E_b} \int_0^1 r\,dr (\mu^2 |u^k|^2 + \mu^2 |v^k|^2 + |w^k|^2) , \qquad (9a)$$

and E_b is the kinetic energy of the basic flow per unit length in the axial direction:

$$E_b = \frac{\pi}{2} \left(\frac{1}{3} + \frac{1}{2} \mu^2 \frac{\Omega^2}{\tilde{R}^2} \right) . \qquad (9b)$$

Similarly, in order to quantify the axial and azimuthal mean-flow components induced by the perturbation disturbance (represented by the $k = 0$ harmonic in (5)), the associated mass flux Φ and Γ, the circulation about the axis averaged over the pipe radius, are defined

$$\Phi = \frac{2\pi}{\Phi_b} \int_0^1 r\,dr\, w^0 , \qquad \Gamma = \frac{2\pi}{\Gamma_b} \int_0^1 r\,dr\, \mu v^0 \qquad (10a)$$

where Φ_b, Γ_b are the corresponding quantities for the basic flow:

$$\Phi_b = \frac{\pi}{2}, \qquad \Gamma_b = \frac{2\pi}{3} \mu \frac{\Omega}{\tilde{R}} . \qquad (10b)$$

Note that in this discussion the mean axial pressure gradient driving the flow is kept fixed and, thus, the disturbance is allowed to modify the mass flux of the basic flow; another choice would be to keep the mass flux fixed and allow the pressure gradient to vary. For finite-amplitude perturbations, these two formulations are not equivalent.

The numerical procedure for computing spiral-wave solutions of the form (5) follows along the lines of the technique developed by

Leonard & Wray.[12] The overall strategy is to expand each velocity Fourier mode $u^k(r)e^{ik\phi}$ in terms of a complete set of divergence-free vector functions $\{\chi_n^k(r)e^{ik\phi}\}$, which behave appropriately as $r \to 0$ and satisfy the no-slip condition at $r = 1$:

$$u^k(r) = \sum_n a_n^k \chi_n^k(r), \qquad (11)$$

$$\nabla \cdot \chi_n^k(r)e^{ik\phi} = 0, \qquad \chi_n^k(r=1) = 0. \qquad (12)$$

The coefficients a_n^k of the spectral expansion (11) are then determined from the momentum equation, which is discretized through a weighted-residual technique, using a set of vector weight functions $\{\xi_m^k(r)e^{-ik\phi}\}$ that are well-behaved at $r = 0$ and satisfy requirements similar to (12). To be more specific, inserting the expansions (5), (11) into the momentum equation, and taking the inner product with each of the weight functions, leads to an infinite set of algebraic equations for the unknown coefficients a_n^k:

$$\left[ikc\, A_{mn}^k + \frac{1}{\tilde{R}} B_{mn}^k + iC_{mn}^k\right] a_n^k + N_m^k = 0, \qquad (13)$$

where A_{mn}^k, B_{mn}^k, C_{mn}^k are known matrices,[17,18] independent of a_n^k, and N_m^k is a vector which is nonlinear in a_m^k. Each term in (13) has a clear interpretation: the first three terms, which are linear in a_n^k, represent, respectively, the temporal acceleration, the viscous term, and the interaction of the basic flow with the perturbation; the last term, which couples the different Fourier modes, represents the nonlinear perturbation interactions.

The success of the spectral technique depends heavily on the specific choice of expansion and weight functions. With the exception of the $k = 0$ family, we use the same sets of expansion and weight vectors as those proposed by Leonard & Wray.[12] The choice of expansion and weight functions for $k = 0$ is made so that the conditions of no mean axial and azimuthal shear stresses at the pipe wall are automatically met; details can be found in Toplosky[17] and in Toplosky & Akylas.[18]

In implementing the spectral technique numerically, the infinite series (5) and (11) are truncated at a finite number of Fourier modes

$k = K$ and expansion functions $n = N$, say, with an equal number ($m \leq N$) of weight functions. Thus, (13) reduces to a finite set of algebraic equations for a finite number of unknowns, which can now be solved iteratively through a multi-dimensional Newton's method to obtain the velocity field and the phase speed for given values of the parameters R, Ω, μ and ℓ; subsequently, the energy, mass flux and average circulation, associated with a spiral wave, can be readily computed from (8), (9) and (10).

3. RESULTS

As explained earlier, the procedure for computing finite-amplitude neutral spiral waves is iterative and requires a sufficiently accurate initial guess in order to achieve convergence. A convenient point of departure for the iteration is provided by the results of linear and weakly nonlinear theories, close to critical conditions for linear instability; these known results are also used as partial checks of the numerical technique.

Finite-amplitude calculations were carried out by continuation in either $\tilde{R}$ or Ω, keeping the rest of the parameters fixed. According to our calculations, all branches of finite-amplitude spiral waves bifurcate supercritically in both the slow and the rapid rotation regime and there is no sign of appearance of limit points. This was found to be the case for $\ell = 1, 2, 3$ and all values of μ in the range $0.1 \leq \mu \leq 1$ examined. For very small values of the disturbance energy, the numerical results agree closely with the predictions of weakly nonlinear theory[17,18] and this provides an additional check of the calculations.

Turning now to a more quantitative discussion of the nonlinear results, Figs. 1(a,b) present plots of the disturbance energy E, induced mass flux Φ, and average circulation Γ, in two particular cases; for convenience, Φ and Γ, which have already been normalized in (10) with respect to the basic flow, are shown in percent in order to bring out the effect of the disturbance on the unperturbed flow more clearly. Figure 1(a) refers to the branch that bifurcates from the point of the linear-neutral-stability curve that requires the minimum amount of rotation for linear instability in the slow-rotation limit ($\Omega = -26.96$, $\tilde{R} = 106.6$, $\ell = 1$, $\mu = 0$). As the rotation is increased (in absolute value) beyond $\Omega = -26.96$ for fixed $\tilde{R} = 106.6$, the induced mass flux Φ, which is always negative implying a mass-flux defect, rises significantly and reaches a value close to 30% of the undisturbed flux; this is in contrast to the induced average circulation, which always remains less than 10% of the undisturbed value and is positive, again indicating a circulation defect (since $\Omega < 0$). Figure 1(b) refers to the rapid-rotation regime and displays the branch that bifurcates from $R = 91.17$, $\tilde{\Omega} = -91.08$ for $\mu = 0.1$, $\ell = 2$; note that here as R is increased, keeping the rest of the parameters fixed, the induced flux defect reaches a value close to 50%, while the disturbance energy and induced circulation still remain relatively very small.

4. DISCUSSION

Based on the results of the present work, it is clear now that there is no immediate connection between nonlinear spiral waves in slowly rotating pipe flow and possible nonlinear neutral states in non-rotating pipe flow; finite-amplitude spiral waves bifurcate supercritically from the linear-neutral-stability curve of slowly rotating pipe flow, and there is not sign (at least in the range of parameters that we examined) that the bifurcation branches fold back towards $\Omega = 0$. This certainly does not imply that the nonlinear states of Smith & Bodonyi[6] do not exist: they were found through an asymptotic analysis, valid as $\tilde{R} \to \infty$, whereas our computations of nonlinear spiral waves were restricted to finite $\tilde{R} \leq 500$; furthermore, even in the limit $\tilde{R} \to \infty$, there is no guarantee that these two classes of nonlinear states are connected. In the range $\tilde{R} \leq 500$, our computations indicate that the states of Smith & Bodonyi[6] (if they exist at such values of $\tilde{R}$) cannot be reached by continuation from slowly rotating pipe flow. At any rate, our attempt to reach the asymptotic nonlinear states of Smith & Bodonyi[6] failed, as spiral-wave disturbances in slowly rotating pipe flow do not appear to be directly connected with nonlinear states in non-rotating pipe flow for $\tilde{R}$ finite. Heuristically speaking, it seems that the stability properties of slowly rotating pipe flow are quite different from those of non-rotating pipe flow both for infinitesimal and finite-amplitude perturbations. It is hoped that a careful experiment would be able to identify such a striking difference in the behaviour of these two seemingly almost identical flows.

An interesting feature of finite-amplitude spiral waves in rotating pipe flow, assuming that the axial pressure gradient is held fixed, is the significant mass flux defect, induced by the perturbation Reynolds stresses. Axial mean-flow distortion is relatively more pronounced in the rapid-rotation regime; for the range of parameters examined, in certain cases it can exceed a defect of 45% of the undisturbed mass flux, while, under the same conditions, the associated relative azimuthal-flow distortion is very small. This can be understood using simple scaling arguments: the linear asymptotic analysis of Pedley[3] shows that, in the rapid-rotation regime (3), all three perturbation velocity components scale with the basic azimuthal velocity at the pipe wall, ωr_0, which is large compared with the centreline velocity W_0 ($W_0/\omega r_0 = O(\mu)$). Thus, in the limit $\mu \to 0$, the induced axial mean flow (normalized with the basic axial flow) is expected to be relatively large compared with the induced azimuthal mean flow (normalized with the basic azimuthal flow). Indeed, as shown in Fig. 1(b), close to the bifurcation point, the induced flux Φ increases much more steeply than the induced circulation Γ. Of course, this qualitative argument cannot reveal that the induced axial mean flow represents a substantial flux defect, a finding that may have some relevance to the phenomenon of vortex breakdown. As already indicated, Ludwieg[15,16] and Pedley[3] noted that linear spiral instability waves in rotating pipe flow, as they twist in the opposite direction to the swirl of the basic flow ($\Omega < 0$ for $\ell > 0$), are reminiscent of similar wave disturbances often observed in vortex breakdown. Our calculations of nonlinear spiral waves, and in particular the substantial flux defect found, lend support to this proposal: the spiral waves observed in vortex breakdown are in fact accompanied with axial-flow stagnation,[13,14] a phenomenon that cannot be explained in terms of a linear instability mechanism. Nevertheless, it should be emphasized that this analogy between finite-amplitude, strictly periodic spiral waves in rotating pipe flow and unsteady, spatially and temporally evolving, spiral-wave disturbances in vortex breakdown, still remains rather loose; further work is necessary in order to establish a possible connection more precisely.

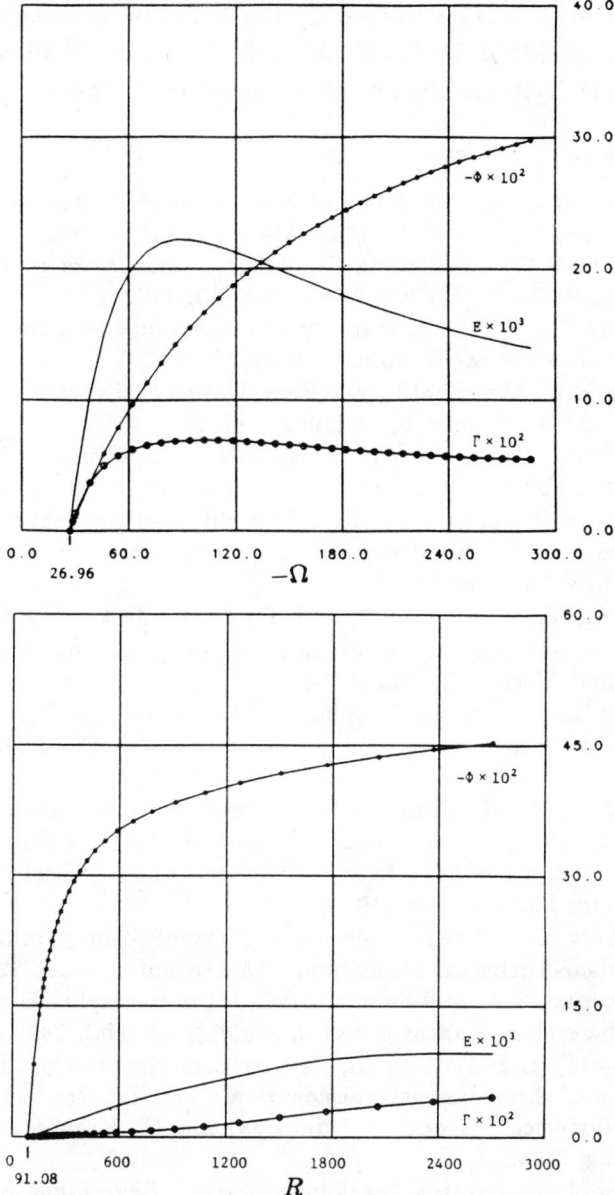

Fig. 1. Disturbance energy E, induced flux Φ, and average circulation Γ.
  (a) slow-rotation regime; $\tilde{R} = 106.6$ and Ω is increased past bifurcation point at $\Omega = -26.96$ ($\ell = 1, \mu = 0; K = 4, N = 10$).
  (b) rapid-rotation regime; $\tilde{\Omega} = -91.08$ and R is increased past bifurcation point at $R = 91.17$ ($\ell = 2, \mu = 0.1; K = 4, N = 10$).

This work was supported by the National Science Foundation under Grant MSM-8451154. Also, N. Toplosky thanks the Naval Underwater Systems Center for his generous support.

References

1. Mackrodt, P.A., "Stability of Hagen-Poiseuille flow with superimposed rigid rotation", *J. Fluid Mech* **73**, 153 (1976).
2. Cotton, F.W. and Salwen, H., "Linear stability of rotating Hagen-Poiseuille flow", *J. Fluid Mech.* **108**, 101 (1981).
3. Pedley, T.J., "On the stability of viscous flow in a rapidly rotating pipe" *J. Fluid Mech.* **35**, 97 (1969).
4. Leite, R.J., "An experimental investigation of the stability of Poiseuille flow" *J. Fluid Mech.* **5**, 81 (1959).
5. Fox, J.A., Lessen, M. and Bhat, W.V., "Experimental investigation of the stability of Hagen-Poiseuille flow", *Phys. Fluids* **11**, 1 (1968).
6. Smith, F.T. and Bodonyi, R.J., "Amplitude-dependent neutral modes in the Hagen-Poiseuille flow through a circular pipe" *Proc. Roy. Soc. Lond.* **A 384**, 463 (1982).
7. Akylas, T.R. and Demurger, J.-P., "The effect of rigid rotation on the finite-amplitude stability of pipe flow at high Reynolds number", *J. Fluid Mech.* **148**, 193 (1984).
8. Zahn, J.-P., Toomre, J., Spiegel, E.A. and Gough, D.O., "Nonlinear cellular motions in Poiseuille channel flow" *J. Fluid Mech.* **62**, 319 (1974).
9. Herbert, T., "Periodic secondary motions in a plane channel", In: *Proc. 5th Intl. Conf. on Numerical Methods in Fluid Dynamics* (A.I. Van de Vooren & P.J. Zandbergen, eds.) *Lecture Notes in Physics* **59**, Springer, (1976).
10. Herbert, T., "Finite-amplitude stability of plane parallel flows", In: *Laminar-Turbulent Transition,*ÃGARDČonf. Proc.ño.224,3-1(1977).
11. Milinazzo, F.A. and Saffman, P.G., "Finite-amplitude steady waves in plane viscous shear flows" *J. Fluid Mech.* **160**, 281 (1985).
12. Leonard, A. and Wray, A., "A new numerical method for the simulation of three-dimensional flow in a pipe". In: *Proc. 8th Intl. Conf. on Numerical Methods in Fluid Dynamics* (E. Krause, ed.), Springer-Verlag, (1982).
13. Hall, M.G., "Vortex breakdown", *Ann. Rev. Fluid Mech.* **4**, 195 (1972).
14. Leibovich, S., "The structure of vortex breakdown" *Ann. Rev. Fluid Mech.* **10**, 221 (1978).
15. Ludwieg, H., "Zur Erklarung der Instabilitat der uber angestellten Deltaflugeln auftretenden freien Wirbelkerne" *Z. Flugwiss.* **10**, 242 (1962).

16. Ludwieg, H., "Erklarung des Wirbelaufplatzens mit Hilfe der Stabilitatstheorie fur Stromungen mit schraubenlinienformigen Stromlinien" *Z. Flugwiss.* **13**, 437 (1965).
17. Toplosky, N., "Finite-amplitude spiral waves in rotating pipe flow", Ph.D. Dissertation, Department of Mechanical Engineering, M.I.T., Cambridge (1987).
18. Toplosky, N. and Akylas, T.R., "Nonlinear spiral waves in rotating pipe flow", *J. Fluid Mech.*, *sub judice*.

BISTABLE CELLULAR FLAMES

A. Bayliss*†, B. Matkowsky*†, and M. Minkoff*

*Mathematics and Computer Science Division
Argonne National Laboratory
Argonne, Illinois 60439-4844

†Department of Engineering Sciences and Applied Mathematics
Northwestern University
Evanston, Illinois 60201

Dedicated to C. C. LIN, pioneer in hydrodynamic stability

1. INTRODUCTION

We consider the problem of a flame stabilized by a point source of fuel in two dimensions. We employ the diffusional thermal model which consists of a reaction, diffusion, convection system of equations for the temperature and concentration of the limiting component of the reaction. In this model, the effect that a given fluid field has on the transport of temperature and concentration is considered, while changes in the underlying flow, due to thermal expansion, are not taken into account. The given fluid velocity field in our problem corresponds to a point source of fuel of strength $2\pi\kappa$.

In addition to κ, the parameters of the problem are the non-dimensional activation energy N, the ratio σ of the unburned temperature T_u to the burned temperature T_b, and the Lewis number L which is the ratio of thermal conductivity to reactant diffusivity. This problem was studied analytically[1] in the limit of $N(1-\sigma) \to \infty$ and $L \approx 1$. That analysis showed that there exists a basic steady solution consisting of a circular flame front of radius $R = \kappa$ with axisymmetric temperature and concentration profiles. It was further shown that this solution was stable for all κ, if $1 > L > L_c$, where the critical Lewis number $L_c = 1 - 2/N(1-\sigma)$ was predicted by the theory. For $L < L_c$ it was shown that there exists a critical value κ_c such that for $\kappa < \kappa_c$ the basic solution was stable while for $\kappa > \kappa_c$ the basic solution was unstable to angular perturbations. In addition, for κ near κ_c, the perturbations were shown to evolve to steady, stable, small-amplitude cellular solutions, which bifurcate supercritically from the basic solution. The cellular solutions were shown to exhibit crests and troughs, with the temperature at the troughs higher than at the crests, as is the case in experimentally observed cellular flames.

The analysis in Ref. 1, employing κ as a bifurcation parameter, is a local analysis valid only in the neighborhood of κ_c. In this paper we extend this analysis into the more fully nonlinear regime. Our numerical results show that as κ increases from κ_c the system exhibits a multistable behavior with at least two distinct stable, steady cellular solutions coexisting for the same parameter values. Each such stable solution has its own domain of attraction, so that the resultant steady solution depends on the initial conditions employed.

Our numerical computations are in polar coordinates (r,ϕ) and use a Chebyshev pseudo-spectral method in r and a Fourier pseudo-spectral method in ϕ.[2] The coordinate system in the r direction is adaptively chosen as described in our paper[3] so as to better resolve the solution. We note that additional calculations of cellular flames, employing L as a bifurcation parameter, are described in our paper.[4] In Section 2 we describe the model and the numerical method, and in Section 3 we discuss our results.

2. MODEL AND NUMERICAL METHOD

We consider the nondimensional reaction, diffusion, convection system for the reduced temperature $\Theta = \dfrac{T-T_u}{T_b-T_u}$ and reactant concentration C, where T_u and T_b represent the temperature in the unburned fresh mixture at the source and the burned temperature at infinity, respectively. In polar coordinates r and ϕ the nondimensional equations are[1]

$$\Theta_t = \Delta\Theta - \frac{\kappa\Theta_r}{r} + \frac{CN^2(1-\sigma)^2}{2L}\exp(N(1-\sigma)(\Theta-1)/(\sigma+(1-\sigma)\Theta))$$

$$C_t = \frac{\Delta C}{L} - \frac{\kappa C_r}{r} - \frac{CN^2(1-\sigma)^2}{2L}\exp(N(1-\sigma)(\Theta-1)/(\sigma+(1-\sigma)\Theta)) \quad (1)$$

Here $\sigma = T_u/T_b$, Δ is the Laplacian, and the terms on the right side of (1), respectively represent diffusion, convection, and a global one-step Arrhenius reaction. The boundary conditions are given by

$$\Theta = 0, C = 1 \quad \text{for } r = 0\,;\quad \Theta\to 1, C\to 0 \quad \text{for } r\to\infty. \quad (2)$$

The analysis in Ref. 1 is valid in the regime $N(1-\sigma)\to\infty$, with $1-\sigma$ small. For large activation energies, the Arrhenius reaction term is negligible except in a thin $O(1/(N(1-\sigma)))$ region (the reaction zone) around the front at $R = \kappa$. Beyond the reaction zone the fuel is essentially exhausted, so that the reaction cannot be maintained. Ahead of the reaction zone the temperature is too low to sustain the reaction. In the limit of $N(1-\sigma)\to\infty$, the reaction zone collapses to the front, with the derivatives of Θ and C experiencing discontinuities there. We remark that this model contains within it the Kuramoto-Sivashinsky equation, which is an evolution equation for the front. It is derived as a long wave theory, in the limit $N(1-\sigma)\to\infty$ and $L\approx L_c$.

In our computations, we employ a finite value of N, so that there no longer exists a sharp front, though the behavior of the solution changes rapidly in the vicinity of the analytically predicted front. To limit the size of the computational domain, we impose the boundary conditions at fixed points r_1, r_2 with $0 < r_1 < \kappa < r_2$. In our computations r_1 and r_2 are sufficiently far from the reaction zone that the artificial boundaries have very little effect. For the computations reported here, $r_1 = 2.4$, $r_2 = 31$.

The numerical method employs a mapping of the coordinate system so that the interval $r_1 \le r \le r_2$ is mapped onto $[-1,1]$. In this new coordinate (say s) the solution is expanded as a sum of Chebyshev polynomials in s and as a sum of trigonometric polynomials in ϕ, that is,

$$\Theta(s,\phi) = \sum_{j=0,l=-M}^{J,M} a_{jl}T_j(s)e^{il\phi} \quad (3)$$

with a similar expansion for C. The coefficients a_{jl} are obtained from collocating at the points $s_j = \cos(j\pi/J)$, $\phi_l = (2\pi l)/M$.[2]

The boundary conditions are

$$\Theta(r_1,\phi) = C(r_2,\phi) = 0$$
$$\Theta(r_2,\phi) = C(r_1,\phi) = 1 \quad (4)$$

These conditions correspond to unburned fuel at $r = r_1$ and complete burning at $r = r_2$.

It is a characteristic feature of the solution that Θ and C are very smooth except in the reaction zone. It is known that the Chebyshev pseudo-spectral method is highly accurate for smooth functions. If the function has large gradients, as is the case here, the method can be inaccurate and can give rise to oscillations due to the Gibbs phenomenon. In order to better resolve the solution in the reaction zone, we introduce a family of coordinate transformations in the radial (i.e., s) direction. Writing $s' = q(s,\underline{\alpha})$, we choose the parameter vector $\underline{\alpha}$ so that the spectral interpolation error is reduced. In practice, $\underline{\alpha}$ is a two-component vector. The specific family of coordinate transformations that we use is given in Ref. 3. As in that paper, the new coordinate system is chosen to minimize a weighted second Sobolev norm of an appropriately chosen linear combination of Θ and C. It is, in fact, desirable to choose $\underline{\alpha}$ as a function of ϕ, since the location of the reaction zone is ϕ dependent. A procedure to do this is currently being developed.

The steady-state solutions are computed by solving the time-dependent system (1) until a steady state is achieved. Thus the solutions we compute are necessarily stable. Two different time-differencing schemes have been used. These were a second-order semi-implicit scheme based on the Crank-Nicolson scheme as described in Ref. 3 and a backward Euler scheme with approximate factorization based on dimensional splitting.[5] In this scheme, for the model equation

$$U_t = U_{rr} + U_{\phi\phi}/r^2 + R(u), \quad (5)$$

we use

$$(U^{n+1}-U^n)/\Delta t = U_{rr}^{n+1} + U_{\phi\phi}^{n+1}/r^2 + R(u^n)$$

or if $U^{n+1}-U^n = \delta$ and D_{rr}, $D_{\phi\phi}$ stand for the second-derivative operators, we have

$$[I-\Delta t D_{rr}-\Delta t D_{\phi\phi}/r^2]\delta = U_{rr}^n + U_{\phi\phi}^n/r^2 + R(U^n). \quad (6)$$

The matrix on the left-hand side of (5) is approximately factored as

$$[I-\Delta t D_{rr}-\Delta t D_{\phi\phi}/r^2] \approx [I-\Delta t D_{\phi\phi}/r^2][I-\Delta t D_{rr}]. \quad (7)$$

In practice (7) permits a larger time step than the Crank-Nicolson scheme and has the advantage that at steady state ($\delta=0$), the steady-state discrete equations are identically satisfied. This scheme is only first-order accurate in time; and since we are interested only in stable solutions, the computed steady states are integrated for a short time with the second-order scheme for different points on the bifurcation curve, to make sure that they are temporally stable. At present, the reaction terms are treated explicitly.

3. RESULTS

We now describe the results of our computations, which were done on the Magnetic Fusion Energy CRAY X-MP computer at Lawrence Livermore National Laboratory. In all cases presented, only the parameter κ was varied. The parameters held fixed were $N = 20$, $\sigma = 0.615$, and $L = 0.44$. In Figures 1a-1f, we plot the reduced temperature Θ as a function of the angle ϕ for a fixed value of r, which is chosen to be in the reaction zone. The value of κ is different in each figure.

In Figure 1a, $\kappa = 8.00$ and we are below the bifurcation point so that the solution is axisymmetric. In Figure 1b, $\kappa = 8.5$. The steady-state solution is a small-amplitude, nearly sinusoidal three cell. The transient decays very slowly, as is characteristic of the behavior near a bifurcation point.[1]

In Figures 1c and 1d, κ has been increased to 9.25 and 11.0, respectively. The amplitude of the cells has grown, and the solution exhibits sharp peaks, which correspond to the pointed crests of the flame front obtained in Ref. 1, by analyzing the sign of the coefficient of the second Fourier harmonic in the analytically derived expansion of the cellular solution. An experimental visualization of this effect, for a flame stabilized on a slot burner, rather than by a point source, can be found in Ref. 6.

In Figures 1c and 1d, the computed solutions were obtained by using the solution corresponding to a lower value of κ, as an initial condition. When we tried to compute the solution at $\kappa = 11.70$ using the solution at $\kappa = 11.0$ as an initial condition, we found growth in the amplitude of the four mode, and a corresponding decay in the three mode, until the solution evolved into a steady four cell. This solution is exhibited in Figure 1f. Continuing this solution to lower values of κ, a four cell solution was found at $\kappa = 11.0$. A cut of this solution is shown in Figure 1e. Figures 1d and 1e demonstrate bistability of cellular flames, that is, the existence of two distinct stable steady states for the same parameter values, each with its own domain of attraction. Plots of the temperature surfaces for these two solutions are shown in Figures 2a and 2b. We have obtained preliminary results that show additional examples of bistability, which will be reported in a subsequent publication. We remark that our computed solutions have been validated by increasing the number of collocation points.

Upon continuing the four cell solution to lower values of κ, we found that the four cell solution lost its stability at a value of κ between 10.1 and 10.2. In fact, a three cell solution at $\kappa = 9.25$ was obtained by integrating the four cell initial data until the four cell decayed and the steady three cell evolved. The region of bistability ranges from κ_1 to κ_2, with κ_1 between 10.1 and 10.2, and κ_2 between 11.2 and 11.25, and a hysteresis phenomenon is observed. Increasing (decreasing) κ beyond (below) κ_2 (κ_1) leads to mode jumping from the three (four) mode to the four (three) mode.

In the actual sequence of our computations, the four cell was found first (see Ref. 4). The three cell was then found by lowering κ and observing a persistent growth in the three mode, with a corresponding decay of the four mode.

We monitor convergence to a steady state by computing the maximum residual—i.e., righthand side of (1)—and normalizing by the initial residual. At a point where a modal transition occurs, the question of convergence can be quite deceptive. For example, for $\kappa = 9.25$ we started with nearly four cell data. The solution at first appears to converge to a steady four cell, but an instability can be observed by noting a slow growth in the three mode. Eventually, the three mode grows to a point where the residual increases and the computation appears to diverge as the three cell develops, and the four cell decays. Finally, the solution converges to a steady three cell.

We remark that this problem also exhibits solutions that oscillate in time about the basic solution. These pulsating solutions occur for $L > L^* > 1$. Our computations, employing L as a bifurcation parameter, reveal that as L is increased through a primary bifurcation point at $L = L^*$, a transition occurs from the basic solution to a solution executing approximately sinusoidal oscillations about the basic solution. As L is increased further, the oscillation develops into a relaxation oscillation, whose peaks become progressively sharper and steeper. Further increasing L leads to a period doubling secondary bifurcation. These results will be reported elsewhere.

Acknowledgments

This research was supported by the Applied Mathematical Sciences subprogram of the Office of Energy Research, U.S. Department of Energy, under contract W-31-109-Eng-38 and grant DEFG02-87ER25027, and by National Science Foundation grant DMS8701543. The results in this paper were reported at a symposium honoring C. C. Lin, held at MIT, Cambridge, Mass., June 22-24, 1987. We thank G. Pieper for her help in preparing the manuscript.

References

1. Matkowsky, B. J., Putnick, L. J., and Sivashinsky, G. I., "A Nonlinear Theory of Cellular Flames", *SIAM J. Appl. Math.* **38**, 489-504 (1980).

2. Gottleib, D. and Orszag, S. A., "Numerical Analysis of Spectral Methods: Theory and Applications", *CBMS-NSF Regional Conference Series in Applied Mathematics* 26, SIAM, Philadelphia (1977).

3. Bayliss, A. and Matkowsky, B. J., "Fronts, Relaxation Oscillations, and Period Doubling in Solid Fuel Combustion", *J. Comp. Phys.* **71**, 147-168 (1987).

4. Bayliss, A., Matkowsky, B. J., and Minkoff, M., "Adaptive Pseudo-Spectral Computation of Cellular Flames Stabilized by a Point Source", *Applied Math. Letters* (1987), to appear.

5. Street, C. L., Zang, T. A., and Hussaini, M. Y., "Spectral Multigrid Methods with Applications to Transonic Potential Flow", *J. Comp. Phys.* **57**, 43-76 (1985).

6. Markstein, G. H., ed., *Nonsteady Flame Propagation*, Macmillan (Pergamon Press), New York, 1964, p. 79.

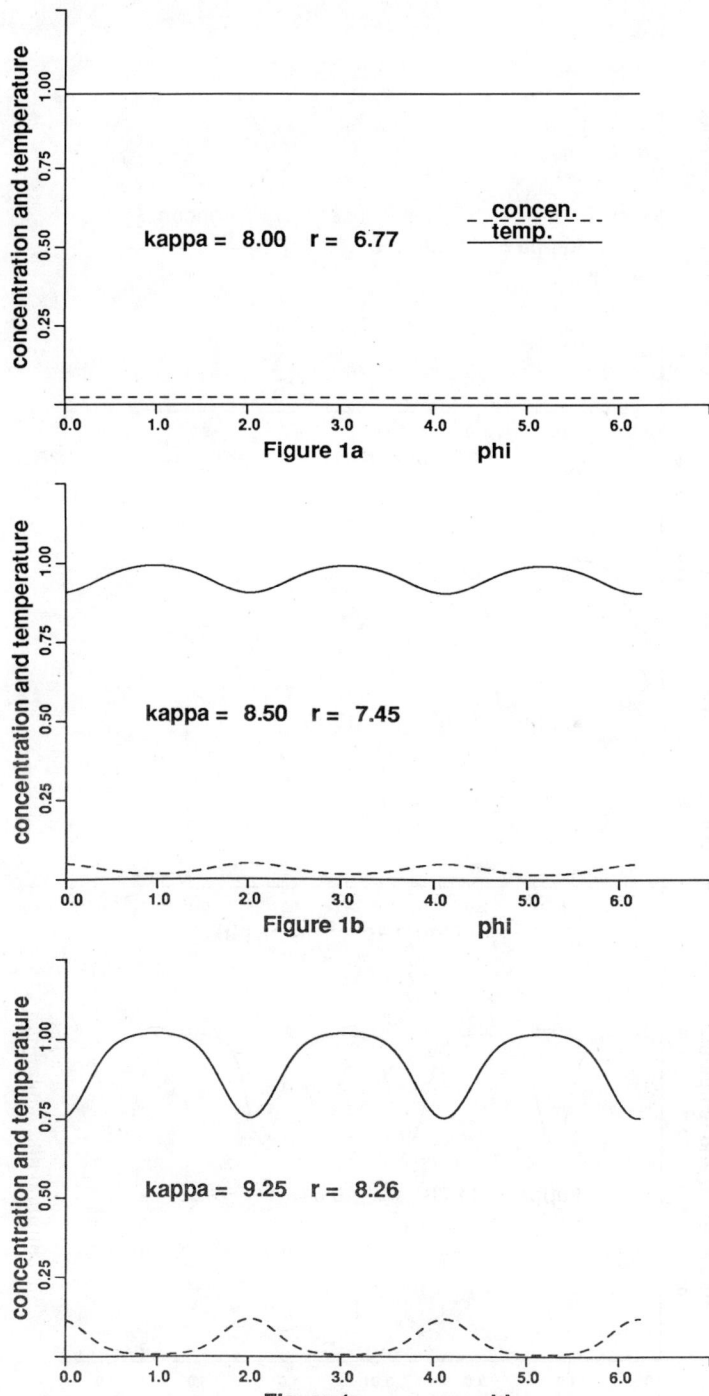

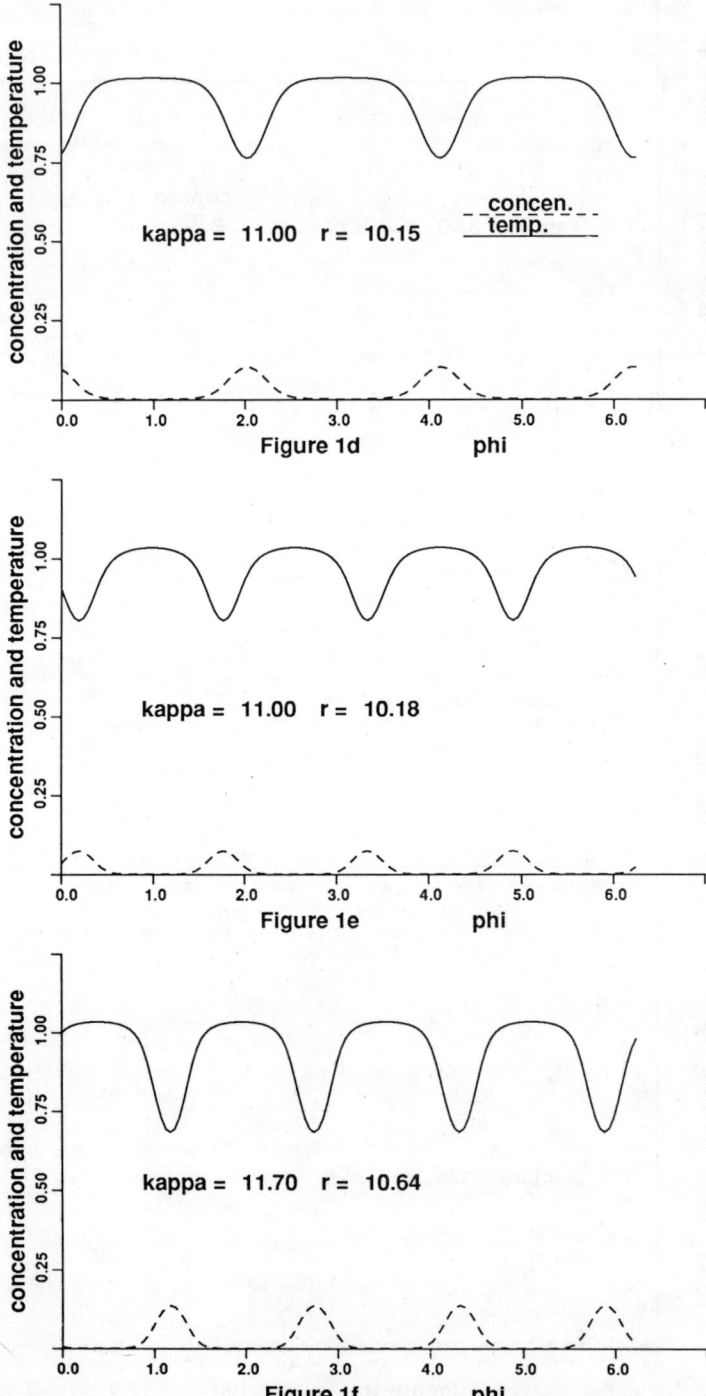

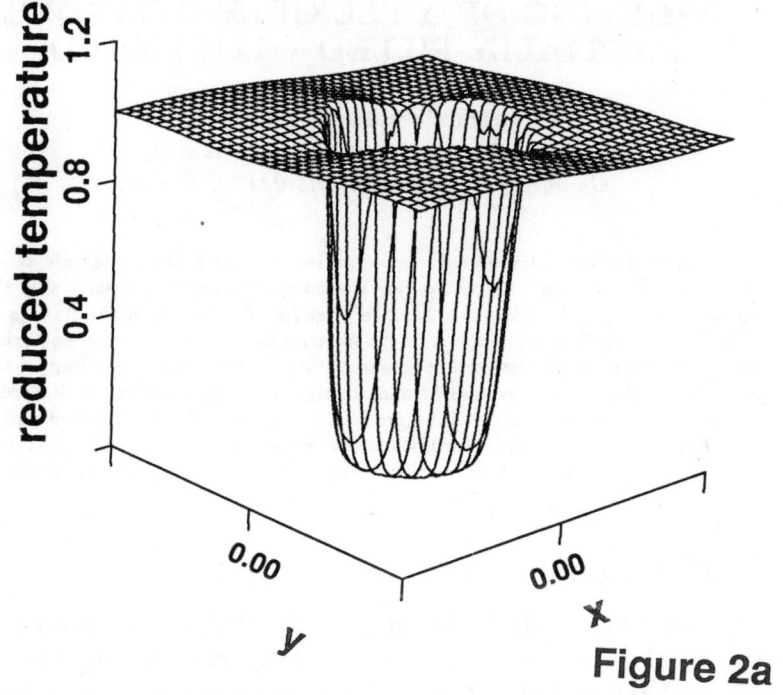

Figure 2a

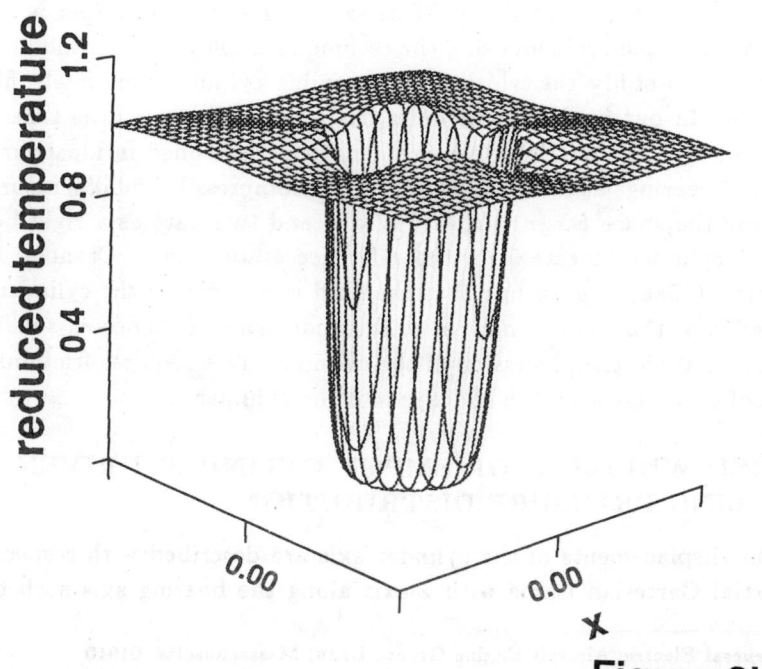

Figure 2b

WHIRLING OF A FLEXIBLE CYLINDER PARTIALLY FILLED WITH LIQUID

S.H. Crandall and J.W. Mroszczyk[*]
Massachusetts Institute of Technology
Cambridge, Massachusetts 02139, U.S.A.

A uniform flexible cylinder with inner radius a rotates about its axis at rotational rate Ω in a gravity-free environment. A liquid partially fills the cylinder with a cylindrical free surface of radius $b < a$ when the fluid is at rest in the rotating reference frame. We consider small dynamic perturbations of this state in the form of steady state asynchronous whirling of the coupled fluid-structure system. A matrix eigenvalue problem of infinite order is derived for the natural whirling modes. Critical speeds are obtained and it is inferred from approximate solutions that supercritical rotation is unstable. Experimental results, which provide support for the analytical predictions, are described.

INTRODUCTION

The whirling of a rigid cylinder partially filled with inviscid liquid was demonstrated to be unstable for a range of supercritical rotational speeds by Kollman[1] and by Wolf[2]. Analytical explanations were given by Kuipers[3] and by Wolf[2]. The extension to a flexible cylinder has been studied by Hendricks[4] and by Crandall and Mroszczyk[5]. The present paper is a reexamination of the problem using the technique employed by Crandall and Mroszczyk[6] to study the whirling of a flexible cylinder completely filled with fluid. In our model the cylinder has inner radius a and is taken to be a uniform Bernoulli-Euler beam of length L suspended in ideal simply supported bearings at its ends. An inviscid incompressible fluid is assumed to occupy the space $b < r < a$, $0 < z < L$ and to rotate as a rigid body with the cylinder at rate Ω in the reference state. As in Crandall and Mroszczyk (1986)[5] we assume that the fluid is retained in the cylinder by rigid end caps that remain normal to the undeformed cylinder axis and do not participate in the precession of the cylinder. This requires frictionless, leakproof joints between the endcaps and the cylinder.

FORCED WHIRLING OF EMPTY CYLINDER DRIVEN BY WHIRLING PRESSURE DISTRIBUTION

The displacements of the cylinder axis are described with respect to an inertial Cartesian frame with z-axis along the bearing axis such that

---

[*] General Electric Aircraft Engine Group, Lynn, Massachusetts 01910

the rotating cylinder extends from one bearing plane at $z = 0$ to the other bearing plane at $z = L$. During steady-state whirl at frequency ω the axis of the cylinder at z is assumed to travel in a circle of radius $R(z)$ in a plane perpendicular to the z-axis with center on the z-axis. By introducing a complex plane with real axis parallel to x and imaginary axis parallel to y the runout of the whirling cylinder axis can be represented by the vector $R(z)e^{i\omega t}$. The excitation which drives this whirl is the resultant force per unit length $F(z)e^{i\omega t}$ exerted on the cylinder by a whirling pressure distribution. The response $R(z)$ to the excitation $F(z)$ is obtained by solving the Bernoulli-Euler beam equation subject to simply supported boundary conditions at the ends. A formal solution[6] is

$$R(z) = \sum_{s=1}^{\infty} B_s \sin s\pi z/L \qquad (1)$$

where

$$B_s = \frac{2}{L} \frac{\int_0^L F(z) \sin s\pi z/L \, dz}{EI(s\pi/L)^4 - \rho_0 A_0 \omega^2} \qquad (2)$$

with EI and $\rho_0 A_0$ representing the flexural modulus and the mass per unit length, respectively, of the cylindrical beam.

FORCED WHIRLING OF ROTATING LIQUID DRIVEN BY CONTAINER MOTION

In the reference state the liquid rotates like a rigid body about the z-axis at a rate Ω, confined by the cylinder wall at $r = a$ and by the end caps at $z = 0$ and $z = L$, and with a free surface at $r = b$. The subsequent fluid motion is described with respect to a cylindrical coordinate system r, θ, z which rotates about the z-axis at rate Ω with respect to the inertial coordinate system. When the cylinder axis whirls with runout $R(z)e^{i\omega t}$ with respect to the non-rotating coordinate system, an observer in the rotating frame will observe a radial displacement wave in the container wall of the form

$$r_{\text{wall}} = a + R(z)\cos(\theta + nt) \qquad (3)$$

where $n = \Omega - \omega$ is the relative whirl rate of the backward wave (3). The pressure in the fluid is a superposition of that due to the static centrifugal field plus that due to the perturbation flow

$$p = \frac{1}{2}\rho r^2 \Omega^2 + \text{Re}\left\{P(r,z)e^{i(\theta + nt)}\right\} \qquad (4)$$

where the perturbation pressure amplitude is[7]

$$P(r,z) = C_0\left(\frac{r}{a} + \nu_0\frac{a}{r}\right) + \sum_{m=1}^{\infty} C_m\left[J_1(k_m r) + \nu_m Y_1(k_m r)\right]\cos m\pi z/L \quad (5)$$

This expression satisfies the boundary conditions of the nonprecessing end caps independently of the parameters k_m, ν_m, and C_m. The linearized momentum and continuity requirements are satisfied if

$$k_m^2 = (m\pi/L)^2(4\Omega^2 - n^2)/n^2 \quad (6)$$

The linearized free-surface boundary condition at $r = b$ is satisfied if

$$\nu_0 = -\frac{2\Omega^2 - 3n\Omega^2 + n^3}{2\Omega^2 - 5n\Omega^2 + n^3}\frac{b^2}{a^2}$$
$$\nu_m = -\frac{(2\Omega^2 - 5n\Omega^2 + n^3)J_1(k_m b) + n\Omega^2 k_m b J_0(k_m b)}{(2\Omega^2 - 5n\Omega^2 + n^3)Y_1(k_m b) + n\Omega^2 k_m b Y_0(k_m b)} \quad (7)$$

and, finally, (5) is consistent with the boundary motion (3) if

$$C_0 = -\left\{\rho(4\Omega^2 - n^2)a/L\int_0^L R(z)\,dz\right\}/\left\{(2\Omega/n + 1) + \nu_0(2\Omega/n - 1)\right\}$$

$$C_m = -\left\{\rho(4\Omega^2 - n^2)2a/L\int_0^L R(z)\cos m\pi z/L\,dz\right\}/$$
$$\left\{[(2\Omega/n - 1)J_1(k_m a) + k_m a J_0(k_m a)]\right\}$$
$$\left\{+\nu_m[(2\Omega/n - 1)Y_1(k_m a) + k_m a Y_0(k_m a)]\right\}$$
$$(8)$$

The force per unit length acting on the cylinder due to the pressure (4) is[7]

$$F(z)e^{i\omega t} = \pi a\left[P(a,z) + \rho\Omega^2 a R(z)\right]e^{i\omega t} \quad (9)$$

with respect to the nonrotating coordinate system.

FREE WHIRLING OF THE COUPLED SYSTEM

In steady-state whirling of the partially filled cylinder the force (9) of the fluid reaction drives the cylinder runout (1) as indicated by (2) and the cylinder runout drives the perturbation pressure (5) as indicated by (8). Analytically this is represented by writing (8) with $R(z)$ replaced by (1) and writing (2) with $F(z)$ replaced by its value in (9) with $P(a,z)$ and $R(z)$

expanded according to (5) and (1), respectively. The result is an infinite partitioned matrix equation

$$\begin{bmatrix} \alpha_{im} & \beta_{is} \\ \gamma_{rm} & \delta_{rs} \end{bmatrix} \begin{Bmatrix} C_m \\ B_s \end{Bmatrix} = 0, \qquad \begin{cases} m, i & =0,1,2,\ldots \\ r, s & =1,2,3,\ldots \end{cases} \qquad (10)$$

in which $[\alpha_{im}]$ and $[\delta_{rs}]$ are diagonal matrices with

$$\begin{aligned} \alpha_{oo} &= (2\Omega/n + 1)\nu_0(2\Omega/n - 1) \\ \alpha_{mm} &= [(2\Omega/n - 1)J_1(k_m a) + k_m a J_0(k_m a)] \\ &\quad + \nu_m [(2\Omega/n - 1)Y_1(k_m a) + k_m a Y_0(k_m a)] \\ \delta_{ss} &= EI(s\pi/L)^4 - \rho_0 A_0 \omega^2 - \pi \rho a^2 \Omega^2 \end{aligned} \qquad (11)$$

and the elements of the coupling matrices are

$$\begin{aligned} \beta_{os} &= \rho a(4\Omega^2 - n^2)\epsilon_{so}/2 \\ \beta_{is} &= \rho A(4\Omega^2 - n^2)\epsilon_{si}, & i &= 1, 2, \ldots \\ \gamma_{ro} &= -\pi a(1 + \nu_0)\epsilon_{ro} \\ \gamma_{rm} &= -\pi a[J_1(k_m a) + \nu_m Y_1(k_m a)]\epsilon_{rm}, & m &= 1, 2, \ldots \end{aligned} \qquad (12)$$

where ϵ_{sm} is zero when s and m have the same parity and equals $(4/\pi)s/(s^2 - m^2)$ when s and m have opposite parity.

A set of exact solutions to (10) is provided by the synchronous whirls with $\omega = \Omega$ or $n = 0$. Here the perturbation pressure amplitude (5) vanishes; i.e., $C_m = 0$ for all m and (10) reduces to $\delta_{ss} = 0$ from which we find the critical speeds Ω_s given by

$$\frac{\Omega_s}{\omega_0} = \frac{s^2}{\sqrt{1+\mu}}, \qquad s = 1, 2, \ldots \qquad (13)$$

where $\omega_0 = \pi^2(EI/\rho_0 A_0)^{1/2}/L^2$ is the fundamental bending frequency of the empty cylinder and $\mu = \pi a^2 \rho / \rho_0 A_0$ is the ratio of the fluid mass required to fill the cylinder to the mass of the cylinder. It is paradoxical that (13) is independent of the actual amount of liquid in the cylinder until it is realized[8] that while the cylinder wall whirls with runout $R(z)$, the free surface remains a cylinder of radius b, concentric with the axis of rotation.

APPROXIMATE SOLUTIONS

Approximate solutions to (10) are obtained by truncating the infinite matrices. Fig. 1 displays the real whirl frequencies ω as a function of the rotation rate Ω for a truncated solution in which the only non-zero coefficients retained are C_0, C_1, C_2 and B_1, B_2, B_3 for the case of a cylinder with $L = 10a$ filled to the point where $b/a = 2/3$ with a fluid for which the mass ratio is $\mu = 0.206$. In Fig. 1 the frequencies of the

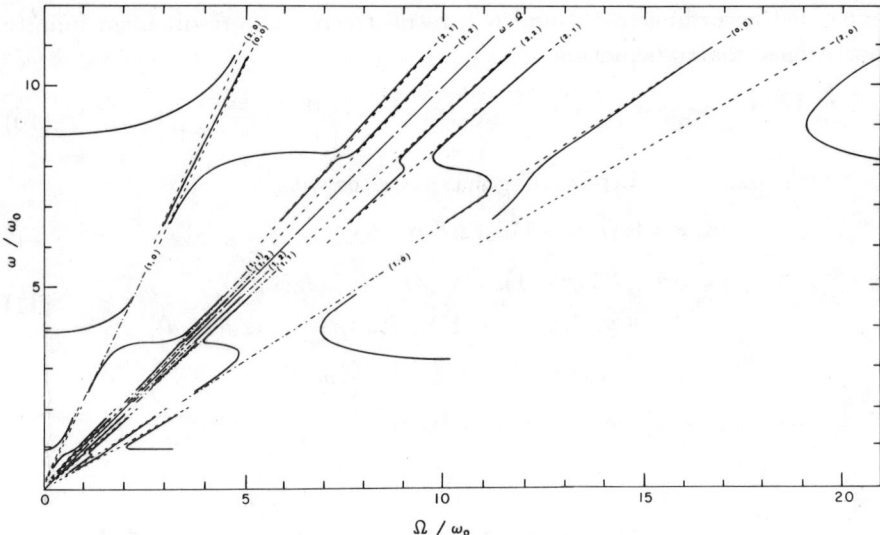

Fig. 1. Natural whirling frequencies ω of a flexible cylinder partially filled with liquid as a function of rotation rate Ω.

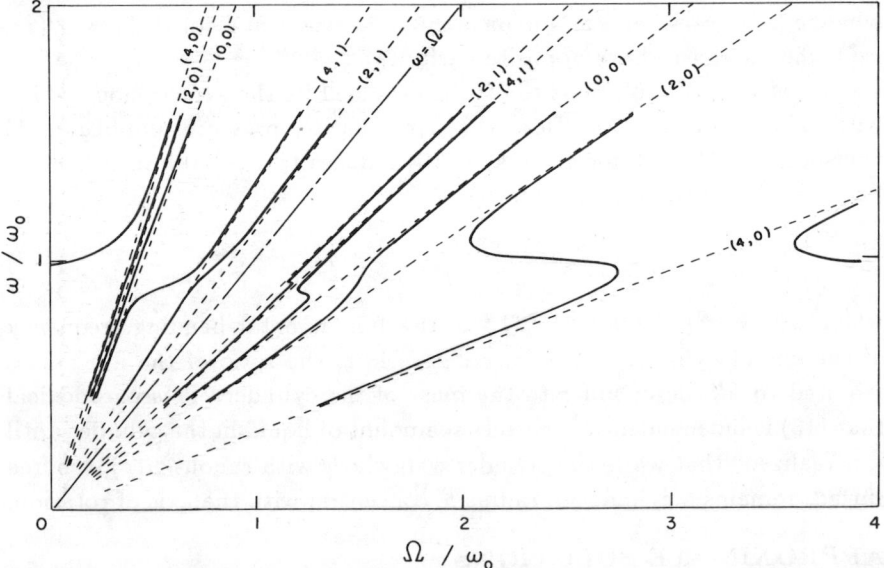

Fig. 2. Natural whirling frequencies of coupled system in neighborhood of fundamental bending frequency ω_0 of empty cylinder.

uncoupled fluid modes[9] are shown as dashed lines labeled (m, p) where m is the number of axial nodal planes and p is the number of nodal cylinders in the radial velocity amplitude. These frequencies correspond to the n-values which make the denominators in (8) vanish. For given m and p there are two distinct fluid modes (m,p), one with subsynchronous frequency $(\omega < \Omega)$ and one with supersynchronous frequency $(\omega > \Omega)$. With m fixed, these frequencies approach Ω from both sides as $p \to \infty$.

The most significant fact displayed in Fig. 1 is that the coupling is conservative for whirl frequencies ω that are supersynchronous, while it is unconservative when $\omega < \Omega$. In the former case the rotation is less than the resonant frequency and the operation is called subcritical: the coupled frequency curves repel one another at the intersection of the uncoupled frequencies. In supercritical operation the coupled frequencies attract one another and join leaving gaps in the speed range where the eigenvalues ω have complex conjugate roots, signalling instability. This effect is visible for the first three cylinder bending modes in Fig. 1. In Fig. 2, the behavior in the vicinity of the first bending mode is shown on an enlarged scale based on a truncation of (10) in which only the coefficients C_0, C_2, C_4, and B_1 are retained. To the left of the synchronous line all whirl frequencies ω are real. To the right of the synchronous line there are gaps between the curves which indicate unstable speed ranges. The precise location of the unstable gaps depends on the truncation selected, but these results strongly suggest that instabilities can be found as close to the synchronous line as desired. This suggests that the first critical speed Ω_1 of (13) is the stability limit for whirling of a flexible cylinder independently of the amount of liquid contained (so long as the entire circumference is wetted).

EXPERIMENTAL VERIFICATION

Experiments were performed[7] using a Lexan plastic cylinder with $L = 10a = 0.707$m. The fluid was water and the mass ratio for the completely filled cylinder was $\mu = 31.2$. Stable whirling modes were excited by applying a transverse vibratory force with frequency ω to a short stub shaft extension of the rotor while it was rotating at rate Ω. The whirling was observed by a pair of non-contacting displacement transducers stationed 90° apart at center span. Results for the case where $b/a = 2/3$ are displayed in Fig. 3. The solid lines are analytical predictions obtained from (10) by truncation, retaining only the coefficients C_0, C_2, and B_1. The circled points represent observed whirling resonance frequencies ω at the indicated rotational rates Ω. In Fig. 3 the experimentally determined stability limit is indicated. It is 5% smaller than the critical speed Ω_1 predicted by (13) based on the measured fundamental bending frequency ω_0 of the empty cylinder. In addition, the experiments showed that the stability limit was substantially independent of the amount of water in the cylinder. The measured stability limit varied less than 2% over the range from $b/a = 0.25$ (nearly full) to $b/a = 0.95$ (nearly empty).

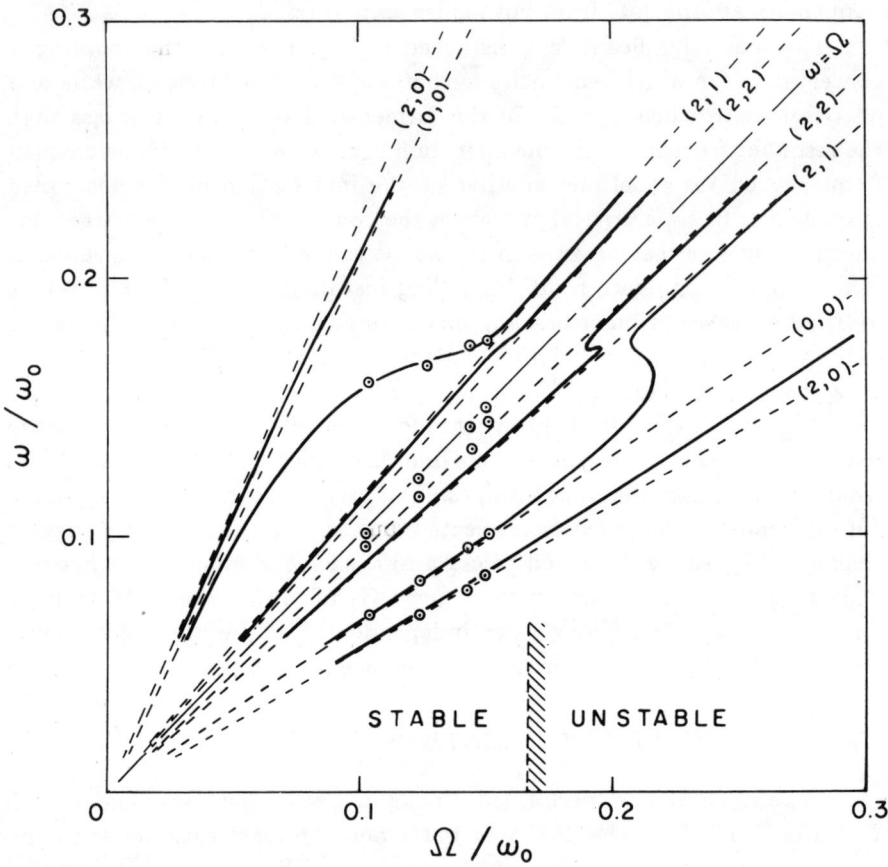

Fig. 3. Comparison of experimental measurements (circled) with analytical predictions (solid lines).

References

1. Kollman, F.G.,"Experimentelle und theoretische Untersuchen uber die kritischen Drehzahlen flussigkeits-gefullter Hohlkorper", *Forschung auf dem Gebiete des Ingenieurwesens* **28**, 115–123 and 147–153 (1962).
2. Wolf, J.A., "Whirl dynamics of a rotor partially filled with liquid, *J. Appl. Mech.* **35**, 676–682 (1968).
3. Kuipers, M., "On the stability of a flexibly mounted rotating cylinder partially filled with liquid", *Appl. Sci. Research* **A13**, 121–137 (1964).

4. Hendricks, S.L., "Dynamics of flexible rotors partially filled with a viscous incompressible fluid, *Ph.D. thesis*, University of Virginia, Charlottesville, VA, May, 1979.
5. Crandall, S.H. and Mroszczyk, J.W., "Stability of a flexible rotor partially filled with liquid", pp. 315–354, *Proc. 6th Intl. Workshop on Gases in Strong Rotation*, Tokyo, Aug. 20-23, 1985.
6. Crandall, S.H. and Mroszczyk, J.W., "Whirling of a flexible cylinder filled with liquid", pp. 449–452, *Proc. Intl. Conf. on Rotordynamics*, Tokyo, Sept. 14-17, 1986.
7. Mroszczyk, J.W., "Instability of dynamic systems based on uncoupled subsystem responses, with applications to rotor dynamics, *Ph.D. thesis*, M.I.T., Cambridge, MA, Nov. 1986.
8. Crandall, S.H., "The physical nature of rotor instability mechanisms", pp. 1–18, *Rotor Dynamical Instability*, ASME, AMD Vol. 55, 1983.
9. Miles, J.W. and Troesch, B.A., "Surface oscillations of a rotating liquid", *J. Appl. Mech.* **28**, 491–496 (1961).

LINEAR STABILITY THEORY FOR A SUBSONIC MIXING LAYER IN THE VISCOUS CRITICAL LAYER REGIME

V. Djordjevic* and L. G. Redekopp

Department of Aerospace Engineering

University of Southern California

Los Angeles, California 90089-1191

The stability of hydrodynamic shear flows has been powerfully impacted by the pioneering work of Professor C. C. Lin. After more than three decades his monograph[1] remains a most valuable resource and provides a clear and concise treatment of, among other subjects, the difficult problem of the asymptotic structure of the flow in the limit of large Reynolds number. His contributions to this subject relates directly to the analysis and results presented in this paper, as does his seminal work[2] on the influence of compressibility on the stability of these flows.

The stability of a parallel free shear flow to infinitesimal perturbations is considered. The basic flow is chosen to have a symmetric vorticity and temperature variation with a single extremum of each occurring along the same streamline $y = y_s$ on which the Lees-Lin criteria

$$\frac{d}{dy}\left[\frac{U_0'(y)}{T_0(y)}\right]_{y=y_s} = 0, \qquad (1)$$

is satisfied. Expressions for the dimensionless velocity $U_0(y)$ and temperature $T(y)$ of the undisturbed state of the shear flow are chosen to be

$$U_0(y) = \tanh y, \quad T_0(y) = 1 + b \operatorname{sech}^{M^2} y \qquad (2)$$

The parameter M is the Mach number of the uniform reference state at $y \to +\infty$ and b specifies the symmetric excess ($b > 0$) or deficit ($-1 < b < 0$) of the temperature along the streamline $y = 0$ where the magnitude of the vorticity is maximum. This specific flow is examined here because the neutral curve in subsonic flow ($M < 1$) is known analytically. In terms of the perturbation pressure expressed as

$$p(x, y, t) = P(y) \exp\{i\alpha(x - ct)\}, \qquad (3)$$

the neutral eigensolution for the inviscid problem is given by[3]

$$P(y) = \operatorname{sech}^{\alpha^2} y, \quad c = c_r + i c_i = 0, \quad \alpha^2 + M^2 = 1. \qquad (4)$$

---

*On leave from Dept. of Mechanical Engineering, University of Belgrade, Yugoslavia

The neutral curve in the (α^2, M^2) plane is the section of a circle of unit radius centered at the origin. For values of (α, M) inside the quarter-circle in the first quadrant, the inviscid flow is linearly unstable. In what follows we examine the diffusive corrections to these results in the limit of large, but not infinite, Reynolds number.

In the limit of large Reynolds number, diffusive effects are first felt in a narrow region, termed the critical layer, around the level $y = 0$ where the Doppler-shifted velocity vanishes. The linear motion within this thin layer and the matching properties for the flow across the critical layer in the limit of large Reynolds number was clarified in a series of very significant papers by Lin[1]. In this report, we assess the effects of Reynolds number Re, Prandtl number Pr, Mach number M, and the viscosity-temperature variation, as manifested through the critical layer dynamics in the linear-viscous limit, on the stability of a compressible free shear flow. The results should be generic to any subsonic critical layer in which the velocity is an odd function and the temperature is an even function in the immediate vicinity of the critical level since the principal results depend only on the first several terms of the Taylor series expansion of the mean profiles and the linear eigenfunctions about the critical level.

The approach we follow is to derive an amplitude equation for the marginal mode in the vicinity of the eigen-loci defined in (4). For this purpose we define a parameter ϵ which is of order the inverse Reynolds number and introduce slow space-time scales as follows:

$$\lambda \epsilon = \frac{1}{\text{Re}}, \quad \xi = \epsilon x, \tau = \epsilon t. \tag{5}$$

λ is a viscous parameter with unit order of magnitude. Furthermore, we take a power-law dependence of the viscosity and thermal conductivity on the temperature which, in dimensionless form, is given by

$$\mu = T^n, \quad k = T^n. \tag{6}$$

Letting $A(\xi, \tau)$ be a unit-order amplitude function, the solution for the perturbation pressure takes the form

$$p(x, y, t) = \left\{ AP(y) + \epsilon \left[\frac{\partial A}{\partial \tau} P_1^{(\tau)}(y) + \frac{\partial A}{\partial \xi} P_1^{(\xi)}(y) + \lambda A P_1^{(\lambda)}(y) \right. \right.$$
$$\left. \left. + B^{\pm}(\xi, \tau) Q(y) \right] + O(\epsilon^2) \right\} \exp\{i\alpha x\}. \tag{7}$$

The function $Q(y)$ is the second, linearly independent solution of the homogenous, eigenvalue equation. It is needed in the solution in order to satisfy the boundary conditions at $y \to \pm\infty$. The $\pm$ superscript designation on the function $B^{\pm}(\xi, \tau)$ denotes different values of the function on either side of the critical layer. In fact, application of the boundary conditions at $y \to \pm\infty$ yields the jump relation

$$B^+ - B^- = -4i(1-M^2)^{3/2}\left[I_1 + bI_2\right]\frac{\partial A}{\partial \xi}, \qquad (8)$$

where I_1 and I_2 are expressions containing gamma functions with arguments which include the Mach number M. The function $P_1^{(\tau)}$ and $P_1^{(\lambda)}$ have $y\ln(y)$ singularities at the critical level, which is evidence of the singular perturbation $\epsilon \sim \mathrm{Re}^{-1} \to 0$.

The construction of a uniformly valid solution through the level $y = 0$ requires the introduction of an inner variable $Y = \epsilon^{-1/3}y$ and solution of the equations of motion in the critical layer $Y = O(1)$ in the limit $\epsilon \to 0$. The details of constructing the inner solution and matching with the outer flow (7) are algebraically involved and cannot be included here. The result, however, is that an additional expression for the jump $(B^+ - B^-)$ is obtained which, when combined with (8), yields an evolution equation for the amplitude function $A(\xi, \tau)$:

$$\frac{\partial A}{\partial \tau} + \lambda r A - \frac{i}{\pi}s\frac{\partial A}{\partial \xi} = 0, \qquad (9)$$

where r and s are lengthy expressions involving the Mach number M, the Prandtl number Pr, the transport coefficient parameter n (viz., equation 6), the ratio of specific heats γ, and the temperature profile parameter $b = 1 - T_0(0)$.

Equation (9) provides a convenient basis for evaluating the effect of the various parameters on the linear stability of a free shear flow. To this end, we look for a solution of the form

$$A(\xi, \tau) = a(\tau)\exp\{-i\Delta\alpha\xi\}, \qquad (10)$$

and evaluate the threshold wave number shift $\Delta\alpha_t$ for neutral stability requiring $\dot{a}(\tau) = 0$. With the specified sign for $\Delta\alpha$, $\Delta\alpha_t > 0$ implies a stabilizing effect and $\Delta\alpha_t < 0$ a destabilizing effect. Results for air (γ=1.4 and Pr=0.716) with n=0.76 are presented in Figures 1 and 2. Curves of constant Mach number are presented since the temperature profile shape (viz., equation 2) is fixed along these curves and only the magnitude of the temperature excess or deficit along the critical-level streamline varies. It is worth noting that the limit b fixed, $M \to 0$ is uniform, providing information on the effect of variable density on the stability of incompressible shear layers. It is clear that the parameter n exerts a strong influence on the stability properties and that strong cooling of the critical layer region has a destabilizing effect. Further studies of compressible free shear flows, including finite-amplitude effects, are in progress and will be reported elsewhere.

L.G.R. was partially supported by the Air Force Office of Scientific Research Grant No. F49620-85-C-0080.

References
1. Lin, C.C., *The Theory of Hydrodynamic Stability*, Cambridge University Press (1955).
2. Less, L. and Lin, C.C., "Investigation of the stability of the laminar boundary layer in a compressible fluid," NACA TN 1115 (1946).
3. Djordjevic, V. and Redekopp, L.G., "Linear Stability Analysis of Inviscid Compressible Flows," submitted to *Physics of Fluids* (1987).

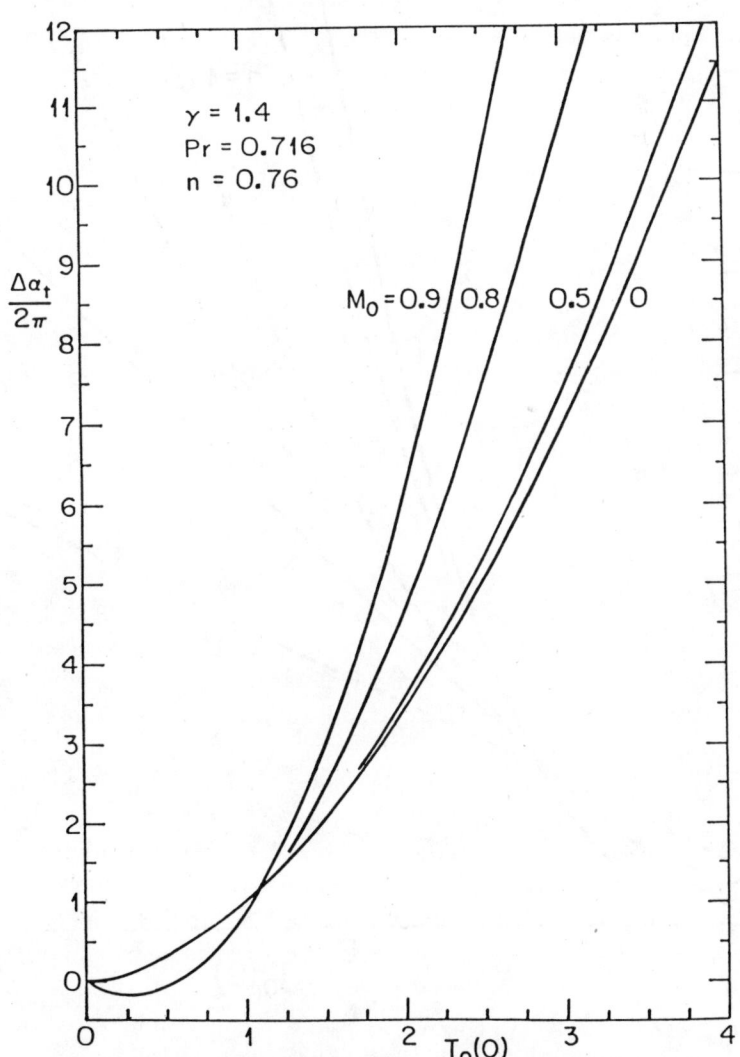

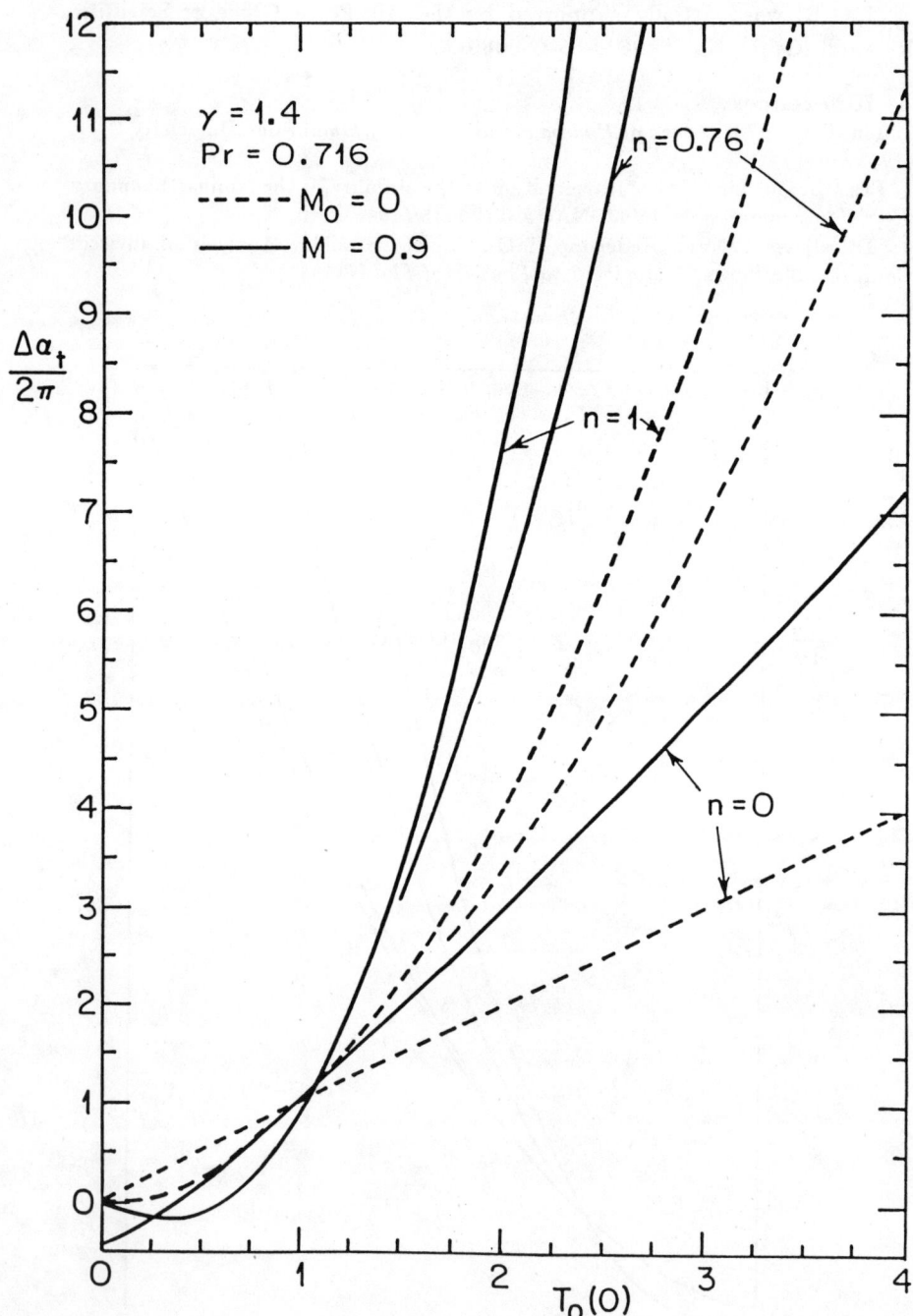

PERTURBATIONS OF JEFFERY-HAMEL FLOWS

Philip G. Drazin
University of Bristol
School of Mathematics, University Walk,
Bristol BS8 1TW, ENGLAND

Small perturbations of a basic Jeffery-Hamel flow (radial flow between two rigid planes) of a uniform incompressible viscous fluid are discussed. Some recent results on spatial and temporal instability, linear and nonlinear, are reported. The angle between the plane boundaries is not assumed to be small, but there is an analogy between the results and those of perturbations of plane parallel flows, perturbations which were elucidated so well by C. C. Lin. The perturbations of Jeffery-Hamel flows are related to flow along a two-dimensional channel with walls of small curvature, much as the perturbations of parallel flows are related to the development of jets, shear layers and the boundary layer on a plate.

1. INTRODUCTION

The aim of this work is to understand flows through a two-dimensional channel of variable width, their instabilities, their bifurcations and their transition to turbulence. To specify these flows, we suppose that the channel has two rigid impermeable walls with equations $z = f(x)$ and $z = -g(x)$ for $-\infty < x < \infty$ in the zx-plane, and that a uniform incompressible fluid of kinematic viscosity ν is driven along the channel at a steady rate Q (volume flux in the x-direction per unit time per unit length spanwise). Then it is convenient to define a flux Reynolds number as $R = Q/2\nu$. Numerical integration of the Navier-Stokes equations has shown recently[1] that instability may occur at quite low values of R and that there is a rich variety of instabilities and bifurcations as R increases.

To understand the mechanisms of these flows, some analysis is desirable. Two asymptotic approximations come to mind. To explain them, we define α as the angle between the centre plane $z = 0$ of the channel and the tangent plane to the upper wall at station x, so that $\tan\alpha(x) = f'(x)$. Similarly, we define the angle β with respect to the lower wall such that $\tan\beta(x) = g'(x)$. Banks *et al.*[2] illustrate this. Then the two approximations are that (P) the walls are nearly parallel, i.e. $|\alpha(x)|, |\beta(x)| \ll 1$ and (R) the walls are nearly plane, i.e., have small curvature, i.e. $|\alpha'(x)|, |\beta'(x)| \ll 1$.

These approximations will be applied, paragraph by paragraph, to channel flows. The nearly parallel approximation, P, is now well understood, not least through the efforts of C. C. Lin,[3,4] so it offers a natural way to express my gratitude to, as well as my admiration for, him, for giving me my first job and wisely guiding my research at that crucial stage of my career. At the same time, the understanding of approximation P helps to describe new applications of the less well-known approximation, R, first proposed by Fraenkel[5] in 1962. The plan of this paper, then, is to review a part of the theory of instability of parallel flows in a paragraph (denoted at the beginning by a P) and immediately to develop the analogous part of the theory of instability of radial flows in the following paragraph (denoted by an R).

2. PERTURBATIONS OF PARALLEL AND RADIAL FLOWS

P1   In seeking a basic flow through a channel with impermeable walls at $z = \pm f$, for constant f, where each wall has prescribed velocity in its own plane, we observe that the problem is steady, two-dimensional and invariant under the group of translations in the x-direction. Accordingly we assume that there is a basic flow which has the same properties, i.e. that the stream function has the form $\psi = \Psi(z)$. In other words, there is a steady plane parallel basic flow with velocity $U = d\Psi/dz$. (Of course, we recognize that there are other flows compatible with the specification of the problem which are unsteady, three-dimensional or not invariant under all translations.) Now the vorticity equation governing the flow gives

$$\frac{d^4 \Psi}{dz^4} = 0 \ .$$

This equation and the boundary conditions give plane Couette-Poiseuille flow, with a solution of the form

$$U(z) = Az^2 + Bz + C$$

for constants A, B and C determined uniquely by the prescribed velocities at the boundaries and the volume flux (or pressure gradient) along the channel.

R1 The analogous problem is due to Jeffery and Hamel. For this we consider flow between two impermeable planes with equations $\theta = \pm \alpha$ driven by a steady line source or sink of strength Q at the intersection $r = 0$ of the two planes, where (r, θ) are plane polar coordinates. We may also suppose that the velocities of the planes are either zero or prescribed such that they are radial and inversely proportional to r. Then the invariances of the problem indicate that we seek basic flows such that the stream function depends only on θ, i.e. that $\psi = \Psi(\theta)$. In other words, there is a steady, plane purely radial basic flow with velocity $U = d\Psi/rd\theta$. Then the vorticity equation governing the basic flow gives

$$\frac{d^4\Psi}{d\theta^4} + 4\frac{d^2\Psi}{d\theta^2} + \frac{2}{\nu}\frac{d\Psi}{d\theta}\frac{d^2\Psi}{d\theta^2} = 0 \ .$$

The solution of this equation and the boundary conditions is not unique, there being an *infinite* number of such flows. The flows were found by Jeffery and by Hamel in terms of Jacobian elliptic functions, and are now well classified.[5] In particular, there is a subcritical pitchfork bifurcation from infinity.

P2 To consider the instability of plane parallel flow, we seek solutions of the form $\psi = \Psi + \psi'$, where the perturbation stream function is $\psi'(x, y, z, t)$, linearizing the equations of motion for small ψ'. The variables x, y and t are separable, so we may consider independent normal modes of the form $\psi'(x, y, z, t) = \text{Re}\{\phi(z)e^{i(kx+ly-ct)}\}$. The perturbations are thereby shown to be governed by the Orr-Sommerfeld problem, which has dimensionless form,

$$\phi^{\text{iv}} - 2k^2\phi'' + k^4\phi = ikR\{(U-c)(\phi'' - k^2\phi) - U''\phi\}$$

and

$$\phi, \phi' = 0 \quad \text{at} \quad z = \pm 1 \ ,$$

where R is a suitable Reynolds number and a prime here denotes differentiation with respect to z. This poses some eigenvalue problems. The first is the one of temporal modes: to find the set of eigenvalues c and corresponding eigenfunctions ϕ for given U, R and all real wavenumbers k. If $\text{Im}(kc) \leq 0$ for all k for all eigenvalues then the basic flow is linearly stable. The second problem is one of spatial modes: to find k, for all real frequencies kc. If $\text{Im}(k) \leq 0$ always then the flow is often said to be stable (provided that the group velocity of the disturbances is positive).

R2 For the linearized problem of instability of a basic radial flow, the variables r and t are not separable together. However, we may separate them individually, taking either temporal modes with a disturbance stream function of the form $\psi' = e^{st}\phi(y, r)$ or *steady* spatial modes of the form $\psi' = r^\lambda \phi(y)$, where $y = \theta/\alpha$. The temporal problem to find the eigenvalues s is a partial-differential one, so only a little progress has been made; more work is needed. The problem for the steady spatial modes is rather narrow, but it seems to be important because the initial bifurcations of flows in a channel are of steady flows, at least when the maximum value of the angle

$\alpha(x)$ is not small. The problem has dimensionless form,

$$\phi^{iv} + \alpha^2\{\lambda^2 + (\lambda-2)^2\}\phi'' + \alpha^4\lambda^2(\lambda-2)^2\phi = \alpha R\{(\lambda-2)\Psi'(\phi'' + \alpha^2\lambda^2\phi) - 2\Psi''\phi' - \lambda\Psi'''\phi\}$$

and

$$\phi, \phi' = 0 \quad \text{at} \quad y = \pm 1,$$

where $R = Q/2\nu$ and a prime here denotes differentiation with respect to y. This determines eigenvalues λ and eigenfunctions ϕ for given Ψ, α and R. A plausible and simplified condition for stability is that disturbances do not grow downstream, i.e. that $\text{Re}(\lambda) \leq 0$.

P3   The abundance of early work on the Orr-Sommerfeld problem, by Lin and others, has been reported by Lin.[4] He described the work of Rayleigh on the special case with $R = 0$, and the calculation of the critical value of R for stability by use of asymptotic analysis as $R \to \infty$. He also described the custom of solving the Orr-Sommerfeld problem for the velocity profile U of a flow which is nearly parallel, even though the problem is posed on the assumption that the basic flow is parallel and an exact solution of the Navier-Stokes equations. Since Lin[4] wrote his book in 1955, many have worked on instability of nearly parallel flows and on weakly nonlinear instability. In particular, Eagles[6] has written some papers on the instability of nearly parallel Jeffery-Hamel flows, and Benney and Lin[7] have described the nonlinear interaction of selected three-dimensional modes of parallel flows. A review of the modern literature has been written recently by Drazin and Reid.[8]

R3   The analogous problem is of perturbations of radial flows. The special case with $R = 0$ is the simplest, the equation becoming biharmonic. This case was mentioned briefly by Rayleigh, but was first treated in detail by Dean and Montagnon. Others have done more work on it, shedding light on the criterion $\text{Re}(\lambda) \leq 0$ for spatial stability. There is no work yet on the asymptotic solution as $R \to \infty$, but it seems that instability arises before R becomes large, at least when α is not small. Indeed, the pitchfork bifurcation of the *basic* Jeffery-Hamel flows gives an explicit 'neutrally stable' eigensolution of the linear problem for all $R : \lambda = 0, \phi = \Psi'$ when $\alpha = \alpha_2(R)$; the function α_2 is that defined by Fraenkel.[5] This solution has been used by Banks et al.[2] to find the weakly nonlinear steady flow in a variable channel where $\alpha(x)$ is close to $\alpha_2(R)$.

3.   RESULTS FOR PERTURBATIONS OF RADIAL FLOWS

Sobey and Drazin[1] and Banks et al.[2] describe various recent results. Perhaps the most important is that all types of Jeffery-Hamel flows except those few with radial velocity U of the same sign for $-\alpha < \theta < \alpha$ are evidently unstable. Further, the condition of instability for diverging flows with positive U, namely $\alpha > \alpha_2(R)$, becomes $\alpha > 4.712/R$ as $R \to \infty$; now it is well-known[4] that plane Poiseuille flow, i.e. a Jeffery-Hamel flow with angle $\alpha = 0$, is unstable if $R > 3848$. (N.B. Defining R as a *flux* Reynolds number, we take the basic plane Poiseuille velocity as $U = 3(1 - z^2)/2$, and 3848 is 2/3 times the well-known critical value 5772 of R.) This suggests that a nearly parallel flow in a diverging channel may become unstable first to a Jeffery-Hamel mode of instability rather than the well-known Orr-Sommerfeld mode if the angle of the channel is such that $\alpha > 4.712/3848$ radians $= 0.07°$. The smallness of this angle emphasizes the practical significance of non-parallelism. The two modes of instability are, however, very different. The Orr-Sommerfeld mode is associated with extraction of energy from the basic flow in the critical layer, and with the subcritical Hopf bifurcation of plane Poiseuille flow. The Jeffery-Hamel mode is associated with the distortion of the basic velocity near the walls, and with the subcritical pitchfork bifurcation of steady radial flows.

References
1. Sobey, I. J. and Drazin, P. G., *J. Fluid Mech.* **171**, 263-287 (1986).
2. Banks, W. H. H., Drazin, P. G. and Zaturska, M. B., *J. Fluid Mech.* (in the press) (1987).
3. Lin, C. C., *Quart. Appl. Math.* **3**, 117-142, 218-234, 277-301 (1945).
4. Lin, C. C., *The Theory of Hydrodynamic Stability*, Cambridge Univ. Press (1955).
5. Fraenkel, L. E., *Proc. Roy. Soc. Lond.* **A267**, 119-138 (1962).
6. Eagles, P. M., *J. Fluid Mech.* **24**, 191-207 (1966).
7. Benney, D. J. and Lin, C. C., *Phys. Fluids* **3**, 656-657 (1960).
8. Drazin, P. G. and Reid, W. H., *Hydrodynamic Stability*, Cambridge Univ. Press (1981).

INSTABILITY DRIVEN BOUNDARY LAYERS ON CURVED WALLS

P. Hall
University of Exeter

W. D. Lakin
Old Dominion University

The fully nonlinear development of small wavelength Görtler vortices in a growing boundary layer on a wall with variable concave curvature is investigated using asymptotic techniques. The vortices considered here have sufficiently large amplitudes that the mean flow correction driven by them is as large as the basic state. A remarkable feature of the analysis is that, even for these large vortices, a relatively simple solution of the nonlinear problem can be obtained asymptotically. The development of the flow field is governed by a pair of nonlinear partial differential evolution equations for the vortex flow and mean flow correction. Through the nonlinear interactions, the fundamental and mean flow correction reinforce each other. Unlike most nonlinear stability problems, the higher harmonics play no role in this interaction. Beyond the position at which the vortices are marginally stable on the basis of linear theory, the vortices spread out across the boundary layer and effectively drive it. With the exception of a thin region near the wall, the vortices come to occupy the entire boundary layer, and even intrude into the free stream. The mean flow develops a "square-root" profile in the vortex region which bears essentially no relationship to the original mean profile. Thus, the mean flow adjusts so as to make these large amplitude vortices locally neutral. Location of the boundaries of the vortex region requires solution of a free boundary value problem involving the boundary layer equations.

1. INTRODUCTION

This work considers large amplitude Görtler vortices in viscous incompressible flows over walls of variable curvature. Recent interest in the Görtler instability stems from practical applications in areas such as Laminar Flow Control. The question of whether this type of instability is likely to induce premature transition because of its effect on the receptivity of the original boundary layer to Tollmien-Schlichting waves or crossflow vortices is still open. However, the present work shows that large amplitude Görtler vortices have the potential to completely alter the basic state both in and downstream from regions of variable convex curvature. Even if the curvature is such that the region of vortex activity terminates at some streamwise position, the downstream mean flow cannot be assumed to simply resume the original boundary layer profile. Consequently, the Görtler vortices and modified mean flow considered here are relevant in any study of the transition process over curved surfaces.

As noted by Hall and Lakin (1988), much of the previous work on the Görtler problem has not accounted for boundary layer growth in a self-consistent manner. Neglect of non-parallel effects in the Görtler problem leads to inconsistent, and in some cases physically absurd, results. Extreme examples are predictions of instability at zero Görtler number or zero wavenumber. Consistent results are obtained, however, at high wavenumber, and in this regime Hall (1982) has shown that an asymptotic solution of the non-parallel problem is possible. The vortices are found to become linearly unstable at a particular downstream location x and to concentrate themselves at a depth in the boundary layer corresponding to the position y where Rayleigh's inviscid instability criterion for the boundary layer is most violated. Hall (1983) has also shown that for $O(1)$ wavenumbers no asymptotic or self-consistent parallel flow caculation is possible, and the neutral stability curve depends on the nature of the initial disturbance.

Asymptotic methods will be used here to describe the fully nonlinear development of small wavelength Görtler vortices in a growing boundary layer. In particular, the vortices have sufficiently large amplitudes that the mean flow correction driven by them is of the same order as the basic state. Further, the depth of the fluid where vortex activity persists is now $O(1)$. Development of the flow field is governed by coupled nonlinear partial differential evolution equations for the vortex flow and mean flow correction. The nonlinear interactions are dominated by a "mean-field" type theory in which the fundamental and mean flow correction reinforce each other. The higher harmonics play no role in this interaction so the Stuart-Watson approach is not applicable.

Downstream of the position where the vortex is neutrally stable on the basis of linear theory, a multi-layer structure develops as shown in Figure 1. The vortices are confined to the core layer labelled region I. The mean flow is driven by the vortices in this layer and indeed is determined as a solvability condition on the equations for the fundamental. The downstream mean velocity component in region I has a simple square root profile which bears essentially no resemblance to the basic state present in the absence of the vortices. The mean flow, in effect, adjusts so as to make the large amplitude vortices locally neutral. This situation is not unlike a scenario postulated by Malkus (1956) in the context of turbulent flows. Malkus argued that the "mean" part of a turbulent flow would organize itself so that any "modes" were marginally stable. In our case, the adjustment is accomplished through centrifugal effects associated with the curved wall.

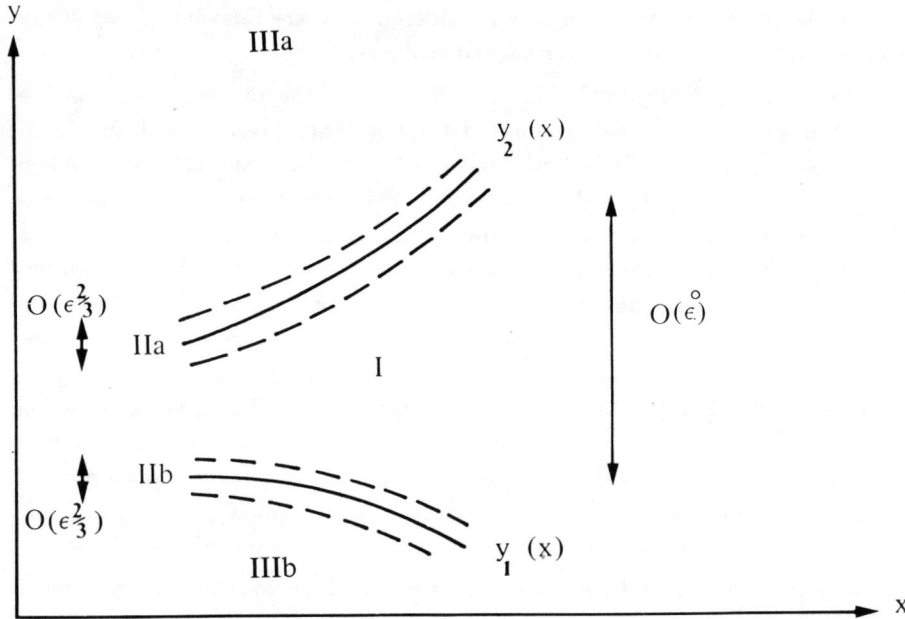

Figure 1. The different regions beyond the downstream position of neutral stability.

Downstream of the onset position, the region of vortex activity remains confined betwen regions IIIa and IIIb. In these two regions, the flow has no spanwise dependence, and the velocity field satisfies the boundary layer equations. If ε is the small dimensionless spanwise wavelength, solutions in region I are matched to solutions in regions IIIa,b through transition regions IIa,b which have depths of order $\varepsilon^{2/3}$. The mean flow is unaltered to lowest order by these transitions regions, but the vortices are reduced to zero as solutions of a nonlinear Airy equation. The locations of regions IIa,b are unknown functions of downstream position. Consequently, a full description of the flow field involves solution of a free boundary value problem.

In section 2 of this paper, the nonlinear Görtler vortex equations are given. An asymptotic solution of these equations for large amplitude small wavelength vortices is discussed in section 3. Section 4 considers a special case where the problem can be reduced to ordinary differential equations. Finally, in section 5 we discuss some implications of our results.

2. THE NONLINEAR GÖRTLER EQUATIONS

Consider the steady flow of a viscous incompressible fluid over a wall of variable concave curvature. With the scalings of Hall (1982), the Navier-Stokes equations for the flow can be written in the form

$$\frac{\partial u}{\partial x} + \frac{\partial v}{\partial y} + \frac{\partial w}{\partial z} = 0 \tag{1}$$

$$\Delta_2 u = u\frac{\partial u}{\partial x} + v\frac{\partial u}{\partial y} + w\frac{\partial u}{\partial z}, \tag{2}$$

$$\Delta_2 v - \frac{1}{2}GKu^2 - \frac{\partial p}{\partial y} = u\frac{\partial v}{\partial x} + v\frac{\partial v}{\partial y} + w\frac{\partial v}{\partial z}, \tag{3}$$

$$\Delta_2 w - \frac{\partial p}{\partial z} = u\frac{\partial w}{\partial x} + v\frac{\partial w}{\partial y} + w\frac{\partial w}{\partial z} \tag{4}$$

where (x, y, z) are the coordinates along the wall, normal to the wall, and in spanwise direction, respectively, (u, v, w) denotes the corresponding velocity vector, p is the pressure, Δ_2 is the two-dimensional Laplacian operator

$$\Delta_2 \equiv \frac{\partial^2}{\partial y^2} + \frac{\partial^2}{\partial z^2}, \tag{5}$$

and terms of relative order $R_e^{-1/2}$ have been neglected. The variable concave wall curvature is given by $K(x)$, and the flow satisfies the boundary conditions

$$u = v = w = 0 \quad \text{at} \quad y = 0, \tag{6}$$

$$u \to 1 \quad \text{as} \quad y \to \infty. \tag{7}$$

In the absence of any vortex, the velocity field is simply $(u, v, w) = (\bar{u}, \bar{v}, 0)$ where $\bar{u}$ and $\bar{v}$ are functions of x and y only satisfying the boundary layer equations

$$\bar{u}\frac{\partial \bar{u}}{\partial x} + \bar{v}\frac{\partial \bar{u}}{\partial y} = \frac{\partial^2 \bar{u}}{\partial y^2}, \tag{8}$$

$$\frac{\partial \bar{u}}{\partial x} + \frac{\partial \bar{v}}{\partial y} = 0. \tag{9}$$

Thus, until a Görtler vortex begins to grow at some downstream location, the mean velocity profile is simply a Blasius boundary layer.

3. EQUATIONS FOR THE LARGE AMPLITUDE GÖRTLER VORTICES

In this section, we describe an asymptotic solution of (1-7) valid in the limit of small vortex wavelength. We suppose that the boundary layer

$(\overline{u}(x,y), \overline{v}(x,y), 0)$ becomes linearly unstable to Görtler vortices at streamwise position $x = x^*$, and that the curvature function $K(x)$ is such that the flow becomes more unstable with increasing x. Consistent with the scaling of the right hand branch of the neutral curve for Görtler vortices, the Görtler number G will be expanded in the form

$$G = G_0 \varepsilon^{-4} + G_1 \varepsilon^{-3} + \ldots . \tag{10}$$

Consider first region I of the flowfield. Appropriate expansions of flow quantities in this core region are discussed in Hall and Lakin (1988). Let $\overline{u}_0$ and $\overline{v}_0$ denote the lowest order streamwise and normal mean velocity terms, respectively, in these expansions, while V_0 denotes the lowest order contribution of the fundamental in the expansion of the normal velocity. Then, the pivotal equations in region I are

$$G_0 K \overline{u}_0 \frac{\partial \overline{u}_0}{\partial y} = 1 \tag{11}$$

$$\frac{\partial \overline{u}_0}{\partial x} + \frac{\partial \overline{v}_0}{\partial y} = 0 , \tag{12}$$

$$\overline{u}_0 \frac{\partial \overline{u}_0}{\partial x} + \overline{v}_0 \frac{\partial \overline{u}_0}{\partial y} - \frac{\partial^2 \overline{u}_0}{\partial y^2} = 2 \frac{\partial}{\partial y} \left\{ \frac{\partial \overline{u}_0}{\partial y} |V_0|^2 \right\} . \tag{13}$$

Thus, the mean streamwise velocity component in the core is determined by (11), the mean normal velocity comes from (12), and (13) then determines $|V_0|^2$. It follows that in the core region, the boundary layer flow is being forced by the vortex which is itself driven by the boundary layer. This is the exact reverse of the roles played by these equations in the linear or weakly nonlinear descriptions of this problem.

Equations (11) and (12) may be integrated to give

$$\overline{u}_0 = \frac{\sqrt{a(x) + 2y}}{\sqrt{G_0 K}} , \tag{14}$$

$$\overline{v}_0 = \frac{-a'(x)\sqrt{a(x) + 2y}}{2\sqrt{G_0 K}} + \frac{K'[a(x) + 2y]^{3/2}}{6K\sqrt{G_0 K}} - b(x) \tag{15}$$

where $a(x)$ and $b(x)$ are arbitrary functions of x and a prime denotes an ordinary derivative with respect to x. Integrating (13) now gives an expression for $|V_0|^2$ which will be omitted, but involves another arbitrary function $B(x)$. The quantity $|V_0|^2$ must be non-negative, and hence setting $V_0 = 0$ specifies the edges $y_1(x)$ and $y_2(x)$ of the core.

The expressions for flow quantities in the core thus contain three undetermined functions of x, and we are not yet able to determine the boundaries of the core region, i.e. the location of the transition regions IIa, b through which the fundamental is reduced to zero. The thickness of layers IIa,b is found to be of order $\varepsilon^{2/3}$. Appropriate expansion of flow quantities in IIa,b and the relevant differential equations are given in Hall and Lakin (1988). Analysis of these equations in IIa,b shows that the mean flow is essentially unaltered by the presence of these layers. Indeed, the first two terms in the expansion of the mean flow in IIa,b are obtained by simply expanding the mean flow of I in terms of the region II stretched variables. Consequently, to lowest order the mean flow in the outer region IIIa must have $\overline{u}, \overline{u}_y$, and $\overline{v}$ match to the coreflow solutions evaluated with $y = y_2$. A similar statement holds for the mean flow in IIIb with $y = y_1$.

The fundamental component $\widehat{V}_0$ is found to satisfy a nonlinear Airy equation in regions IIa,b. Appropriate solutions of this equation are a particular form of the second Painlevé transcendent. Consequently, both the fundamental and the higher harmonics decay to zero in IIa,b, so the Görtler vortex is trapped between regions IIIa and IIIb. Thus, in IIIa,b, there is only a mean velocity field which has no z-dependence.

In regions IIIa,b, $w = p = 0$ while $\overline{u}$ and $\overline{v}$ satisfy equations (8) and (9) subject to the boundary conditions (6) in IIIb and condition (7) in IIIa. As noted above, solutions for $\overline{u}$ and $\overline{v}$ must match to the core solutions across $y_1(x)$ and $y_2(x)$, and this condition together with (14) and (15) defines a free boundary value problem for y_1, y_2, and the functions $a(x), b(x)$, and $B(x)$ in the core expressions.

Clearly, no analytic solution of this free boundary value problem is available. In fact, to solve the problem at a given position x larger than x^* it is necessary to know something about the "upstream" nature of the instability. This amounts to finding an asymptotic form for the solution close to the position x^* where the original boundary layer becomes linearly unstable.

For general curvature functions $K(x)$, the free boundary value problem has been considered by Hall and Lakin (1988) using asymptotic-numerical methods. As shown symbolically in Figure 2, solution involves asymptotic representations of the solution for $x - x^*$ small, numerical integration of the equations until $x - x^*$ is "large", and additional large x asymptotics. Rather than give details of this procedure, it is more useful here to consider a special case involving $K(x)$ proportional to the square root of x with the Görtler number large. For curvatures of this type, the sets of partial differential equations can be reduced to ordinary differenial equations through use of a similarity variable. For large Görtler number, these equations can be completely analyzed using asymptotic methods alone. Indeed, the large Görtler number asymptotics provides the motivation for the scalings adopted by Hall and Lakin (1988) in the case of general curvatures and unspecified values of G.

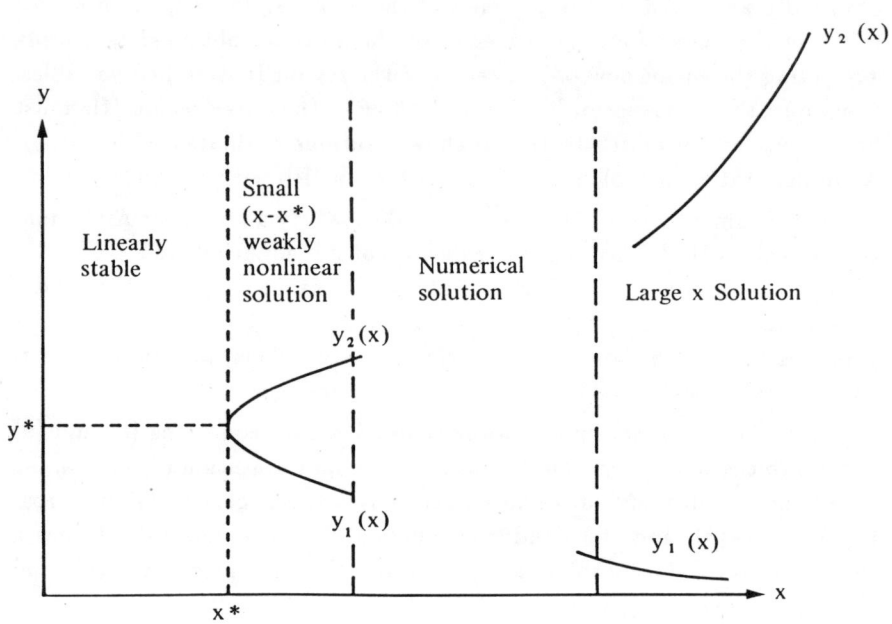

Figure 2. The different regions of vortex activity as the vortices develop in the downstream direction for $K \sim x^M$, $M > \frac{1}{2}$.

4. THE SIMILARITY CASE WITH G LARGE

Consider now the case where $K(x) = \kappa\sqrt{x}$ with κ a constant. We define a similarity variable η by

$$\eta = y/\sqrt{2x} \tag{16}$$

and seek solutions for the mean flow of the form

$$\bar{u} = f'(\eta), \quad \bar{v} = \frac{1}{\sqrt{2x}}(\eta f' - f) . \tag{17}$$

In the core, equations (11) through (13) reduce to the ordinary differential equations

$$\kappa G_0 f' f'' = 1 , \tag{18}$$

$$f''' + f f'' = -2\frac{d}{d\eta}\{f''|V_0|^2\} \tag{19}$$

whose solutions are

$$f(\eta) = \frac{2}{3}\left(\frac{2}{\kappa G_0}\right)^{1/2}(\eta+a)^{3/2} + b, \tag{20}$$

$$|V_0|^2 = -\frac{1}{2} - \frac{1}{6}\left(\frac{2}{\kappa G_0}\right)^{1/2}(\eta+a)^{5/2} - b(\eta+a) + 2B\left(\frac{2}{\kappa G_0}\right)^{-1/2}(\eta+a)^{1/2}. \tag{21}$$

In these expressions, a, b, and B are now simply undetermined constants. Again, as $|V_0|^2$ must remain non-negative, the boundaries η_1 and η_2 of region I are given by $|V_0|^2 = 0$. Examination of the possible balances which can reduce $|V_0|^2$ to zero now shows that for large G, η_1 will be of order G_0^{-3} while η_2 will be of order G_0.

As the mean flow is essentially unaltered by the presence of IIa,b, these transition regions will not be explicitly analyzed here. Consider now, region IIIb which lies between the core region and the wall. In this region, the mean flow satisfies the boundary layer equations and the no-slip boundary conditions. Consequently, in IIIb, $f(\eta)$ is a solution of the Blasius equation

$$f''' + ff'' = 0 \tag{22}$$

with the boundary conditions

$$f(0) = f'(0) = 0. \tag{23}$$

Consistent with the magnitude of η_1, we define the new variable $\varsigma = G_0^3 \eta$ and expand $f(\eta)$ in IIIb as

$$f(\eta) = G_0^5 F_0(\varsigma) + \cdots. \tag{24}$$

To lowest order, we thus have

$$F_0''' = 0, \quad F(0) = F'(0) = 0 \tag{25}$$

and hence

$$F_0(\varsigma) = c\varsigma^2 \tag{26}$$

where c is another undetermined constant.

As mean quantities in regions I and IIIb match across η_1, the value of η_1 now follows from (26) and (11). The matching of $f(\eta)$ and $f'(\eta)$ across η_1 also gives a and b, while the condition that $V(\eta_1) = 0$ determines B.

Results are

$$\eta_1 = \frac{1}{4\kappa} c^{-2} G_0^{-3},$$

$$a = -\frac{1}{8\kappa} c^{-2} G_0^{-3},$$

$$b = \frac{1}{48\kappa^2} c^{-2} G_0^{-5}, \qquad (27)$$

$$B = cG_0.$$

All of the expressions above involve the unknown constant c which must be determined by considering region IIIa. At the outer boundary of the core region, $V_0(\eta_2)$ in (21) must vanish. This condition, and the relations in (27), now show that

$$\eta_2 = (6c\kappa)^{1/2} G_0. \qquad (28)$$

Above η_2 in region IIIa, $f(\eta)$ will again satisfy the unforced Blasius equation and the outer boundary condition $f' \to 1$ as $\eta \to \infty$. As η_2 is large, this implies that $f'(\eta)$ in IIIa differs from unity by terms which are exponentially small. Matching the solutions for f' across η_2 now gives that

$$c = \frac{\kappa}{24} \qquad (29)$$

completing the asymptotic solution of the problem for large Görtler number. We note that (29) and (26) now allow determination of the skin friction. In particular, these relations show that

$$f''(0) = \frac{\kappa G_0}{12}. \qquad (30)$$

To assess the accuracy of the large Görtler number asymptotics, the present results were compared with results obtained by direct numerical solution of the boundary layer equations for a curvature function $K(x) = \sqrt{x}/10$ at a Görtler number $G = 146.0955$. As shown in Table 1, excellent agreement was obtained between the asymptotic and numerical values of η_1, η_2 and $f''(0)$.

Table 1. Comparison of asymptotic results for $G_0 \gg 1$ and numerical results for $K(x) = \sqrt{x}/10$ and $G_0 = 146.0955$.

	Asymptotic	Numerical
η_1	0.13	0.13
η_2	5.16	4.99
$f''(0)$	0.861	0.866

5. DISCUSSION

If the curvature function $K(x)$ increases at least as quickly as $\sqrt{x}$, the flow is linearly unstable for all values of x larger than x^*. The mean flow downstream of x^* is now driven by the Görtler vortex and bears essentially no relation to the Blasius flow present in the absence of the vortex. Indeed, in this case Hall and Lakin (1988) have shown that the region of vortex activity expands into the free stream and thus thickens the undisturbed boundary layer. Thus, for growing boundary layers over curved surfaces, the relevant mean flow in a secondary instability problem is unlikely to be the undisturbed mean flow, and may depend critically on downstream position.

This conclusion still holds even if $K(x)$ increases less quickly than $\sqrt{x}$. The basic state will now become linearly stable at some finite position beyond x^*, and the core region will terminate at a downstream position $\hat{x}$. Between x^* and $\hat{x}$, the mean flow is again modified at lowest order by the presence of the vortices. However, the influence of the vortices on the boundary layer does not necessarily cease at $\hat{x}$. This is because the initial velocity distribution at $\hat{x}$ for the subsequent boundary layer development will in general be quite different from that appropriate to the undisturbed flow. It thus cannot be assumed that the decay of the vortices beyond $\hat{x}$ implies that the boundary layer will simply return to its undisturbed state.

References
1. Hall, P. *J. I. M. A.*, **29**, p. 173 (1982).
2. Hall, P. *J. Fluid Mech.*, **130**, p. 41 (1983).
3. Hall, P. and Lakin, W. D., *Proc. Roy. Soc. London*, **A415**, p. 421, (1988).
4. Malkus, W., *J. Fluid Mech.*, **1**, p. 521 (1956).

KELVIN-HELMHOLTZ STABILITY AND TWO-PHASE FLOW

D.Y. Hsieh
Division of Applied Mathematics
Brown University
Providence, RI 02912

The ill-posedness of the two-phase flow problems is explicitly related to the Kelvin-Helmholtz instability of two compressible fluid layers. A long wavelength formulation of the Kelvin-Helmholtz problem is derived.

I. THE KELVIN-HELMHOLTZ STABILITY PROBLEM

The stability of two fluid layers in relative motion, known as Kelvin-Helmholtz stability, has been a subject of extensive studies[1-4]. Most of these studies deal with incompressible fluids. Landau[5] first considered the Kelvin-Helmholtz stability of compressible fluids. He treated the stability problem of two fluid layers with infinite depths, and showed that when the relative speed of the fluid layers are supersonic, the system can be stabilized.

The configuration of the fluid system is shown in Fig. 1. We use subscripts 1 and 2 to denote variables in these two fluids, and subscripts 0 to denote the equilibrium value. Fluid 1 occupies the region $-h_{10} < y < \eta$, while fulid 2 occupies the region $\eta < y < h_{20}$.

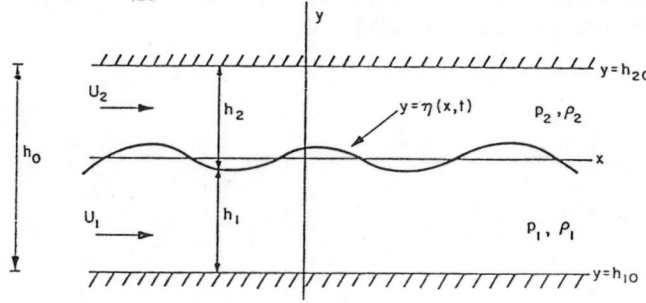

Fig. 1 The Schematic configuration of the fluid system.

Let ρ_i be the density, $\underline{u}_i$ the velocity, p_i the pressure, a_i the sound speed, and g the gravitation constant. Taking the perturbed interface as $\eta e^{i(\underline{k}\cdot\underline{x}+nt)}$, and considering the problem of two fluid layers of infinte depths, Landau obtained the following dispersion relation:

$$\left\{ \frac{\rho_{10}(n+kU_1)^2}{\left[k^2 - \frac{1}{a_1^2}(n+kU_1)^2\right]^{1/2}} \right\} = -\{1 \to 2\}, \tag{1}$$

where $\{1\to 2\}$ denotes the same expression as the one on the left hand side except that the subscript 1 is changed to 2. Based on (1), Landau showed that the system is stable if $|U_1 - U_2|$ is sufficiently supersonic, (e.g $|U_1 - U_2| > 2\sqrt{2}\, a_1$, if $a_1 = a_2$ and $\rho_{10} = \rho_{20}$).

A straightforward generalization of Landau's analysis to the problem of two layers of finite depths lead to the following dispersion relation:

$$\left\{ \frac{\rho_{10}(n+kU_1)^2 \cosh\left[k^2 - \frac{1}{a_1^2}(n+kU_1)^2\right]^{1/2} h_{10}}{\left[k^2 - \frac{1}{a_1^2}(n+kU_1)^2\right]^{1/2} \sinh\left[k^2 - \frac{1}{a_1^2}(n+kU_1)^2\right]^{1/2} h_{10}} \right\}$$

$$= -\{1 \to 2\}, \tag{2}$$

For the limiting case $h_{10} \to 0$ and $h_{20} \to 0$, the dispersion relation (2) becomes

$$\rho_{10} h_{20} (n + kU_1)^2 \left[k^2 - \frac{1}{a_2^2} (n + kU_2)^2 \right]$$

$$= - \rho_{20} h_{10} (n + kU_2)^2 \left[k^2 - \frac{1}{a_1^2} (n + kU_1)^2 \right] \tag{3}$$

II. INSTABILITY OF TWO PHASE FLOWS

Consider the flow of a mixture of two compressible, inviscid fluids, which will be designated by the subscripts 1 and 2. For simplicity, we shall consider one dimensional flow problems. The simplest or basic model of such two phase flow has the following governing equations[6,7]:

$$\frac{\partial}{\partial t} (\alpha_i \rho_i) + \frac{\partial}{\partial x} (\alpha_i \rho_i u_i) = 0 \quad, \quad i = 1,2, \tag{4}$$

$$\frac{\partial u_i}{\partial t} + u_i \frac{\partial u_i}{\partial x} = - \frac{1}{\rho_i} \frac{\partial p}{\partial x} = - \frac{c_i^2}{\rho_i} \frac{\partial \rho_i}{\partial x} \quad, \quad i = 1,2 \quad, \tag{5}$$

Where α_i is the volume fraction of ith phase, and $c_i = \left[\frac{dp_i}{d\rho_i} \right]$. is the sound speed of the ith phase, (4) are the continuity equations and (5) the momentum equations. It is obvious that

$$\alpha_1 + \alpha_2 = 1 \:. \tag{6}$$

In this basic model, we have taken $p_1 = p_2 = p$, and assume the fluids to be barotropic, thus

$$c_1^2 d\rho_1 = c_2^2 d\rho_2 = dp. \tag{7}$$

The stability of this nonlinear system may be investigated by studying the characteristics of the partial differential equations (4) and (5). If all the characteristics are real, then the system is hyperbolic, or we say the initial problem is well-posed. In other words, the system is stable. Otherwise, the problem is ill-posed, or the system is unstable. Stewart and Wendroff [6] have carried out such analysis and found the criterion of stability. So far as the stability criterion is concerned, identical results can be obtained by a linear analysis. The dispension relation for the linear problem can be readily found to be:

$$\left[\frac{\alpha_{10}\rho_{20}}{c_1^2} + \frac{\alpha_{20}\rho_{10}}{c_2^2}\right](n+kU_1)^2(n+kU_2)^2$$
$$- \alpha_{20}\rho_{10}k^2(n+kU_1) - \alpha_{10}\rho_{20}k^2(n+kU_2) = 0. \quad (8)$$

If we make the identification of α_{i0} with h_{i0} and c_i with a_i, equation (8) is exactly the same as the equation (3).

Denote

$$U_c^2 = \left[(\alpha_{10}\rho_{20})^{1/3} + (\alpha_{20}\rho_{10})^{1/3}\right]^3 / \left[\frac{\alpha_{10}\rho_{20}}{c_1^2} + \frac{\alpha_{20}\rho_{10}}{c_2^2}\right]. \quad (9)$$

Then it can be shown[6] all four roots for $\left[\frac{n}{k}\right]$ of equation (8) are real if

$$(U_1 - U_2)^2 \geq U_c^2, \quad (10)$$

and there are two complex roots for $\left[\frac{n}{k}\right]$ if

$$0 < (U_1 - U_2)^2 < U_c^2. \quad (11)$$

U_c is clearly of the order of c_1 or c_2. Thus the flow system can be stabilized if the relative speed is sufficiently supersonic. The identity of equations (8) and (3) demonstrates that the underlying physical mechanism of the so-called ill-posedness of the two phase flow problems is the Kelvin-Helmholtz instability. Although this point has been repeatedly mentioned in the two-phase flow literature, we believe this is the first time an explicit connection is established between the rigorous Kelvin-Helmholtz stability analysis and the two phase problems.

III. THE LONG WAVE LENGTH KELVIN-HELMHOLTZ PROBLEM

Let us now make a long wavelength approximate analysis of the rigorous Kelvin-Helmholtz problem, in a manner which is standard in the shallow water theory. Then we may include the effect of gravitation and resistive forces, which would cause great difficulty in the rigorous analysis of Kelvin-Helmholtz problem of compressible fluids.

Without going into details, we found[8] that the governing equations are

$$\frac{\partial}{\partial t}(\rho_1 h_1) + \frac{\partial}{\partial x}(\rho_1 h_1 u_1) = 0 , \tag{12}$$

$$\frac{\partial}{\partial t}(\rho_2 h_2) + \frac{\partial}{\partial x}(\rho_2 h_2 u_2) = 0 , \tag{13}$$

$$\rho_1 \left[\frac{\partial u_1}{\partial t} + u_1 \frac{\partial u_1}{\partial x} + g \frac{\partial h_1}{\partial x}\right] = -\frac{\partial p}{\partial x} + \mu_1 \frac{\partial^2 u_1}{\partial x^2} , \tag{14}$$

$$\rho_2 \left[\frac{\partial u_2}{\partial t} + u_2 \frac{\partial u_2}{\partial x} + g \frac{\partial h_1}{\partial x}\right] = -\frac{\partial p}{\partial x} + \mu_2 \frac{\partial^2 u_2}{\partial x^2} , \tag{15}$$

and

$$h_1 + h_2 = h_0. \tag{16}$$

If we set $g = \mu_1 = \mu_2 = 0$, and identify h_i with α_i, we see that the equations (12)-(15) are identical to equations (4) and (5). It is no wonder that we obtain the same stability criterion for these two cases.

The long wavelength formulation (12)-(16) offers some new perspectives to the nonlinear study of Kelvin-Helmholtz problem of compressible fluid with the inclusion of gravitational effect and resistive forces. Some aspect of the problem has been treated elsewhere[8], but a more comprehensive and fuller study is still to be carried out.

REFERENCES

1. Helmholtz, H., Philos. Mag., Ser. 4, **36**, 337, (1968).
2. Lord Kelvin, **"Mathematical and Physical Papers"** (Cambridge U.P., Cambridge 1910), Vol. **IV**, pp. 69-85.
3. Chandrasekhar, S., **"Hydrodynamic and Hydromagnetic Stability"** (Oxford., U.P., Oxford, 1961), Chap. XI.
4. Hsieh, D.Y., and Chen, F., Phys. Fluids, **28**, 1253, (1985).
5. Landau, L.D., Compt. Rend. (Doklady) Acad. Sci. URSS **44**, 139, (1944).
6. Stewart, H.B., and Wendroff, B., J. Comp. Phys. **56**, 363, (1984).
7. Hsieh, D.Y., **"On Dynamics of Bubbly Liquids"** To be published.
8. Hsieh, D.Y., **"Kelvin-Helmholtz Stability and Two-Phase Flow"**, Report of Applied Mathematics, Brown University, January 1987.

HIGH TAYLOR NUMBER COUETTE FLOW BETWEEN CONCENTRIC CYLINDERS AND LONG ECCENTRIC CYLINDERS

Martin Lessen, *Consulting Engineer, 12 Country Club Dr.,Rochester, NY 14618*
Yates Memorial Professor of Engineering Emeritus,University of Rochester

High Taylor number Couette flow between concentric cylinders with an axial pressure gradient is examined using the previously justified postulate of marginal instability of the large scale Taylor vortices in the gap as well as the small scale Goertler vortices against the inner and outer cylindrical walls. Simple, closed form expressions are obtained for the torque on the inner cylinder, the spin up of the core flow in the gap, and the axial flow rate. The bifurcation of the wall mode from the Goertler to the Tollmien-Schlichting disturbance is then examined.

The case of long, eccentric cylinders is then similarly studied and simple, closed form expressions are obtained for the force on the inner cylinder.

1. Introduction. G.I. Taylor (1935) studied "turbulent" circular Couette flow after his earlier pioneering study of the stability of laminar circular Couette flow. He found, in this turbulent case, that the average flow field in the annulus was essentially of constant circulation (potential vortex) except at the walls where in a thin-boundary layer, the average flow circulation quickly went to the value corresponding to that of the wall. In Barcilon, Brindley, Lessen and Mobbs (1979), it was then found that the postulate of marginal instability of the outer wall boundary layer with respect to a Goertler disturbance yielded flow characteristics that compared favorably with observation. (see fig. 1).

The basic transport mechanism operating at high Taylor number with regard to the flow field is that standing Taylor vortices transport angular momentum between the walls of the annulus; since there is little dissipation away from the walls because of the small effect of viscosity at such Taylor numbers, the angular momenta of the fluid convected to the outer wall and that convected to the inner wall are preserved and hence the average circulation in the gap between the wall boundary layers is essentially constant. At each wall, however, the circulation varies from that of the gap to that of the wall in a thin boundary layer. This variation of circulation is accompanied by Goertler instabilities at both walls. The Goertler (wall) instabilities are each centrifugally driven, of finite amplitude, and fixed wave number. Since the source and sink of angular momentum at the edges of the boundary layer are essentially unlimited, the effect of the finite amplitude disturbance is to thin down the boundary layers until they are marginally unstable with respect to the Goertler disturbance. Although the Goertler disturbance is of finite amplitude, it is so small that linear stability theory is used to obtain the critical Goertler number to good approximation. The boundary layer average flow configuration is assumed to be similar to that of Blasius flow; in fact, because the mechanisms involved are different, the actual flow differs somewhat from the Blasius distribution but the differences probably have only a small effect on the Goertler instability.

Since the mechanism of axial momentum transport is the same as that for angular momentum transport, the flow due to an axial pressure gradient can be easily inferred. The flow regime studied in particular corresponds to that of reasonably high Taylor number and fairly low axial Reynolds number. For this case, the turbulence driven by the centrifugal instability dominates and renders the average flow field easily soluable. (see fig. 1).

2. Analysis. We will now consider flow in a cylindrical annulus where the outer wall is fixed and the inner wall is rotating about its axis of symmetry. We will assume that the flow field is driven by both the relative motion between the walls and axial pressure gradient. If δ_i and δ_o are the boundary layer thick- nesses at the inner wall (of radius of curvature R_i) and the outer wall (of radius of curvature R_o) respectively, ν is the kinematic viscosity of the fluid flowing in the annulus, and U_{rit} and U_{rot} are the tangential components of the flow velocity relative to the inner and outer walls respectively, the Goertler numbers G_i and G_o pertaining to the inner and outer walls respectively can be written as

$$G_i = \frac{U_{rit}\delta_i}{\nu}\left(\frac{\delta_i}{R_i}\right)^{\frac{1}{2}} \quad ; \quad G_o = \frac{U_{rot}\delta_o}{\nu}\left(\frac{\delta_o}{R_o}\right)^{\frac{1}{2}} \tag{1}$$

If it is assumed that

$$\delta_i, \delta_o << (R_o - R_i),$$

then, with constant circulation gap flow, the tangential flow velocities at the edges of the outer and inner wall boundary layers relative to the walls are given by

$$U_{rot} = U_{mot} = U_{mit}\frac{R_i}{R_o}$$

$$U_{rit} = U_{it} - U_{mit} = U_{it}(1 - \frac{U_{mit}}{U_{it}}) = U_{it}(1-\lambda) \qquad (2)$$

where U_{mot} and U_{mit} are the tangential components of the absolute flow velocity at the edge of the boundary layer on the outer wall and inner wall respectively, U_{it} is the tangential velocity of the inner wall, and $\lambda = U_{mit}/U_{it}$.

At this point, the tangential components of the shear stresses on the outer and inner walls will be evaluated. From (1) and (2)

$$\delta_o = \left(\frac{\nu G_o}{U_{mit} R_i}\right)^{\frac{2}{3}} R_o \; ; \quad \delta_i = \left(\frac{\nu G_i}{(1-\lambda) U_{it} R_i}\right)^{\frac{2}{3}} R_i \qquad (3)$$

If δ_i and δ_o are taken to be lengths representing one unit of the Blasius variable (assuming the boundary layers are approximated by the Blasius velocity distribution), the tangential components of the shearing stresses at the outer and inner walls (τ_{ot} and τ_{it} respectively) can be written as

$$\tau_{ot} = \rho\nu \frac{U_{mit}}{\delta_o} \frac{R_i}{R_o} K \; ; \quad \tau_{it} = \rho\nu \frac{(1-\lambda) U_{it}}{\delta_i} K \qquad (4)$$

where ρ=density of fluid, and for the Blasius velocity distribution, $K = .33206$.

3. <u>Equilibrium Swirl Solution</u>. Far away from the entry end of a long annulus (equilibrium average velocity distribution) and G_o and G_i at their <u>critical</u> values for marginal instability

$$\tau_o R_o = \tau_{it} R_i \;.$$

Substituting for τ_{ot}, τ_{it}, from (4) and δ_o, δ_i from (3) one obtains

$$\lambda = \frac{1}{\left(\frac{G_i}{G_o}\right)^{\frac{2}{5}} \left(\frac{R_i}{R_o}\right)^{\frac{3}{5}} + 1} \qquad (5)$$

The Goertler instability characteristics of the Blasius boundary layer for flow along a concave wall and flow along a convex wall (for various λ) have been calculated and the results are given in Fig. 2. By investigating the Goertler instability of the inner boundary layer flow and obtaining the critical G_i as $G_i(\lambda)$, Eq. (5) can be made to yield

$$\lambda = \lambda\left(\frac{R_i}{R_o}\right).$$

It is seen that for $\lambda = 1/2$, the <u>minimum</u> critical $G_i = G_o = .68$. Therefore, for $R_i/R_o = 1$; $\lambda = .5$ satisfies Eq. (5). The torque T per unit length in the axial direction necessary to drive the inner wall is therefore

$$T = 2\pi \tau_{it} R_i^2 = .664\, \pi\rho\nu\, \frac{(1-\lambda)}{\delta_i}\, U_{it} R_i^2.$$

For $\lambda = .5$, $U_{it} = \Omega R_i$, $\Omega = $ angular velocity of inner wall, $G_i = .68$,

$$T = .85 \rho \nu^{\frac{1}{3}} \Omega^{\frac{5}{3}} R_i^{\frac{10}{3}} \tag{6}$$

Though the result in Eq. (6) is compatible with that in Barcilon and Brindley (1984), the derivation follows more in the spirit of Barcilon, Brindley, Lessen and Mobbs (1979) and is, in fact, a priori independent of the large-scale Taylor vortex problem. In addition, the Goertler instability problem of the inner boundary layer flow was not considered in Barcilon and Brindley.

The axial flow in the annulus due to an axial pressure gradient may be calculated in a manner similar to the foregoing because the momentum transport in both the axial and tangential directions is accomplished in similar manner by the Taylor and Goertler vortices. The axial boundary layers on the inner and outer walls therefore have the same thicknesses as the tangential ones respectively. For a narrow annulus where $R_i/R_o = 1$, the axial components of the shear stresses on the inner and outer walls are therefore the same. In the gap between the boundary layers, the axial flow field is essentially plug flow because of the efficient momentum transport due to the large scale Taylor vortices. If $R_o - R_i = h$, U_{ma} is mean flow velocity in the axial direction, τ_a is axial component of shearing stress at the walls, $R_o = R_i = R$, $\delta_o = \delta_i = \delta$

$$\tau_a = \rho\, \frac{\nu U_{ma}}{\delta}\, K \quad (K = .33206).$$

Since by equilibrium, if $p = $ pressure in fluid

$$-\frac{h}{2}\frac{dp}{dx} = \tau_a$$

then

$$U_{ma} = -\frac{h\delta}{.66\rho v}\frac{dp}{dx}$$

and

$$Vol\ flow\ rate = +2\pi R h U_{ma} = -\frac{2\pi R \delta h^2}{.66\rho v}\frac{dp}{dx}$$

Since

$$\delta = \left(\frac{2vG}{\Omega R^2}\right)^{\frac{2}{3}} R,$$

then for $G = .68$

$$Vol\ flow\ rate = -11.7\frac{R^{\frac{2}{3}}h^2}{\rho v^{\frac{1}{3}}\Omega^{\frac{2}{3}}}\frac{dp}{dx} \qquad (7)$$

4. **Effect of Inlet Swirl.** In the usual rotating shaft seal, fluid enters the pressure side of the seal with a swirl that does not correspond to that of fully developed flow in a long seal. This problem will now be addressed.

As before, transport of momentum in the gap between the wall boundary layers via Taylor vortices will cause the flow in the gap to have a uniform circulation at any particular axial coordinate. For inlet swirl corresponding to a tangential velocity $U_{mit} < U_{it}$, the following analysis applies.

The net torque, T_N per unit axial length on the fluid may be written using Eq. (4) as

$$-T_N = 2\pi(\tau_{ot}R_o^2 - \tau_{it}R_i^2) = 2\pi\rho v K R_i U_{mit}\left[\frac{R_o}{\delta_o} - \frac{(1-\lambda)R_i}{\lambda\delta_i}\right] \qquad (8)$$

and introducing δ_o and δ_i from Eq. (3a) and (3b) with G_o and G_i at their minimum critical values (marginally unstable) and approximating $G_o = G_i = G$; $R_o = R_i = R$

$$-T_N = \frac{\rho v^{\frac{1}{3}} K}{(2\pi G)^{\frac{2}{3}}} \left[\Gamma^{\frac{5}{3}} - (\Gamma_i - \Gamma)^{\frac{5}{3}} \right] \tag{9}$$

where $\Gamma = 2\pi U_{mit} R_i$; $\Gamma_i = 2\pi U_{it} R_i$.

The rate of change in circulation with axial distance can be calculated through

$$+T_N = \rho R h U_{ma} \frac{d\Gamma}{dx} \tag{10}$$

or, if

$$\frac{(2\pi G)^{\frac{2}{3}} h R U_{ma}}{v^{\frac{1}{3}} K} = \gamma = 7.9 \frac{h R U_{ma}}{v^{\frac{1}{3}}}$$

and G is assumed to be essentially constant, then

$$\int \frac{d\Gamma}{\left[\Gamma^{\frac{5}{3}} - (\Gamma_1 - \Gamma)^{\frac{5}{3}} \right]} = -\frac{x}{\gamma} \tag{11}$$

The asymptotic properties of Eq. (11) may be examined for small initial swirl $\Gamma << \Gamma_1$ and adjustment of swirl to the equilibrium swirl Γ-$\Gamma_1/2$.

For $\Gamma << \Gamma_1$ (Inlet)

$$\Gamma = \frac{\Gamma_1^{\frac{5}{3}}}{\gamma} x. \tag{12}$$

For $\Gamma \sim \Gamma_1/2$ (adjustment to equilibrium swirl) or

$$\frac{\Gamma_1}{2} - \Gamma = Ce^{\frac{10}{3\gamma}\left(\frac{\Gamma_1}{2}\right)^{-\frac{2}{3}} x} \tag{13}$$

The torque per unit axial length supplied by the inner cylinder or shaft is simply

$$T = \frac{\rho v^{\frac{1}{3}} K}{(2nG)^{\frac{2}{3}}} (\Gamma_1 - \Gamma)^{\frac{5}{3}} \tag{14}$$

and the total torque supplied is

$$T_{tot} = \int T dx = \frac{\rho v^{\frac{1}{3}} K}{(2nG)^{\frac{2}{3}}} \int (\Gamma_1 - \Gamma)^{\frac{5}{3}} dx \tag{15}$$

From Eq. (11) Γ can be obtained as $\Gamma(x)$ therefore T_{tot} can be evaluated.

The pressure drop in the axial direction can be obtained using

$$-h \frac{dp}{dx} = \tau_{ia} + \tau_{oa} = \left(\frac{\rho v U_{ma}}{\delta_1} + \frac{\rho v U_{ma}}{\delta_0} \right) K$$

or

$$p = -.126 \frac{\rho v^{\frac{1}{3}} U_{ma}}{hR} \int \left[\Gamma^{\frac{2}{3}} + (\Gamma_1 - \Gamma)^{\frac{2}{3}} \right] dx \tag{16}$$

5. Eccentric Couette Flow. The case of eccentric inner and outer cylindrical boundaries with parallel axes of symmetry is now discussed for the two-dimensional, equilibrium swirl case. Reference is made to Fig. 3a. For the small gap approximation it follows by definition that

$$h = \bar{h}(1 + \varepsilon \cos\theta)$$

and from continuity that

$$U_{mt} = U_{mt}\frac{\bar{h}}{h} = \frac{\bar{U}_{mt}}{1+\varepsilon\cos\theta}.$$

By definition,

$$\lambda = \frac{U_{mt}}{U_{it}}; \quad \bar{\lambda} = \frac{\bar{U}_{mt}}{U_{it}}$$

and

where

$$U_{rit} = U_{it} - U_{mt} = U_{it}\left(1 - \frac{\bar{\lambda}}{1+\varepsilon\cos\theta}\right)$$

$$(\bar{}) = \frac{1}{2\pi}\int_0^{2\pi}(\ )d\theta.$$

From momentum conservation to first order for a narrow gap (see Fig. 3b)

$$\frac{dp}{d\theta} = \left(\tau_{it} - \tau_{\theta t}\right)\frac{R}{h} - \frac{d}{d\theta}\left(\rho U_{mt}^2\right)$$

or

$$p + \frac{\rho U_{mt}^2}{2} = R\int\frac{\tau_{it} - \tau_{ot}}{h}d\theta \qquad (17)$$

which is Bernoullis' Equation modified by wall stresses.

The marginally unstable outer and inner boundary layers are given as in Eq. (3) extended to the eccentric case and specialized to the narrow-gap case, yielding

$$\delta_0 = \left[\frac{vG_o(1+\varepsilon\cos\theta)}{\bar{U}_{mt}}\right]^{\frac{2}{3}} R^{\frac{1}{3}} \qquad (18a)$$

$$\delta_i = \left[\frac{vG_i}{U_{it}(1 - \frac{\bar{\lambda}}{1 + \varepsilon\cos\theta})} \right]^{\frac{2}{3}} R^{\frac{1}{3}} \tag{18b}$$

where δ_o, δ_i are now functions of θ.

Using (18a) and (18b), the stresses on fluid at the outer and inner boundaries respectively are, <u>to first order in ε</u>

$$\tau_{ot} = \rho v \frac{U_{rit}}{\delta_o} K = \frac{\rho v^{\frac{1}{3}} \bar{U}_{mt}^{\frac{5}{3}} K}{G_o^{\frac{2}{3}} R^{\frac{1}{3}}} (1 - \frac{5}{3} \varepsilon\cos\theta ...) \tag{19a}$$

$$\tau_{it} = \rho v \frac{U_{rit}}{\delta_i} K = \frac{\rho v^{\frac{1}{3}} U_{it}^{\frac{5}{3}} K}{G_i^{\frac{2}{3}} R^{\frac{1}{3}}} (1 - \lambda)^{\frac{5}{3}} (1 + \frac{5}{3} \varepsilon \frac{\bar{\lambda}}{1 - \bar{\lambda}} \cos\theta ...) \tag{19b}$$

Since, for equilibrium swirl,

$$\int_0^{2\pi} (\tau_{it} - \tau_{ot}) d\theta = 0$$

then for Go = Gi, and to first order in ε

$$\frac{1 - \bar{\lambda}}{\bar{\lambda}} = 1 \quad \text{or} \quad \bar{\lambda} = \frac{1}{2} \tag{20}$$

$$p = \left[\frac{5}{3(2)^{\frac{2}{3}}} \frac{\rho v^{\frac{1}{3}} R^{\frac{2}{3}} U_{it}^{\frac{5}{3}} K}{G^{\frac{2}{3}} \bar{h}} \sin\theta + 2\rho \bar{\lambda}^2 U_{it}^2 \cos\theta \right] \varepsilon + const \tag{21a}$$

It then follows from (17) that

$$\tau_{it} = \frac{1}{2^{\frac{1}{5}}} \frac{\rho v^{\frac{1}{3}} U_{it}^{\frac{5}{3}} K}{G^{\frac{2}{3}} R^{\frac{1}{3}}} \left(1 + \frac{5}{3} \varepsilon \cos\theta\right) \qquad (21b)$$

The net force in the x and y directions, F_x and F_y respectively, per unit length in the axial direction, exerted by the inner cylinder on the fluid are given by

$$F_x = \int pR\,d\theta\cos\theta - \int \tau_{it} R\,d\theta\sin\theta$$

$$F_y = \int pR\,d\theta\sin\theta + \int \tau_{it} R\,d\theta\cos\theta \,.$$

Since the reaction forces by the fluid on the inner cylinder are opposite to the forces of the inner cylinder on the fluid, the reaction forces on the cylinder per unit length in the axial direction are

$$X_x = -F_x = -\frac{\pi\rho\Omega^2 R^3 \varepsilon}{2} \qquad (22a)$$

$$X_y = -F_y = -\frac{5}{3}\frac{\pi}{2^{\frac{2}{3}}} \frac{\rho v^{\frac{1}{3}} R^{\frac{7}{3}} \Omega^{\frac{5}{3}} K}{G^{\frac{2}{3}}} \left(\frac{R}{h} + \frac{1}{2}\right)\varepsilon \qquad (22b)$$

7. **Results and Discussion.** The results of this investigation are contained in Eqs. (5), (6), (7), (11), (15), (16), (20), (22a), and (22b). All of the results are in closed form. For the concentric case, Eq. (5) gives the ratio of the potential core flow circulation to the circulation of the driving cylinder. Eq. (6) gives the torque on the inner driving cylinder in turbulent Taylor-Couette flow with a fixed outer cylinder in the small gap approximation; the torque is seen to be independent of gap width and hence Taylor number. In Eq. (7), the solution for the volume rate of axial flow for turbulent Taylor-Couette flow with an axial pressure gradient is presented and it is seen that the volume flow rate is a function of the gap width. For turbulent Taylor-Couette flow with an axial pressure gradient but with non-equilibrium swirl, the adjustment of the swirl toward equilibrium is presented in Eq. (11); the torque applied by the inner cylinder for this case is presented in Eq. (15) and the pressure variation with axial position is given in Eq. (16).

For the two-dimensional eccentric case, Eq. (20) gives the ratio of the potential core flow circulation to the circulation of the driving cylinder and to first order in the eccentricity parameter ε, the result is the same as that in Eq. 5 for the concentric seal. Eqs. (22a) and (22b) give the reaction forces on the inner cylinder due to its eccentricity with respect to the outer cylinder and it is seen that the reaction force in the x direction is the same direction as the eccentricity thus indicating the possibility of a mechanical instability.

For an inner cylinder displacement in the negative x direction, the reacting force on the inner cylinder in the x direction contains no viscous terms and is therefore solely due to inertial effects; the resulting reaction in the y direction, however, is seen to be solely due to viscous forces.

In Smith and Townsend (1982) it was reported that in Taylor Couette flow at sufficiently high rotational velocity after the Goertler vortices appear, the Goertler vortices themselves are disrupted. Such behavior was also observed by Aihara, Tomita and Ito (1985) in connection with Goertler vortices in boundary layer flow over a concave plate. Unstable shearing layers formed in between adjacent vortices were alluded to by Barcilon and Brindley as causing the disruption. Such shearing layers between adjacent vortices in the second stage instability of Blasius flow were calculated by Lessen and Koh (1985). However, in order for the character of the turbulent flow at the boundary of the outer cylinder in Taylor-Couette flow to resemble the state of affairs in the flat plate (Blasius) boundary layer, another type of transition is suggested.

Since

$$G_o = \frac{U_{rot} \delta_o}{\nu} \left(\frac{\delta_o}{R_o}\right)^{\frac{1}{2}} = Re_t \left(\frac{\delta_o}{R_o}\right)^{\frac{1}{2}}$$

where Re_t is the Reynolds number of the boundary layer flow against the outer cylinder in the tangential direction,

$$Re_t = G_o \left(\frac{R_o}{\delta_o}\right)^{\frac{1}{2}}.$$

Because with increase of U_{rot}, δ_o must decrease for fixed G_o at marginal instability, then it follows that Re_t must also increase. For no axial flow, if Re_t exceeds the critical value Re_{crit} for a Tollmien-Schlichting instability, the Goertler disturbance will be superseded by Tollmien-Schlichting waves and the character of turbulence will change to resemble that of turbulent boundary layer flow over a flat plate.

Also, for the case of axial flow, there will be an axial boundary layer Reynolds number, Re_a.

There is also the possibility that for $Re_t < Re_{crit}$,

$$Re_{tot}^2 = Re_t^2 + Re_a^2 > Re_{crit}^2$$

and that the addition of axial flow will drastically change the turbulent nature of the boundary layer flow as in the purely axial flow case.

Bibliography

1. Aihara, Y.; Tomita, Y.; Ito, A (1985) Generation, Development and Distortion of Longitudinal Vortices in Boundary Layers along Concave and Flat Plates, Proc. IUTAM Symp. on Laminar-Turbulent Transition, Novosibirsk 1984, Ed. V.V. Kozlov, Springer Verlag, Berlin, 447.
2. Barcilon, A.; Brindley, J.; Lessen, M.; Mobbs, F.R. (1979) Marginal Instability in Flows at Very High Taylor Number, J.F.M., 94, 453.
3. Barcilon, A.; Brindley, J. (1984) Organized Structures in Turbulent Taylor Couette Flow, J.F.M., 143, 429.
4. Lessen, M.; Koh, P.-H. (1985) Instability and Turbulent Bursting in the Boundary Layer, Proc. IUTAM Symp. on Laminar-Turbulent Transition, Novosibirsk 1984, Ed. V.V. Kozlov, Springer Verlag, Berlin, 39.
5. Smith, G.P.; Townsend, A.A. (1982) Turbulent Couette Flow Between Rotating Cylinders, J.F.M., 123, 187.
6. Taylor, G.I. (1935) Distribution of Velocity and Temperature Between Concentric Rotating Cylinders, Proc. Roy. Soc. A151, 494-512.

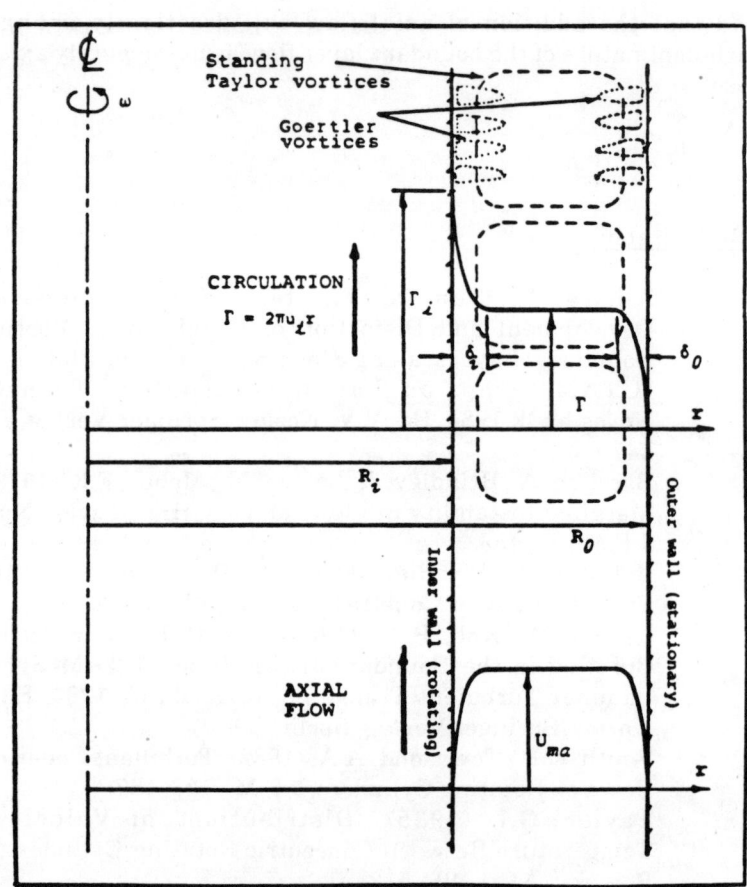

Figure 1 Simple, Rotating Shaft Seal.

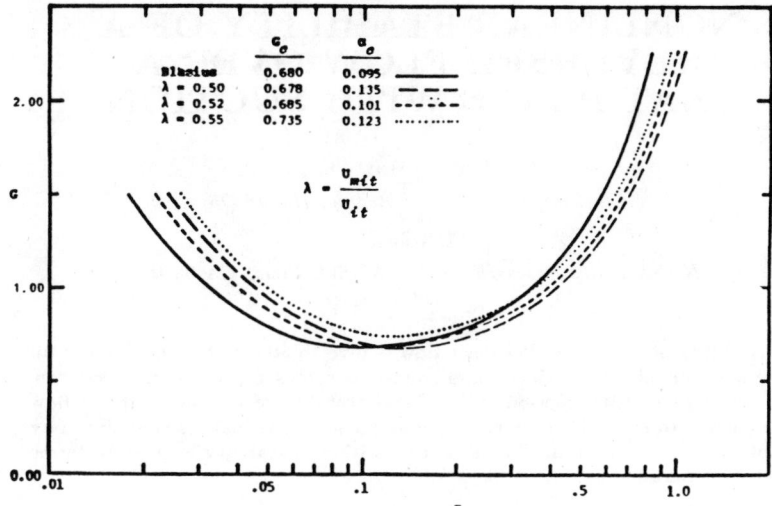

Figure 2 **Neutral Curves for Goertler Instability**

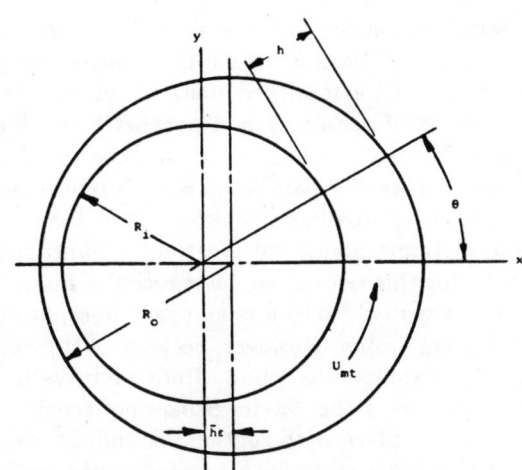

Fig. 3(a). The Eccentric Shaft Seal or Bearing

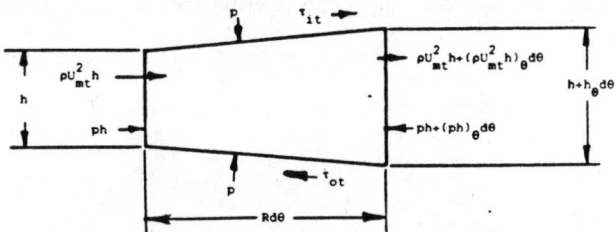

Fig. 3(b). An Element of the Annulus of Fig. 3(a).

NONLINEAR STABILITY OF A REVERSED FLOW OVER A FLAT PLATE WITH SUCTION

S.P. Lin and H.B. Chen
Clarkson University, Potsdam, NY 13676

M. Tobak
NASA Ames Research Center, Moffett Field, CA 94035

The stability of a non-parallel basic flow is investigated. The basic flow is an exact solution of the Navier-Stokes equations representing the reversed flow over a flat plate with suction. The global stability of this non-parallel flow with respect to three-dimensional finite amplitude disturbances is analyzed by use of the energy method. A sufficient condition of stability is given in terms of relevant parameters.

1. INTRODUCTION

Stability of weakly non-parallel flows, such as boundary-layer flows without flow reversal, have been analyzed quite successfully by use of various extensions of the well established method for parallel flows[1,2]. The linear stability analysis of genuinely non-parallel flows does not have the advantage of being allowable to Fourier decompose the disturbances as in the case of parallel and weakly parallel flows. Moreover, most of the non-parallel basic flows are approximate solutions, and their stability analysis suffers from the uncertainty about the effect of the approximate description of the basic states. For this reason, we have recently analyzed the stability of the Taylor's vortex array[3,4] which is an exact non-parallel flow solution of the Navier-Stokes equations. However, because of the decaying nature of the basic state, no instability was found. Here we investigate the stability of another exact solution of the Navier-Stokes equations representing the reversed flow over a flat plate with suction. A sufficient condition for the global stability of this basic state with respect to three-dimensional finite amplitude disturbances is obtained with the energy method. It is shown that suction can be destabilizing when it enhances the flow reversal.

2. GLOBAL STABILITY

Consider the flow of an incompressible Newtonian fluid. The governing equations are the Navier-Stokes equations

$$\partial_t \boldsymbol{v} - \boldsymbol{v} \cdot \nabla \boldsymbol{v} = -(1/\rho)\nabla P + \nu \nabla^2 \boldsymbol{v}$$
$$\nabla \cdot \boldsymbol{v} = 0, \tag{1}$$

where t is time, $\boldsymbol{v}$ is the velocity, ∇ is the gradient operator, ρ and ν are respectively the density and kinematic viscosity of the fluid, P is the pressure and ∇^2 is the Laplacian. Lin and Tobak[5] found the following exact solution of (1),

$$\begin{aligned} U &= U_o\{1 - nB\exp[(U_o/\nu)(mX - nZ)]\}, \\ V &= -U_o mB\exp[(U_o/\nu)(mX - nZ)], \\ m &= 0.5\left[1 \pm (1-4n^2)^{0.5}\right], B \neq 0, \end{aligned} \tag{2}$$

where (U, V) are the (X, Z) components of the velocity in Cartesian coordiantes, and B and n are constants. Equation (2) describes the flow over a flat plate $-\infty \leq X \leq 0$, if $0 \leq n \leq 0.5$. The case $B > 0$ corresponds to suction and $B < 0$ corresponds to blowing at the bottom plate.

To study the stability of this basic flow, we perturb the flow and write (1) as

$$\partial_t(\boldsymbol{U}+\boldsymbol{u}) + (\boldsymbol{U}+\boldsymbol{u})\cdot\nabla(\boldsymbol{U}+\boldsymbol{u}) = -(1/\rho)\nabla(P+p) + \nu\nabla^2(\boldsymbol{U}+\boldsymbol{u}),$$
$$\nabla \cdot (\boldsymbol{U}+\boldsymbol{u}) = 0, \tag{3}$$

where $\boldsymbol{U}$ is the basic state velocity vector field, P is the basic flow pressure field, and $\boldsymbol{u}$ and p are respectively the velocity and pressure perturbations. Subtracting the basic flow from (3), forming the inner product of the resulting equation with $\boldsymbol{u}$ and using the fact that the velocity field is divergence free, we have

$$[\partial_t + (\boldsymbol{U}+\boldsymbol{u})\cdot\nabla]e(t) = -\boldsymbol{u}\cdot(\boldsymbol{u}\cdot\nabla)\boldsymbol{U} - (1/\rho)\nabla\cdot(\boldsymbol{u}p) + \nu[\nabla\cdot(\boldsymbol{u}\cdot\nabla)\boldsymbol{u} - \nabla\boldsymbol{u}:\nabla\boldsymbol{u}], \tag{4}$$

where $e(t) = (\boldsymbol{u}\cdot\boldsymbol{u})/2$ is the kinetic energy of the disturbance per unit mass of fluid.

The flow domain is $-\infty \leq X \leq 0, 0 \leq Z \leq \infty$. Integrating (4) over the entire flow domain using the divergence theorem and the boundary condition

$$\boldsymbol{u}\mid_s = \boldsymbol{0}, \tag{5}$$

where the subscript s denotes that $\boldsymbol{u}$ is to be evaluated at the boundary s of the flow domain, we have

$$\frac{d\langle E\rangle}{dt} = -\nu\langle\nabla\boldsymbol{u}:\nabla\boldsymbol{u}\rangle - \langle\boldsymbol{u}\cdot(\boldsymbol{u}\cdot\nabla)\boldsymbol{U}\rangle, \tag{6}$$

where $\langle \cdot \rangle$ denotes integration over the flow domain. For two-dimensional basic flows such as the one considered, we have

$$\langle \boldsymbol{u} \cdot (\boldsymbol{u} \cdot \nabla) U \rangle = \langle u^2 \partial_X U + v^2 \partial_Z V + uv(\partial_X V + \partial_Z U) \rangle, \tag{7}$$

where (u, v) are the (X, Z) components of velocity perturbation. It should be pointed out that $\boldsymbol{u}$ was not assumed to be two dimensional in (7). It can be seen from (2) that the shear rate of deformation in the basic flow is

$$\partial_X V + \partial_Z U = (U_o^2 B/\nu)(n^2 - m^2)\exp[(U_o/\nu)(mX - nZ)],$$

and

$$\partial_X U = -\partial_Z V = -(U_o^2 B/\nu) nm \exp[(U_o/\nu)(mX - nZ)].$$

Since $(n^2 - m^2)_{\min} = -1$, thus

$$(\partial_X V + \partial_Z U)_{\min} = -U_o^2 B/\nu, \tag{8}$$

and

$$(\partial_X U)_{\min} = -(U_o^2 B/\nu)nm; \; nm \geq 0,$$
$$(\partial_Z V)_{\min} = 0. \tag{9}$$

On the other hand, it can be shown by use of Schwarz's inequality that[6]

$$\langle \nabla \boldsymbol{u} : \nabla \boldsymbol{u} \rangle \geq \frac{2}{D^2} \langle \boldsymbol{u} \cdot \boldsymbol{u} \rangle, \tag{10}$$

where D is the distance in the Z-direction above the flat plate, where condition (5) is invoked. It should be pointed out that (10) is not valid when $D \to \infty$, which is the case in the flow given by (2). Fortunately, physical and numerical simulation of the present flow can only be achieved in a finite domain. Hence the conclusions to be drawn from the following analysis remain meaningful when the present exact solution is applied to practically realizable siutations. It follows from (6), (7), (8), (9) and (10) that

$$\frac{d\langle E \rangle}{dt} \leq -\frac{2\nu}{D^2} \langle \boldsymbol{u} \cdot \boldsymbol{u} \rangle + (U_o^2 B/\nu)(nmu^2 + uv). \tag{11}$$

It can be shown that

$$(nm)_{\max} = 3\sqrt{3}/16,$$
$$u^2 + v^2 > 2uv, \quad u^2 + v^2 + w^2 > u^2 + v^2 > u^2.$$

It then follows from (11) that

$$\frac{d\langle E \rangle}{dt} \leq \langle E \rangle [-(4\nu/D^2) + U_o^2 B/\nu].$$

Integration of this inequality yields

$$\langle E \rangle \leq \langle E(0) \rangle \exp\left\{ [-(4\nu/D^2) + U_o^2 B/\nu] t \right\}, \tag{12}$$

where $\langle E(0) \rangle$ is the energy integral of any kinematically admissible initial disturbance. Hence the sufficient condition for global nonlinear stability is

$$B < (2\nu/DU_o)^2 \tag{13}$$

3. DISCUSSION

The sufficient condition of stability (13) is always satisfied by negative B which corresponds to blowing. Hence blowing which creates no flow reversal as described by the exact solution (2) is always stable. However the basic flow is stable with respect to periodic disturbances of spatial period D for any finite suction ($B > 0$) if the free stream velocity is sufficiently small and the kinematic viscosity of the fluid is sufficiently large such that inequality (13) is satisfied. Statement (13) also states that the flow is more unstable to spatially periodic disturbances of larger period in the direction normal to the flat plate. It is seen from (2) that a larger positive B creates a larger region of flow reversal. It follows from (13) that the flow reversal is destabilizing. Finally, we point out that since all of the conditions used to arrive at (13) hold for both two- and three-dimensional finite disturbances, the sufficient stability condition (13) and the above conclusions hold for both types of disturbances.

This work was supported in part by a NASA grant NCC2-280.

References

1. Lin, C.C., *The Theory of Hydrodynamic Stability*, Cambridge University Press, Cambridge, England (1955).
2. Drazin, P.G. and Reid, W.H., *Hydrodynamic Stability*, Cambridge University Press, Cambridge, England (1985).
3. Lin, S.P. and Tobak, M., "Spectral Stability of Taylor's Vortex Array," *Phys. Fluids* **29**, 3477 (1987).
4. Lin, S.P. and Tobak, M., "Nonlinear Stability of Taylor's Vortex Array," *Phys. Fluids* **30**, 605 (1987).
5. Lin, S.P. and Tobak, M., "Reversed Flow Above a Plate with Suction," *AIAA Journal* **24**, 334 (1986).
6. Joseph, D.D., *Stability of Fluid Motions*, Springer-Verlag, Berlin (1976).

BAROTROPIC INSTABILITY OF THE BICKLEY JET

S.A. Maslowe
McGill University
Montreal, Quebec
H3A 2K6 CANADA

The linear stability of the zonal shear flow $\bar{u} = -\text{sech}^2 y$ is investigated in the framework of the β-plane approximation. It turns out that this retrograde jet is more unstable than its eastward propagating counterpart and has some surprising characteristics. First, this is a rare example of a flow in which barotropically unstable modes occur that do not have a critical point. Secondly, there is evidence that for some wavenumbers and values of β the varicose mode is more unstable than the sinuous mode; when $\beta = 0$, by contrast, this is never the case.

1. INTRODUCTION

An important early study of the linear stability of jets both with and without the influence of planetary rotation is described in a paper by Foote and Lin[1]. This paper, by considering the inviscid limit of the Orr–Sommerfeld equation, illustrated how the path of integration should be deformed in the complex plane around critical points so that solutions of the Rayleigh and Kuo equations are valid limits of the viscous problem. Other useful results were obtained directly from the inviscid equations by considering the variation of the Reynolds stress. These ideas apply, of course, to the Bickley jet for the most part. However, it will also be seen below that certain exceptional cases arise which fall outside the scope of the Foote and Lin analysis. Specifically, there are unstable modes which have no critical point (defined as a value of y such that the velocity $\bar{u} = c$, the phase speed) and a singular neutral mode exists for which $\bar{u}'_c = 0$, i.e. the critical point is at the jet maximum. For perturbations whose stream function is of the form $\hat{\psi} = \phi(y) exp\{i\alpha(x-ct)\}$, the linear, inviscid theory revolves around the Rayleigh-Kuo equation

$$(\bar{u} - c)(\phi'' - \alpha^2 \phi) + (\beta - \bar{u}'')\phi = 0, \qquad (1)$$

where α is the wavenumber, β is the derivative of the Coriolis parameter (assumed constant) and c is a complex constant whose real part is the phase speed. As noted by Howard and Drazin[2], although β is always positive, changing its sign is mathematically equivalent to reversing the flow direction; hence, it is to be understood that the results presented below for $\beta < 0$ and $\bar{u} = \text{sech}^2 y$ correspond, in reality, to a retrograde jet with $\beta < 0$.

Two neutral modes for the Bickley jet were found by Lipps[3], namely, the sinuous mode

$$c = \tfrac{1}{6}\alpha^2, \beta = \tfrac{1}{6}\alpha^2(4-\alpha^2), \phi = \text{sech}^2 y \qquad (2)$$

and the varicose mode

$$c = \tfrac{1}{6}(3+\alpha^2), \quad \beta = \tfrac{1}{6}(1-\alpha^2)(3+\alpha^2), \quad \phi = \text{sech} y \ \text{tanh} y. \qquad (3)$$

For a given positive value of β, the maximimum amplification factor αc_i is obtained by considering the sinuous mode and therefore most numerical computations have been done for that mode. However, for long waves with β nonzero the varicose mode is generally more unstable, so that latter would appear to merit further attention. Howard and Drazin found, in addition, the singular neutral mode

$$c = 1, \quad \beta = -\tfrac{1}{9}\alpha^2(9-\alpha^2), \quad \phi = (\text{sech} y)^{\alpha^2/3}(\text{tanh} y)^{2-\alpha^2/3}. \qquad (4)$$

The interpretation of this mode is not clear and it will be discussed further in Section 3. First, however, the stability properties of the sinuous mode are elucidated in the following section.

2. STABILITY CHARACTERISTICS OF THE SINUOUS MODE

From a generalisation of Rayleigh's inflexion point theorem, it follows that the quantity $(\beta - \bar{u}'')$ must change sign somewhere for instability to occur; stability is therefore guaranteed for $\beta < -2$. However, because this is not a sufficient condition, it cannot be concluded that $\beta = -2$ forms part of the stability boundary. Lipps, on the other hand, used Lin's perturbation formula to show that for $\beta > -2$ the neutral modes (2) and (3) do constitute stability boundaries. The missing portion from $\beta = -2, \alpha^2 = 6$ (or $\alpha^2 = 3$ in the case of the varicose mode) must be computed numerically. In the rest of this section, the term "neutral curve" refers to this part.

In a survey article treating the general topic of barotropic instability, Kuo[4] has presented numerical calculations for the unstable sinuous modes. These calculations are accurate for moderately unstable waves, but seem to be in error for weakly amplified waves when β is negative. Kuo also appears to have assumed that the singular mode (4) forms part of the stability boundary, but it will be seen below that such is not the case.

Subsequently, in a more thorough numerical investigation, Deblonde[5] computed a portion of the missing stability boundary. It turns out, however, that at least part of this neutral curve is not a stability boundary and there is instability on either side of it. Some of her results are illustrated in Figure 1, below.

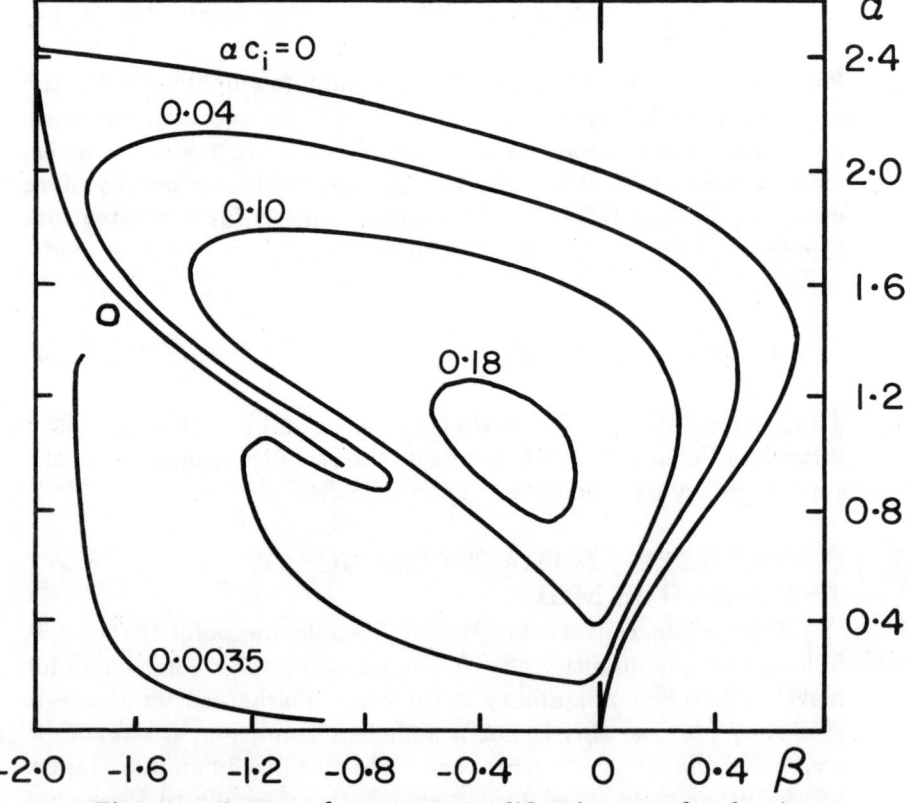

Figure 1. Curves of constant amplification rate for the sinuous instability mode.

The curves of constant αc_i were found by both Kuo and Deblonde to be smooth for $\alpha c_i \geq 0.10$. However, Deblonde observed that a sort of kink reversal begins to form when αc_i is smaller than 0.10. This is evident, for example, on the curve $\alpha c_i = 0.04$, shown on Figure 1. Computation of the neutral curve is made difficult by several factors. These include the existence of a continuous spectrum of modified Rossby waves to the left and the fact that c is not far from unity along the curve. It is in fact the nonanalytic behavior of the dispersion relation along the neutral curve that can be used to locate it precisely. Essentially, two modified Rossby waves coalesce as β is increased to become a single unstable mode once the neutral curve is crossed. A procedure exploiting this observation is to compute c as a function of β at constant α (or as a function of α with β constant) for the lowest Rossby mode and, if this curve has a minimum, it corresponds to a point on the stability boundary. This procedure, whose rationale is outlined by Drazin et al.[6], was used to compute the neutral curve in Fig. 1, which seems to end at about $\alpha^2 = 3$. At that point, the associated curve for the modified Rossby wave no longer has a minimum and further analysis is required to determine the continuation of the neutral mode.

A somewhat surprising result found by Deblonde was that weakly amplified modes without critical layers occur in a region toward the upper left of the diagram in Figure 1. The existence of such modes is permitted by a modified version of the semicircle theorem due to Pedlosky[7], but their occurrence is more prevalent in the context of baroclinic instability. Her results, however, are supported by the arguments and modified Rossby wave calculations of Drazin et al. (cf. their §6.4).

Given that certain results obtained by Deblonde were expected and that there was some disagreement with those of Kuo, a few words about her numerical procedures are appropriate. Both finite-difference and initial–value (shooting) methods were employed, as well as coordinate transformations to make the integration domain finite. When c_i was small, the contour of integration was deformed into the complex plane in accordance with the analysis of Foote and Lin in so that it was usually not necessary to pass near a singularity. (However, there are regions in parameter space where the two critical points approach each other and a very small step size must then be utilized.) It was verified that programs employing different methods gave the same results and agreement with the long wave, small-β expansions of Howard and Drazin was very good in the domain where the latter are valid.

3. THE VARICOSE MODE AND THE SINGULAR MODE

The varicose mode has received little attention because it has lower amplification rates when $\beta = 0$ and it is unstable for a smaller range of wavenumbers when $\beta \neq 0$. However, it is more unstable to long waves for positive β and, according to the analysis of Howard and Drazin, it has some intriguing properties when β is negative. Specifically, in the limit $\alpha \to 0$ with β/α^2 fixed, there is a neutral curve corresponding to $\beta = \alpha^2$. On either side of this netural curve, there is instability and the unstable solutions for $-\beta > \alpha^2$ have $c_r > 1$, i.e. there is no critical point. These conclusions depend upon accepting both signs in front of certain square roots but, in any case, the varicose modes seem worthy of numerical study and the present author is presently engaged in such computations. Further motivation is provided by some recent experimental results reported by Swinney et al.[8] which indicate that for a retrograde jet the varicose mode may be the most unstable.

This brings us finally to the singular model given by (4). Some analysis is required to determine how the solution should be continued across the branch point at $y = 0$. The singularity is of the same form as that encountered in stratified shear flows and a phase change is therefore possible without a jump in Reynolds stress (because $\phi(0) = 0$). Interpretation of this solution by considering the large-Reynolds number limit of the associated Orr–Sommerfeld problem seems feasible. Because $\bar{u}'_c = 0$ now, the critical layer thickness is $(\alpha \mathrm{Re})^{-1/4}$ instead of the usual $(\alpha \mathrm{Re})^{-1/3}$. Introducing the inner variables

$$\eta = (\alpha \mathrm{Re})^{1/4} y \quad \text{and} \quad \chi(\eta) = \phi(y),$$

the governing equation in the critical layer is now found to be

$$\chi^{IV} + i\eta^2 \chi'' - i(\beta + 2)\chi = 0. \tag{5}$$

Series solutions of (5) in powers of η have been obtained and, following a procedure due to Watson[9], these four solutions can be expressed as contour integrals. The saddle-point method can then be used to find the asymptotic behavior χ for large $|\eta|$ and, in this way, the analytic continuation of the eigenfunction ϕ in (4) is determined. These results will be published elsewhere once the saddle–point analysis has been completed.

References

1. Foote, J.R. and Lin, C.C., "Some Recent Investigations in the Theory of Hydrodynamic Stability", *Q. Appl. Math.* **8**, 265-280 (1950).
2. Howard, L.N. and Drazin, P.G., "On Instability of Parallel Flow of Inviscid Fluid in a Rotating System with Variable Coriolis Parameter", *J. Math. Phys.* **43**, 83-99 (1964).
3. Lipps, F.B., "The Barotropic Stability of the Mean Winds in the Atmosphere", *J. Fluid Mech.* **12**, 397-407 (1962).
4. Kuo, H.L., "Dynamics of Quasigeostrophic Flows and Instability Theory", *Adv. Appl. Mech.* **13**, 247-330 (1973).
5. Deblonde, G., "Instabilité Barotrope du Jet de Bickley", M. Sc. dissertation, McGill University (1981).
6. Drazin, P.G., Beaumont, D.N. and Coaker, S.A., "On Rossby Waves Modified by Basic Shear and Barotropic Instability", *J. Fluid Mech.* **124**, 439-456 (1982).
7. Pedlosky, J., " The Stability of Currents in the Atmostphere and Ocean: Part I", *J. Atmos. Sci.* **21**, 201-219 (1982).
8. Sommeria, J.L. Meyers, S. and Swinney, H., "Coherent Structures in Quasi-Geostrophic Jets", Symp. on Turbulent Shear Flows, Toulouse, France (Sept., 1987).
9. Watson, G.N. "The Diffraction of Electric Waves by the Earth", *Proc. Ray. Soc.* **A 95**, 83-99 (1918).

Instability of Stratified Shear Flow with Ri > 0.25 Everyplace

Richard S. Lindzen
Department of Earth, Atmospheric and Planetary Sciences
M.I.T.
Cambridge, MA 02139

As a graduate student at Harvard, I cross-registered at M.I.T. in 1960-61 in order to take C.C. Lin's course in hydrodynamics. It was in this course that I first learned that viscous Poiseuille flow could be unstable.

To help us understand this mathematically difficult seeming result, C.C. presented us the mathematically simpler result that standing water waves which had no Reynolds stresses in the absence of viscosity, developed such stresses when viscosity was present. This result was indeed simpler and readily understandable; unfortunately, it didn't really help me to understand the Orr Sommerfeld problem!

This problem has continued to bother me for the last 26 years. As a meteorologist, however, I have had to do most of my worrying and thinking on this problem in the "background mode."

The talk presented, briefly described my present understanding of shear instability – an understanding significantly influenced by C.C. and his colleagues in a series of seminars that we jointly held at M.I.T. during the years 1982-83. These seminars arose from the mutual recognition that there was much in common between the wave overreflection approach that I was taking to shear instability and the wave overreflection approach C.C. was taking to galactic spiral instabilities. This understanding has been applied to the viscous Orr-Sommerfeld problem[1]; it has also allowed us to anticipate the viscous instability of inviscidly stable stratified shear flow[2] (i.e., flow for which the Richardson Number is everywhere greater than 0.25) – an instability which a student of mine, Ron Miller, and I have computationally confirmed[3]. Our overall approach to shear instability is given in a comprehensive review[4].

Insofar as the contents of my talk are completely contained in papers in print or in press, there seems to be little point in repeating them here in an overly compressed form. However, I welcome this opportunity to acknowledge the very important and generous influence of C.C. on this work.

References

1. Lindzen, R.S., and Rambaldi,S., *J. Fluid Mech.*, 165, 355 (1986).
2. Lindzen, R.S., and Barker, J., *J. Fluid Mech.*, 151, 189 (1985).
3. Miller, R., and Lindzen, R.S., *Geophys. and Astrophys. Fl. Dyn.*, in press.
4. Lindzen, R.S., *Pure Appl. Geophys.*, , in press.

NONLINEAR DEVELOPMENT OF GÖRTLER VORTICES AND THE GENERATION OF HIGH SHEAR LAYERS IN THE BOUNDARY LAYER

A. S. Sabry & J. T. C. Liu

Laboratory for Fluid Mechanics, Turbulence and Computation
and
The Division of Engineering, Brown University
Providence, Rhode Island 02912, U.S.A.

Preface

C. C. Lin's initial work on hydrodynamic stability,[1] his monograph[2] on this subject, and his sustaining work in fluid dynamics[3] and astrophysics[4] have, indeed, most significantly stimulated and enriched his scientific colleagues and have led to the development of fertile research in subject areas that are of enormous scientific and technological importance.

We dedicate the present modest work to C. C. Lin, the presentation is given by the second author, based on the doctoral thesis work of the first author.

Abstract

In general, three-dimensional longitudinal vortices in the boundary layer are responsible for the generation of high inflectional point shear layers that, in turn, break down into other scales of instabilities and turbulence. This process occurs in transitional and also in turbulent boundary layers. As a prototype of such longitudinal vortices, the nonlinear development of Görtler vortices is studied numerically and detailed comparisons of results with and physical interpretations of observations are made.

1. INTRODUCTION

In his article on the mechanics of viscous fluids in Durand-Volume III, Prandtl[5] anticipated much of the present scenario in the transition to turbulence in laminar boundary layers as well as the fine-grained turbulence generation mechanisms in turbulent boundary layers:

"The beginning of turbulence in flows through pipes or along smooth walls can be often regarded as caused by three-dimensional disturbances of the following form. Let us assume the existence of feebly disturbing vortex motions with axes parallel to the main direction of flow. The effect of these vortices is to shift sideways the boundary layer which develops along the walls, to thicken it at some points and also occasionally to drag part of it into the unretarded flow. A band of retarded fluid brought by such means in between portions of fluid moving with greater velocities is however unstable and very soon breaks up into separate vortices. Apparently events of this kind are responsible in most cases for the production of turbulence. In practice it is observed that vortices produced by the disintegration of an unstable flow multiply, as a rule, very quickly, so that once such vortices are present, turbulence soon spreads throughout the whole flow. The appearance of fully developed turbulence is properly attributed to this phenomenon of which, however, no theoretical explanation has hitherto been possible... ."

Prandtl's[5] remarkable foresight in transition is borne out by the now classical experiments[6-8] on the development of three-dimensionalities in boundary layer transition. This takes the form of alternate rows of longitudinal vorticity that advects low momentum fluid upwards, forming high, inflectional shear layers in the streamwise velocity, called "peaks", where subsequent breakdown into other coherent scales and fine-grained turbulence occurs. The discussion of these experiments, in the light of nonlinear development from small, plane disturbances upstream[1] is given by Stuart,[9] who constructed a kinematic model to elucidate the formation of intense shear layers by vortex stretching and advection in absence of streamwise modulation. More recently Russell and Landahl[10] extended these ideas to a finite-spanwise "flat eddy". What Prandtl had not foreseen at that time is that his perceived role of longitudinal vorticity elements and the formation of high shear layers is now found to be omnipresent in turbulent boundary layers.[11-14] There are, in fact, strong analogies between the longitudinal vorticity elements and their effects in transitional and turbulent boundary layers.[15] From a theoretical point of view, it would be most desirable to extend the original kinematical consideration of the problem[9] to one involving the full, but approximate, dynamical equations. The insight and perspective gained could possibly lead to the control of the formation of high shear layers and hence the implicit control of the generation of turbulence in transitional and in turbulent boundary layers. To this end, Görtler vortices[16] serve as a convenient as well as appropriate[15] prototype of longitudinal vorticity elements in boundary layers with the simplicity of zero streamwise modulation as in the original problem discussed by Stuart.[9]

2. FORMULATION AND BASIC EQUATIONS

For a review of the development of the linearized theory of Görtler vortices,[16] we refer to Herbert[17] and Floryan and Saric.[18] More recently, spatially developing Görtler vortices in a growing boundary layer have been given considerable attention by Hall and coworkers.[19-24] Even for the linear problem, from appropriate scaling, Hall concludes that the problem is governed by parabolic partial differential equations with the streamwise variable playing the role of "time". In this situation, the search for the neutral curve is no longer as meaningful as in the case of the idealized parallel flow problem. Instead, both the linear and nonlinear Görtler problems in developing boundary layers are necessarily "initial value" problems, the disturbance amplitude develops and the "neutral" solution is but one stage, or eventual stage, in the transient solution in terms of the "timelike" variable. Experiments[25–27] indicate that the Görtler vortices develop in the vicinity of amplified modes with dimensionless wave number of order unity, for which asymptotic methods for large wave numbers can only serve as an indirect guide.

In what follows, we shall formulate the nonlinear Görtler problem, with emphasis on the generation of high, intense shear layers in the boundary layer, in terms of temporal development. It is considerably simpler than the spatial problem as a comparison between the mean flow problems in absence of disturbances already illustrates: the temporal problem for the mean flow is governed by the linear heat diffusion equation (the Rayleigh-Stokes problem) while that for the spatial problem is the nonlinear boundary layer equations (the Blasius problem). Although there is no one-to-one transformation available, the much simpler temporal problem serves to unmask many of the physical features of the spatial problem.

The Navier-Stokes equations are written in terms of a coordinate system placed on the surface of a stationary, slightly curved concave wall with radius of curvature R. The equations of motion are nondimensionalized using the following definitions: $t = T/(X_0/U_0), x = X/X_0, y = Y/\delta_0, z = Z/\delta_0, p = (P/\rho U_0^2)\mathrm{Re}, u = U/U_0, v = (V/U_0)\mathrm{Re}^{1/2}, w = (W/U_0)\mathrm{Re}^{1/2}$, where $t, p, (x, y, z)$ and (u, v, w) are the dimensionless quantities for the time T, pressure P, mutually orthogonal coordinates (X, Y, Z) and their coorresponding velocities (U, V, W) respectively. Here X is measured in the streamwise direction, Y normal to the wall, and Z in the spanwise direction. Also X_0 is a typical length scale in the streamwise direction and $\delta_0 = (\nu X_0/U_0)^{1/2}$ is that for both the normal and spanwise coordinates, U_0 is the free stream velocity, ν the kinematic viscosity, ρ the fluid density and $\mathrm{Re} = U_0 X_0/\nu$, is the Reynolds number. For the temporal problem we neglect the streamwise derivatives and terms of order (δ_0/R) thus the equations of continuity and motion become:

$$\frac{\partial v}{\partial y} + \frac{\partial w}{\partial z} = 0 \qquad (1)$$

$$\frac{\partial u}{\partial t} + v\frac{\partial u}{\partial y} + w\frac{\partial u}{\partial z} = \nabla_c^2 u \qquad (2)$$

$$\frac{\partial v}{\partial t} + v\frac{\partial v}{\partial y} + w\frac{\partial v}{\partial z} + G^2 u^2 = -\frac{\partial p}{\partial y} + \nabla_c^2 v \qquad (3)$$

$$\frac{\partial w}{\partial t} + v\frac{\partial w}{\partial y} + w\frac{\partial w}{\partial z} = -\frac{\partial p}{\partial z} + \nabla_c^2 w \qquad (4)$$

where ∇_c^2 is the Laplacian in the (y,z) plane and $G = (U_0 \delta_0/\nu)(\delta_0/R)^{1/2}$, is the Görtler number. The linearized form of the above equations corresponds to the original temporal problem of Görtler[16] with the parallel flow approximation. Hall[19] has shown that in terms of spatially developing boundary layers, this approximation is justified only for $G \gg 1$.

In the temporal problem, although the mean flow is parallel at each instant, the boundary layer is developing in time and a set of nonlinear partial differential equations is solved. Our intention is to study the nonlinear evolution of initially amplified Görtler vortices which occur at large values of the Görtler number and at dimensionless wave numbers of order unity, corresponding to observations.[27]

The boundary conditions are: at $y = 0, u = v = w = 0$; at $y = \infty, u = 1, v = w = 0$; at $z = 0, \pi/\alpha, \partial u/\partial z = 0, v = w = 0$; where α is the dimensionless wave number in the spanwise direction. The initial conditions are obtained from the solution of the linearized version of the above system of equations corresponding to the linear Görtler problem in a Blasius boundary layer with a given amplitude.

3. NUMERICAL PROCEDURE

In solving the system of equations (1-4) we used the streamfunction-vorticity formulation and followed a similar procedure suggested by Pearson[28] and Aziz and Hellum.[29] Both streamwise momentum and vorticity equations are solved using the alternating direction implicit (ADI) method, while the Poisson equation relating the yz-section streamfunction and streamwise vorticity is solved using the successive over relaxation (SOR) method.

The linear solution, which is presumed known, is obtained by using the computer code developed by Scott and Watt[30] for solving boundary value ordinary differential equations. The mesh used for the nonlinear problem has sixteen equally spaced cells in the spanwise direction covering a half spanwise wave length. The normal direction is divided into two hundred nonequally spaced cells covering a length of fifty δ_0, with cell sizes widening further away from the wall.

4. COMPARISON WITH EXPERIMENTS AND DISCUSSION

We have, in §1, embedded the nonlinear Görtler vortices among broader perspectives in transitional and turbulent boundary layers, focusing on longitudinal vorticity elements and the generation of high intense shear layers as the common issue. In order to progress meaningfully in future work with our much simplified nonlinear temporal problem, we shall present the results of a computation directed towards the comparison with recent experiments,[27] particularly with emphasis on the recoverableness of detailed flow structures. Our computations begin in the linear stability region of Swearingen's observations[27] which corresponds to $X = X_0 = 60$ cm in his measurements, for which $G \simeq 9.22, \alpha \simeq 0.462$, with $\delta_0 \simeq 0.132$ cm, $U_0 \simeq 5$ m/s, Re $\simeq 2.055 \times 10^5$, and the maximum value of $u'_{G\ RMS}/U_0 \simeq 0.12$. The initial disturbance energy content across the boundary layer was about $0.018\delta_0 U_0^2$. We find that a good comparison with the experiment is obtained by using a convection velocity of about $U_c \sim 0.644 U_0$.

In Figure 1 we show a comparison between the experimental streamwise growth of the maximum streamwise turbulent velocity fluctuations $u'_{T\ RMS}$, the growth of the maximum Görtler streamwise velocity $u'_{G\ RMS}$, and the computed maximum $u'_{G\ RMS}$, where RMS stands for root mean square. A very good agreement with the experiment is found up to $X \sim 110$ cm, subsequent to which the turbulence, which is not accounted for in the formulation, becomes dominant all over the flow field.

We show in Figure 2 a comparison of the measured values of the displacement thickness and the computed ones for both the low speed regions (peaks) and the high speed regions (valleys). A reasonable agreement with the experiment is observed up to $X \sim 100$ cm. The deviation between computation and experiment for larger X is again due to the presence of turbulence.

The comparison between computed and measured values of the velocity derivative (or shear stress) at the wall for both the values at peaks and valleys is shown in Figure 3. Also shown are the values for the Blasius and empirical turbulent boundary layer over a flat plate, as well as the computed spanwise-averaged values. There is the tendency of the latter "bridging" the laminar and (almost) the turbulent boundary layer over most of the transition region.

In Figure 4 we show the evolution of the streamwise velocity profile at the "peak region", with normal coordinates scaled by the wall parameters. A very good agreement between the measurements and the computation of the development of inflectional profile is noticed up to about $X \sim 100$ cm. At about $X = 110$ cm one may conclude that the presence of turbulence and other instabilities become apparent by the smoothing of the highly unstable velocity profile created by the streamwise Görtler vortices as well as a steepening of the velocity gradient at the wall. The streamwise vortices themselves do not give the transition to intense turbulent flows directly, but instead, are a means of setting up localized intense free shear layer instabilities that ultimately breakdown into the turbulence.

Our future intention, relying on our numerical model, would be to study the effect of the nonlinear interaction between different spanwise modes for the ultimate aim of the active control of the streamwise vorticity elements crucial to transitional and turbulent boundary layers as envisioned much earlier by Prandtl.[5]

ACKNOWLEDGEMENTS

This work is partially supported by the NSF Fluid Dynamics and Hydraulics Program Grant MSM-83-20307 and by the DARPA/Applied and Computational Mathematics Program through its University Research Initiative. We are thankful to J. T. Stuart for his continued interest. One of us (JTCL) acknowledges a Visiting Fellowship, supported by the U.K. S.E.R.C. at the Department of Mathematics, Imperial College, London during 1987/88.

References
1. Lin, C. C. *Quart. Appl. Math.* **3**, 117, 218, 277.
2. Lin, C. C. **The Theory of Hydrodynamic Stability**, Cambridge (1955).
3. Lin, C. C. **Selected Papers of C. C. Lin Vol. 1: Fluid Mechanics**, World Scientific (1987).
4. Lin, C. C. **Selected Papers of C. C. Lin Vol. 2: Astrophysics**, World Scientific (1987).
5. Prandtl, L. in **Aerodynamic Theory 3**, 34. Springer (1935).
6. Klebanoff, P. S., Tidstrom, K. D. and Sargent, L. M. *J. Fluid Mech.* **12**, 1 (1962).
7. Kovasznay, L. S. G., Komoda, H. and Vasudera, B. R. in **Proc. 1962 Heat Trans. Fluid Mech. Inst.**, 1. Stanford Univ. Press (1962).
8. Hama, F. R. and Nutant, J. in **Proc. 1963 Heat Trans. Fluid Mech. Inst.**, 77. Stanford University Press (1963).
9. Stuart, J. T. NPL Aero. Res. Rep. no. 1147; also NATO AGARD Rep. no. 514 (1965).
10. Russell, J. M. and Landahl, M. T. *Phys. Fluid* **27**, 557 (1984).
11. Kline, S. J., Reynolds, W. C., Schraub, F. A. and Runstaldler, P. W. *J. Fluid Mech.* **30**, 741 (1967).
12. Corino, E. R. and Brodkey, R. S. *J. Fluid Mech.* **37**, 1 (1969).
13. Willmarth, W. W. *Adv. Appl. Mech.* **15**, 159 (1975).
14. Blackwelder, R. F. and Kaplan, R. E. *J. Fluid Mech.* **76**, 89 (1976).
15. Blackwelder, R. F. *Phys. Fluids* **26**, 2807 (1983).
16. Görtler, H. NACA Tech. Memo. no. 1375 (1954).

17. Herbert, T. *Arch. Mech.* **28**, 1039 (1976).
18. Floryan, J. M. and Saric, W. S. *AIAA J.* **20**, 316 (1982).
19. Hall, P. *J. Fluid Mech.* **124**, 475 (1982).
20. Hall, P. *J.I.M.A.* **29**, 173 (1982).
21. Hall, P. *J. Fluid Mech.* **130**, 41 (1983).
22. Hall, P. Proc. R. Soc. Lond. **A399**, 135 (1985).
23. Hall, P. and Bennett, J. *J. Fluid Mech.* **171**, 441 (1986).
24. Hall, P. and Lakin, W. D. ICASE Rept. no. 87-16 (1987).
25. Tani, I. *J. Geophys, Res.* **67**, 3075 (1962).
26. Bippes, H. NASA TM no. 75243 (1978).
27. Swearingen, J. D. Ph.D. Thesis, Univ. Southern California (1985).
28. Pearson, C. A. *J. Fluid Mech.* **21**, 611 (1965).
29. Aziz, K. and Hellums, C. *Phys. Fluids* **10**, 314 (1967).
30. Scott, M. R. and Watts, H. A. *SIAM J. Numer. Anal.* **14**, 40 (1977).

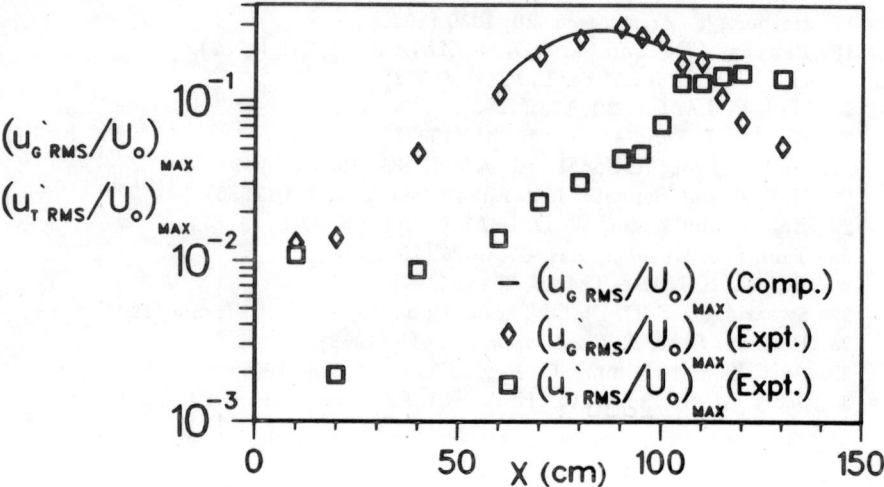

Figure 1. Streamwise velocity fluctuations: (Comp): present computations, (Expt.): Swearingen.[27]

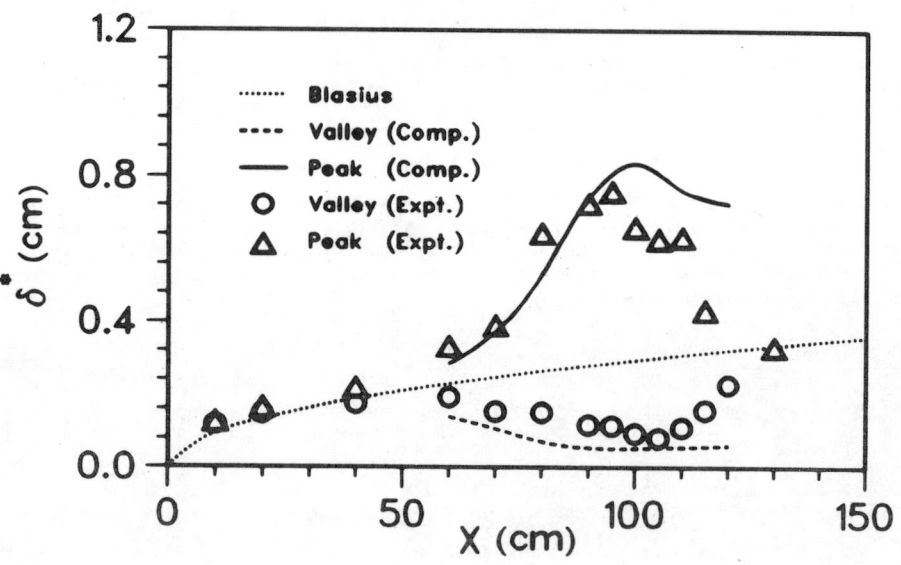

Figure 2. Displacement thickness: (Comp.): present computations, (Expt.): Swearingen.[27]

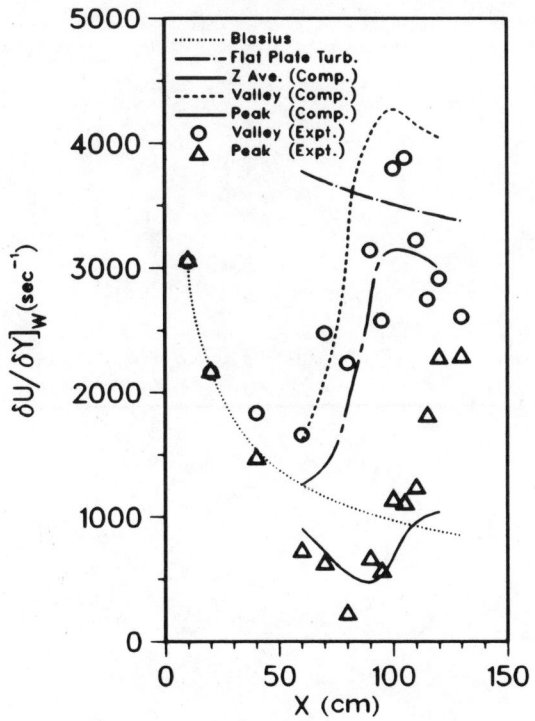

Figure 3. Streamwise velocity gradient at the wall, (Comp.): present computations, (Expt.): Swearingen.[27]

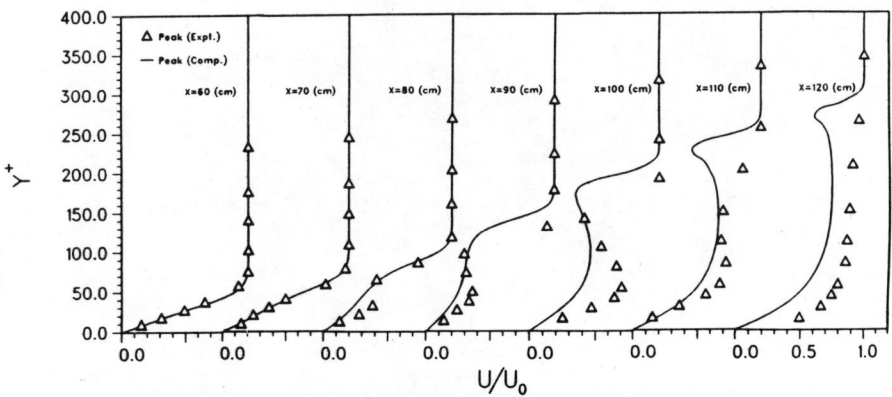

Figure 4. Development of high intense shear layers in streamwise velocity: (Comp.): present computations, (Expt.): Swearingen.[27]

III. GENERAL FLUID MECHANICS

SOME HELE SHAW CELL FLOWS AND THEIR VISUALIZATION

Louis N. Howard
Florida State University

1. The Hele Shaw cell, an arragement in which a liquid flows between closely spaced parallel plates, is familiar as a way of physically illustrating irrotational two dimensional flows, and was so used by Hele Shaw himself [1]. Henry Selby Hele Shaw, referred to in the author index of Lamb's treatise as "Shaw, H.J.S. Hele", was one person (with plenty of names). Professor Drazin tells me that he was Professor of Engineering at Bristol from 1882 to 1885, and did not use a hyphen. The elementary theory of such flows is given on p. 582 of [2], in Lamb's typically terse style; he also refers there to a paper of Stokes. If the flow domain is taken to be $|z^*| < h$ and (x^*, y^*) in a region of characteristic size L (which is also taken as a characteristic scale of horizontal variation of the flow variables), and if a characteristic horizontal velocity is taken as V, one is led to introduce the dimensionless variables $(x, y) = (x^*, y^*)/L, \zeta = z^*/h, t = t^*V/L, (u, v) = (u^*, v^*)/V, w = w^*L/hV, p = p^*\mu VL/h^2$ and the parameters $\epsilon = h/L, R = UL/\nu$ (in which μ and ν are the dynamic and kinematic viscosity coefficients). The flow equations then become

$$\begin{aligned} \epsilon^2 R(u_t + uu_x + vu_y + wu_\zeta) + p_x &= u_{\zeta\zeta} + \epsilon^2 \nabla_1^2 u \\ \epsilon^2 R(v_t + uv_x + vv_y + wv_\zeta) + p_y &= v_{\zeta\zeta} + \epsilon^2 \nabla_1^2 v \\ \epsilon^2 R(w_t + uw_x + vw_y + ww_\zeta) + \epsilon^{-2} p_\zeta &= w_{\zeta\zeta} + \epsilon^2 \nabla_1^2 w \\ u_x + v_y + w_\zeta &= 0. \end{aligned} \quad (1)$$

Here ∇_1^2 is the horizontal Laplace operator, and the density has been assumed constant, with gravity neglected. [Also of interest is the case of two immiscible fluids of different densities and viscosities, in which case gravity is also usually significant, but this will not be written down explicitly here.]

For small ϵ and moderate R an appropriate expansion has the form $u = u_0 + \epsilon^2 u_1 + ...$, (with similar expansions for $v, w,$ and p). With this equations (1) and the boundary conditions $u = v = w = 0$ on $\zeta = \pm 1$

readily show that
$$p_0 = p_0(x,y)$$
$$u_0 = U_0(x,y)(1-\zeta^2)$$
$$v_0 = V_0(x,y)(1-\zeta^2) \qquad (2)$$
$$w_0 = 0,$$

and
$$U_0 = -\frac{1}{2}p_{0x}$$
$$V_0 = -\frac{1}{2}p_{0y} \qquad (3)$$
$$U_{0x} + V_{0y} = 0$$

(Time may also enter as an independent variable, but at this order only parametrically.)

Thus U_0, V_0, the velocity components at the center plane $\zeta = 0$, are an ideal irrotational and two dimensional flow. (U_0 and V_0 are also proportional to the vertically averaged flow).

However, not <u>all</u> irrotational two dimensional flows can be simulated this way – (3) shows that the velocity potential is $-p_0/2$ which is of necessity single valued, and flows in multiply connected regions which have non-zero circulation about circuits enclosing obstacles do not occur in Hele Shaw cells, at least when the assumptions of the elementary theory are applicable. Hele Shaw cell flows are really dominated by viscosity, and the mathematical analogy between them and irrotational inviscid flow must not be extended carelessly to other physical concepts. Two examples of this are

(1) Pressure in an ideal flow is related to the velocity potential through Bernoulli's equation $\phi_t + \frac{1}{2}|\nabla\phi|^2 + p = 0$; this "pressure" has nothing to do with the real physical pressure p_0 in a Hele Shaw cell, which is in fact essentially the velocity potential itself. A buoyant object or bubble in an inviscid fluid would rise (if stable) with a constant acceleration related to the virtual mass, but in a Hele Shaw cell would rise (again, if stable) with a constant <u>velocity</u>.

(2) The principle of Galilean relativity does not apply to Hele Shaw cell flows the way it does in inviscid irrotational flow: the presence of the walls at $\zeta = \pm 1$ is in some ways always felt, and for instance the force on a cylindrical object towed through a Hele Shaw cell at a certain

velocity is not necessarily the same as that on the same object fixed relative to the walls and immersed in a flow which at great distances is uniform (in x and y) with the same relative velocity (either on $\zeta = 0$ or in ζ-average).

Some corrections to the elementary theory can be obtained by continuing the expansion in ϵ^2. The effects of non-zero R can be estimated this way, and may well be of some significance in typical demonstrations. While these corrections correspond to including nonlinear terms in (1), they do not correspond to taking account of inertia in the analogous inviscid flow which is in this sense already "inertial" – the Laplace equation for such flows is linear, but not a linearization. Other corrections are associated with the horizontal viscous terms; these are perhaps of greatest importance near the boundaries of the (horizontal) domain, where boundary layers (whose thickness is of the order of h, as noted already by Lamb) intervene. These will be mentioned again below.

In order to visualize flows in Hele Shaw cells, one may utilize dye streaks, which are helpful to locate streamlines in steady flow. Particularly intriguing patterns are often seen however by using instead a "rheoscopic fluid" – a suspension of small flakes of plastic or aluminum which tend to become oriented in some parts of the flow and so affect the light relfected from the fluid. This phenomenon is exploited for aesthetic purposes in the "Kalliroscope", a device invented by Paul Matisse, which is essentially a Hele Shaw cell in a vertical position, warmed form below so that thermal convection produces changing patterns of light reflection. Aluminum, though readily available, is less satisfactory because of its fairly rapid settling. Some shampoos contain such flakes – they also settle but not so rapidly as aluminum. Better water-based rheoscopic fluids, and a very well balanced one based on Freon 113, can be had (though not for a song) from the Kalliroscope Corporation. By looking closely one can see the motions of the individual tracer particles. Since they are at different distances from the faces of the cell, even those at about the same (x, y) position will have varying velocities, but to the extent that the elementary theory is correct, the <u>direction</u> of the averaged or $\zeta = 0$ velocity should be correctly indicated. However the interpretation of the patters of light relfection is more subtle, and I should like to discuss this next.

2. Small rigid particles immersed in a larger scale flow to first approx-

imation "see" only the velocity of the larger scale flow at the place where they are located – if they are not buoyant nor attached to anything they essentially just move with the flow. To a higher approximation they are also aware, so to speak, of the gradients in the flow. The effect of the flow on the particles may be estimated by treating them as isolated rigid bodies in a Stokes flow which at infinity has a linear form, say $\mathbf{u_0} + \mathbf{r_0} \cdot \mathbf{G}$, where $\mathbf{u_0}$ is the value of the large scale velocity at a point fixed in the particle, $\mathbf{r_0}$ is the position vector with this point as origin, and G is the tensor $\nabla \mathbf{u}$, also evaluated at this point. This is approximately the case if

a) The particles are small enough that the Reynolds number based on their size is very small;

b) They are not <u>so</u> small that their Brownian motion is dominant;

c) They are not too densely distributed, i.e. their typical separation is much larger than their size;

d) They are not too close to walls, again in terms of their size.

a) is almost certainly true for the kind of flow visualization particles considered here; the others are perhaps less certain, but will be assumed anyway.

We describe the motion of the particle (a rigid body) by means of the velocity $\underline{U}$ of a point O fixed in it, and by its angular velocity $\boldsymbol{\omega}$. The gradient G of the Stokes flow are infinity may be described by its symmetric part $S = \frac{1}{2}(G + G^*) = \frac{1}{2}$ (deformation tensor of $\mathbf{u}$) and by the fluid angular velocity $\boldsymbol{\omega}_f = \frac{1}{2}\nabla \times \mathbf{u}$, whose components are related to the nontrivial components of the antisymmetric part of G in the usual way (G^* is the transpose of G). Then the force and torque (about O) on the particle are given by the formulas (cf. [3])

$$\begin{aligned} \mathbf{F} &= -\mu[K_t \cdot (\mathbf{U} - \mathbf{u_0}) + C^* \cdot (\boldsymbol{\omega} - \boldsymbol{\omega}_f) + \Phi : S] \\ \mathbf{T} &= -\mu[C \cdot (\mathbf{U} - \mathbf{U}_0) + K_r \cdot (\boldsymbol{\omega} - \boldsymbol{\omega}_f) + \tau : S] \end{aligned} \quad (4)$$

in which K_t and K_r are the translational and rotational <u>resistance tensors</u> and C is the <u>coupling</u> (pseudo) <u>tensor</u> (all of second order) for the body in question, and Φ and τ are (third order) shear force and shear torque tensors. These tensors depend only on the shape and orientation of the body and (with the exception of K_t) on the choice of the point O. For a non-chiral body the coupling tensors can be taken to be zero by a suitable choice of O. Let us in fact assume that all the flow visualization particles are <u>circular disks</u> of radius a, place the point O at the center of the disk in question, and describe its orientation by its unit normal vector $\mathbf{i}_1$, extending this vector to an orthonormal right handed set by vectors $\mathbf{i}_2$ and $\mathbf{i}_3$ in the plane of the disk. In this case the coupling tensor C and the shear force tensor Φ are zero, and the others are given by ([3], [4]):

$$K_t = \frac{16a}{3}[2I + \mathbf{i}_1\mathbf{i}_1], \quad K_r = \frac{32a^3}{3}I$$
$$\tau = \frac{16a^3}{3}[\mathbf{i}_3\mathbf{i}_1\mathbf{i}_2 - \mathbf{i}_2\mathbf{i}_1\mathbf{i}_3 + \mathbf{i}_3\mathbf{i}_2\mathbf{i}_1 - \mathbf{i}_2\mathbf{i}_3\mathbf{i}_1] \quad (5)$$

where I is the identity tensor.

Assuming the particles to be neutrally buoyant, the net force and torque on each of them must be zero; (5) and (4) then show that $\mathbf{U} = \mathbf{u}_0$, i.e. the disks move with the flow, and

$$\boldsymbol{\omega} = \boldsymbol{\omega}_f - (\mathbf{i}_1 \times S) \cdot \mathbf{i}_1 \quad (6)$$

Since the unit vector $\mathbf{i}_1$ is attached to the particle, its rate of change is described by the usual rigid body formula

$$\frac{d\mathbf{i}_1}{dt} = \boldsymbol{\omega} \times \mathbf{i}_1$$

Using 6) in this we obtain

$$\frac{d\mathbf{i}_1}{dt} = \boldsymbol{\omega}_f \times \mathbf{i}_1 - S \cdot \mathbf{i}_1 + \mathbf{i}_1(\mathbf{i}_1 \cdot S \cdot \mathbf{i}_1) = -G \cdot \mathbf{i}_1 + \mathbf{i}_1(\mathbf{i}_1 \cdot S \cdot \mathbf{i}_1) \quad (7)$$

since $G \cdot \mathbf{i}_1 = (\nabla \mathbf{u}) \cdot \mathbf{i}_1 = \frac{1}{2}(\nabla \mathbf{u} + (\nabla \mathbf{u}^*)) \cdot \mathbf{i}_1 + \frac{1}{2}(\nabla \mathbf{u} - (\nabla \mathbf{u})^*) \cdot \mathbf{i}_1 = S \cdot \mathbf{i}_1 - \frac{1}{2}(\nabla \times \mathbf{u}) \times \mathbf{i}_1 = S \cdot \mathbf{i}_1 - \boldsymbol{\omega}_f \times \mathbf{i}_1$. Now the nonlinear equation (7) for $\mathbf{i}_1$ can be solved in terms of the solution of a linear equation in a rather simple way. Represent G by the matrix of its components with respect to some fixed basis, also calling this matrix G, and let L be a column vector satisfying

$$\frac{dL}{dt} = -GL \quad (8)$$

From this we find

$$a) \frac{dL^T}{dt} = -L^T G^T$$
$$b) \frac{d}{dt}(L^T L) = -L^T G^T L - L^T G L = -2L^T S L$$

(where the matrix $S = (G + G^T)/2$) and then

$$c) \frac{d}{dt}(L/(L^T L)^{\frac{1}{2}}) = -GL/(L^T L)^{\frac{1}{2}} - \frac{1}{2}(L^T L)^{-\frac{3}{2}} L(-2L^T SL)$$
$$= -GL/(L^T L)^{\frac{1}{2}} + \qquad (9)$$
$$(L/(L^T L)^{\frac{1}{2}})^T S(L/(L^T L)^{\frac{1}{2}}) L/(L^T L)^{\frac{1}{2}}$$

Comparing (9) with the matrix interpretation of (7) we see that the unit vector $L/(L^T L)^{\frac{1}{2}}$ satisfies the same equation as $\mathbf{i_1}$. Thus if (8) is solved with L initially equal to the column of components of $\mathbf{i_1}$, we can get $\mathbf{i_1}$ thereafter as $L/(L^T L)^{\frac{1}{2}}$.

Now in general G is not a constant matrix, even if the flow is steady, because the particle moves with the flow and G (evaluated at the particle's center) will usually change along the particle's path. But if this change is not too rapid, G is approximately constant and the solution of (8) is $L = \exp(-tG)L_0$. If the eigenvalue of G furthest to the left in the complex plane is a simple real eigenvalue, this shows that L tends to the direction of the corresponding eigenvector. In the case of pure shear, where $\boldsymbol{\omega}_f = 0$ and $G = S$, this means that the disks will tend to orient so that their normals are in the direction of <u>greatest compression</u>, if there is a single such direction as is in general the case. If $\boldsymbol{\omega}_f$ is not zero, G may have two complex conjugate eigenvalues, and if these have negative real part they are furthest to the left because the trace of G is zero (incompressible flow). In this case $\mathbf{i_1}$ will usually rotate, in general somewhat non-uniformly, in the plane of the real and imaginary parts of the corresponding eigenvector, and hence about the line perpendicular to this plane, which is in the direction of the <u>real</u> left eigenvector.

The distinction between the cases of approach to a limiting direction and rotation can be made in physically more familiar terms as follows. In a (right handed) system of principal axes for the deformation tensor the matrix G is given by

$$G = \frac{1}{2} \begin{bmatrix} D_1 & \Omega_3 & -\Omega_2 \\ -\Omega_3 & D_2 & \Omega_1 \\ \Omega_2 & -\Omega_1 & D_3 \end{bmatrix}$$

where the D_i are the principal deformations and the Ω_i are the components of vorticity in the principal axis system. In this form it is fairly easy to see that the distinction can be made in terms of the invariants

$$a = D_1 D_2 + D_2 D_3 + D_3 D_1 + \sum \Omega_i^2 = -\frac{1}{2} \sum D_i^2 + \sum \Omega_i^2$$

(the second equality because $\sum D_i = 0$), which is a kind of measure of the relative magnitudes of vorticity and deformation, and

$$b = -D_1 D_2 D_3 - \sum \Omega_i^2 D_i$$

which takes some account of the orientation of the vorticity vector with respect to the principal axes of deformation. (The characteristic polynomial of G is $\lambda^3 + a\lambda + b$.) It turns out that if $a > 0$ we get rotation (of the normal to the disk) when $b < 0$, but if $a < 0$ only when $b < -2(-a/3)^{\frac{3}{2}}$; otherwise a limiting orientation is approached. Of course, in the latter case the disk may, and generally does, rotate <u>about</u> its axis of symmetry.

3. Now in a Hele Shaw cell the velocity components are given approximately by the formulas (2), from which the gradient matrix G is readily found to be (in a system of coordinates with third axis normal to the plane of the cell)

$$G = \begin{bmatrix} U_{0x}(1-\zeta^2) & V_{0x}(1-\zeta^2) & 0 \\ U_{0y}(1-\zeta^2) & V_{0y}(1-\zeta^2) & 0 \\ -2\zeta U_0/\epsilon & -2\zeta V_0/\epsilon & 0 \end{bmatrix} \qquad (10)$$

Equations (8) thus show that the first two components of the vector L behave essentially as they would in the two dimensional flow (U_0, V_0), and since this flow is irrotational they tend to become oriented along its direction of <u>maximal compression</u>. Since this direction is not usually orthogonal to (U_0, V_0), the third of equations (8) then shows that (except right on $\zeta = 0$) the third component of L will grow rapidly in magnitude, which means that $\mathbf{i}_1$ will rapidly assume a direction normal to the plane of the cell. As a general rule, then, small disks in a Hele Shaw cell flow would be expected to be oriented parallel to the cell, and to reflect light pretty much uniformly over the cell. Two significant exceptions to this, however, are:

a) Along a particle path which happens to be orthogonal to the local direction of maximal compression of the (U_0, V_0) flow, the strong tendency for disk orientation parallel to the plane of the cell is not present. Such paths, if any, should thus be visible as darker lines in reflected light.

b) In places where velocity gradients in the x or y directions are comparable to those in the z direction, the argument given above must be modified (because equation (2) is no longer the correct description of the flow, and also because G would no longer have just one dominant row as it does in (10)), and the limiting orientation of the disks might well not be parallel to the plane of the cell. It is also conceivable that in such places the disk might not assume a definite orientation, but continually tumble. In either case the light reflection would be expected to be altered.

Case a) occurs for instance if a more or less uniform (in x and y) flow is produced in a Hele Shaw cell and caused to flow around an obstacle like a solid circular disk placed between the plates. The (U_0, V_0) flow then resembles ideal flow without circulation past a circle, and along the dividing streamline on the downstream side, the direction of maximal compression is normal to the streamline. This dividing streamline becomes visible as a dark line by reflected light. On the upstream side, however, the normal to the dividing streamline is a direction of maximal expansion rather than compression, and no dark line is seen. A related and somewhat more easily produced example is obtained with a slightly inclined Hele Shaw cell containing an air bubble in the rheoscopic fluid. The slowly rising bubble leaves behind itself a slender track which appears dark by reflected light. Under certain conditions rising bubbles do not rise steadily in a straight line, but oscillate from side to side. This produces patterns that have a considerable resemblance to alternating vortex trails in high Reynolds number flow past a cylinder. This analogy should not be pushed too far; once the bubble has passed, the fluid comes rapidly to rest, but the pattern remains in the light reflection – in the absence of flow, some time is required for the disks to become randomly oriented.

[In the lecture some demonstrations of flows produced by rising bubbles were given, using Hele Shaw cells placed on an overhead projector and slightly inclined. A change in the amount of light reflected is of course accompanied by a complementary change in the amount transmitted. Some photographs of similar flows are reproduced in the figures.]

Case b) always occurs near the edges of obstructions, or the edges of the cell, or of rising buoyant bubbles of another fluid. There are also sometimes regions of strong shear inside the flow which become visible in this way. This seems to be the case with the Kalliroscope. Indeed, it was in attempting to understand some experiments on thermal convection in Hele Shaw cells, in which rising and moving plumes are seen clearly with this kind of visualization, that I became interested in the matters discussed here. [In the lecture, a movie film of such an experiment which was made some time ago by my colleague R. Krishnamurti was shown.]

4. The nature of the flow in the boundary layer near a solid obstacle in a Hele Shaw cell is fairly well indicated by the simplest special case – a wall on $y = 0$ extending over all x and $|\zeta| < 1$, with a constant Poiseuille flow $u_\infty = \frac{1}{2}(1-\zeta^2)$ (driven by a uniform pressure gradient in the x-direction) being approached as y becomes large compared to the plate separation. We set $y = \epsilon\eta$, and then the only non-trivial equation of the set (1) becomes

$$p_x = u_{\zeta\zeta} + u_{\eta\eta} \tag{11}$$

in which u, the only non-zero velocity component, is a function of η and ζ and p_x is a constant. From the flow u_∞ as $\eta \to \infty$ we see that in fact $p_x = -1$. u is thus $u_\infty + \psi(\eta, \zeta)$ where ψ is harmonic, zero for $\eta \to \infty$ and for $\zeta = \pm 1$, and on $\eta = 0$ is $-\frac{1}{2}(1-\zeta^2)$. This problem for ψ is easily solved; one representation of the solution is

$$\psi = \sum c_n \sin n\pi(\zeta+1)/2 \exp(-n\pi\eta/2) \tag{12}$$

in which the summation here (and hereafter) is for <u>odd</u> n, and the c_n are found (from the condition on $\eta = 0$) to be:

$$c_n = -16/(n\pi)^3 \tag{13}$$

We thus have $u_x = 0$, and

$$\begin{aligned} u_y &= \frac{8}{\epsilon\pi^2} \sum \sin n\pi(\zeta+1)/2 \exp(-n\pi\eta/2)/n^2 \\ u_z &= \frac{1}{\epsilon}\zeta - \frac{8}{\epsilon\pi^2} \sum \cos n\pi(\zeta+1)/2 \exp(-n\pi\eta/2)/n^2 \end{aligned} \tag{14}$$

Equations (8) then become

$$\frac{dL_1}{dt} = 0, \quad \frac{dL_2}{dt} = -u_y L_1, \quad \frac{dL_3}{dt} = -u_z L_1,$$

so, usually, L_2 and L_3 rapidly grow large and their ratio approaches the limit u_y/u_z, a function of η and ζ which can be computed fairly easily from (14). The last figure indicates the orientation of the disks calculated in this way for the region $0 < \eta, \zeta < 1$. It is evident that reflection of normally incident light should fade out as the wall on $\eta = 0$ is approached. It appears to be essentially this phenomenon that makes the edges of rising plumes in the convection experiment visible with a rheoscopic fluid.

REFERENCES

Shaw, H.J.S. Hele Trans. Inst. Nav. Arch. 40 (1898)
Lamb, H. "Hydrodynamics", Dover Publ. (1945)
Happel, J. and H. Brenner "Low Reynolds number hydrodynamics", Martinus Nijhoff Publ. (1986)
Brenner, H. "The Stokes resistance of an arbitrary particle. Part III. Shear fields", Chem. Engrg. Sci. 19, 631-651 (1964)

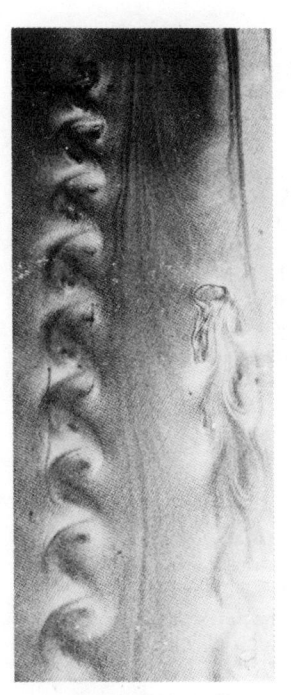
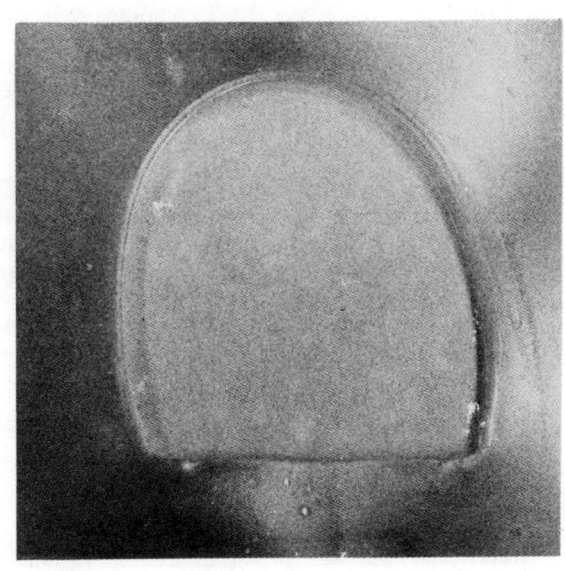

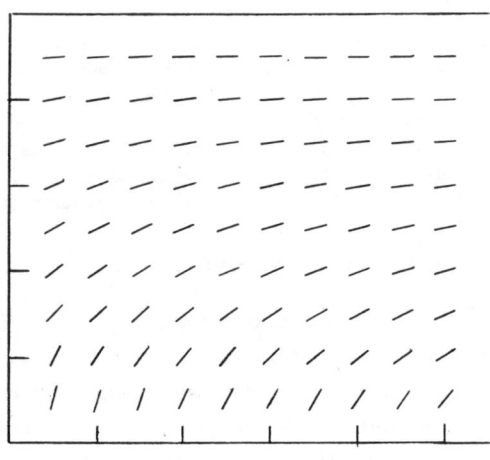

FORCED GENERATION OF SOLITARY WAVES

T. Yaotsu Wu

California Institute of Technology
Pasadena, CA 91125 USA

This is a preliminary study on the determination of the basic mechanism underlying the phenomenon whereby a forcing distribution moving uniformly with a constant transcritical velocity in shallow water can generate, continuously and periodically, a succession of solitary waves, advancing in procession ahead of the disturbance, while a train of weakly dispersive waves develop behind the steadily progressing disturbance. This phenomenon was reported first numerically by Wu & Wu[1] based on the generalized Bousinesq model with a two-dimensional surface pressure and by Lee[2] using the forced KdV model with two kinds of external forcing disturbances, a free surface pressure and a moving submerged topography. These models were found by Lee[2] in broad agreement with experiment employing a two-dimensional cambered topography towed along the bottom of a layer of shallow water.

This remarkable phenomenon shows that the response of a dynamical system to steady forcing need not asymptotically tend to a steady state, but instead the response can be conspicuously periodic when the system is being forced at resonance. Conversely, not all of the transcritical disturbances will necessarily result in a response having the precursor solitons periodically generated. In fact, a family of forced steady solitary waves, all of a permanent form of finite amplitude, was found to exist according to the fKdV model, each of which is accompanied by a characteristic forcing function moving with a constant near critical velocity. With reference to each forced steady solitary wave, a nonlinear perturbation equation is derived; it depends on the depth Froude number and the initial condition as two control parameters and it admits a trivial solution. However, if the forced steady solution is sufficiently perturbed, it will evolve to generate precursor solitons periodically in the same manner as by non-characteristic forcing functions. The preliminary result obtained here suggests that the periodically perturbed motion thus induced appears to be a bifurcation of the solution whose type remains to be determined.

1. INTRODUCTION

Studies of generation of solitary waves (in a soliton-bearing system) by external forcing agencies have emerged to become a rather new subject, though recorded chance observations of certain events obviously related to this phenomenon can be traced back to a history of about fifty years (for an overall review see Wu[3]). With some imagination, it could even be dated back to the time of John Scott Russell who reported observing in 1834 the "large solitary wave" that "suddenly rolled forward with great velocity" from a boat which had been "rapidly drawn along a narrow channel by a pair of horses" and was then suddenly stopped, with no reason given or sought. The water then "accumulated round the prow of the vessel in a state of violent agitation, suddenly leaving it behind, rolling forward ..." If the vessel were given an additional pair of strong and well-behaved horses to make certain of the rapid motion of the boat was maintained, could Russell then possibly have observed the same fascinating phenomenon as encountered by Huang et al.[4] and Sun[5] during their ship model tests? They discovered, independently, that as the ship model (about 1.5m long) was towed on a very shallow water (nearly 0.15 m deep) in a towing tank, at a constant transcritical velocity U, corresponding to the depth Froude number $0.8 < F < 1.2$ ($F = U/(gh_o)^{\frac{1}{2}}$, g being the gravitational constant, h_o the undisturbed water depth) unsteady waves were continuously generated, one after another, that were singularly different from the familiar ship waves, but periodically detached themselves from the ship model, quickly stretched straight across the channel and rolled away in procession ahead of the the advancing model. Even further, one might wonder that if this phenomenon were discovered some 150 years earlier, couldn't it have changed the history of hydrodynamics that followed?

Back to reality, we note that there are several intriguing features of this "singular and beautiful" phenomenon that are thought provoking. First, such a phenomenon can occur in any soliton-bearing system. That is to say, for a system possessing the same nonlinear and dispersive effects as those in shallow water waves, steady forcing disturbances can result in periodic responses. Such systems are known to occur in various physical, chemical, biological, geological and astrophysical disciplines in addition to mechanics where the phenomenon was first discovered. We further note that even for three-dimensional forcing distributions, such as that given by a ship hull, the resonant responses invariably would settle to a two-dimensional formation if the medium is transversely bounded, as in a towing tank. Equally interesting is the new finding[3] (see also Section 3) that there exist eigen-distributions of resonant forcing which can result in waves of finite amplitude in permanent form without any unsteady waves being radiated. But, when sufficiently perturbed, the steady local effects can become unstable and will evolve into an unsteady form of periodic responses, generating the same upstream advancing solitons as just described.

To facilitate investigations of these new problems, we first have to extend our scope from one for physically closed systems to that for open systems and to extend space dimension from one to more than one. By an open system we mean a system having exchanges of mass, momentum and energy with some external agencies. With such generalizations, recent research activities have been directed to investigate the fundamental properties of the mathematical models, to develop effective methods of solution, and to establish new concepts and understanding about the phenomenon in question. To describe the position where we now stand in this new field with respect to the current soliton theory, the latter being already highly developed for closed systems with one space dimension, the following aspects are worthy of note.

We signify the difference between physically closed and physically open systems by expressing their conservation laws in the following evolution form:

$$E_t^{(n)} + F_x^{(n)} = 0 \qquad \text{(for a closed system)} \qquad (1a)$$

$$E_t^{(n)} + F_x^{(n)} = Q^{(n)}(x,t) \qquad \text{(for an open system)} \qquad (1b)$$

$n = 1, 2, \ldots, N(\leq \infty)$, where $E^{(n)}(u;x,t)$ is a density function of order n, which generally depends on the state function $u(x,t)$ of the system, $F^{(n)}(u;x,t)$ is the corresponding flux function and $Q^{(n)}(x,t)$ is the associated source or forcing function, which is prescribed over a specific range of the space coordinate x and the time t (with obvious extensions for the case of more than one space dimensions). This set of equations is supposed to contain the basic equation(s) of motion, such as the KdV equation, after which the system in question is modelled.

Using the classical KdV equation, for example, to analyze a problem of a closed system (which we call a CS-problem) and the forced KdV (fKdV) equation for a problem pertaining to an open system (called an OS-problem), we observe the following contrasting points. The solution to a CS-problem, with the boundary fixed as required, is uniquely determined by the initial data, whereas the solution to an OS-problem depends on initial and boundary conditions as well as the specified forcing functions. A closed system of the KdV family has a countably infinite number of conservation laws, rendering the system completely integrable[6,7] (in asymptotic sense). In contrast, the open system may have several basic conservation laws, is not known to be integrable, and its solutions can be obtained only by numerical and asymptotic expansion methods. The CS-problems usually deal with asymptotic states, leaving the transient processes either out of the question or considered by numerical means, whereas the OS-problems are invariably transient in nature. The free soliton solutions of a closed system are known to be robust, very stable to moderate disturbances, whereas the stability criteria of the solutions of an open system are yet to be established.

It gives me a great pleasure to have this privileged opportunity to present to Professor C.C. Lin and his friends this progress report on some preliminary studies of this fascinating problem which has only a very short history but may promise a future of vigorous growth. It is fortunate there are expert pioneers in the audience who have made significant contributions to this new field, including C.C. himself[8], and we have already heard earlier in this Symposium from Dr. Ru-lin Chou and Prof. C.K. Chu about their interesting study on the generation of periodic solitons in plasma by external forcing in the form of a moving boundary. This is a token of our warmest tribute to Prof. Lin from my research group at Caltech, also C.C's Alma Mater, where, as everywhere else, his outstanding performance as a leader of our profession has been winning increasingly high pride and esteem. We hope that this subject may also appeal to him and receive the guidance he has so generously been giving to the other subjects. With this we wish him a most gratifying and fulfilling era of new creativity.

2. THE THEORETICAL MODELS

There are several theoretical models that can admit external forcing disturbances in the form of a surface pressure moving over the top free surface and a topography moving along the floor of a water layer. The classical Boussinesq equations have been generalized[9,10] to include forcing functions, applicable to an inhomogeneous medium so that the water depth may be slowly varying in two horizontal dimensions, and time. This generalized Boussinesq (gB) model has been used by Schember[11] and Lepelletier[12] to investigate some problems related to tsunamis and harbor oscillations. It was also used by Wu & Wu[1] to discover numerically this phenomenon of runaway solitons. A different approach based on the concept of a directed-sheet model[13] was adopted by Ertekin, Wehausen & Webster[14-17] to study this class of problems. More limited to unidirectional motions is the forced Korteweg-de Vries (fKdV) model which has been employed with regular and singular forcing representations for various studies[2,3,18,19]. For applications to internal solitary waves in a density-stratified fluid, the fKdV model has also been used to investigate the near resonant motion of a specific normal mode[20-26]. Comparisons between the gB model and the fKdV model and with experiment has recently been carried out by Lee & coworkers[2,27], who found a broad agreement between experiment and the two theoretical models at critical and near critical speeds ($0.9 < F < 1.1$).

We shall adopt the fKdV model for the subsequent investigation, partly for its established validity and partly for the appeal to mathematical simplicity. For left-going forcings the fKdV equation reads

$$\varsigma_t + [(F-1) - \frac{3}{2}\varsigma]\varsigma_x - \frac{1}{6}\varsigma_{xxx} = \frac{1}{2}(p_a + b)_x, \qquad (2)$$

where ς is the free surface elevation of a water layer initially of uniform depth, the subscripts denote partial differentiation, and the external forcing is given in the form of the surface pressure $p_a(x,t)$ and the bottom topography $b(x,t)$. In (2) and the sequel, the length, the time and the mass are given in units of the constant water depth h_o, $(h_o/g)^{\frac{1}{2}}$, and ρh_o^3 respectively, where g is the gravitational constant and ρ is the uniform density of the fluid. The x-coordinate is fixed with respect to the disturbance moving to the left with constant velocity U with respect to the fluid at rest at infinity, corresponding to the depth Froude number F ($F = U/(gh_o)^{\frac{1}{2}}$). As is implied by the underlying assumptions, this model is for weakly nonlinear, weakly dispersive and weakly forced waves at near resonance such that

$$\epsilon = \frac{h_o^2}{\lambda^2} \ll 1, \quad \alpha = \frac{a}{h_o} \ll 1, \quad \alpha = O(\epsilon), \tag{3a}$$

$$|p_a + b| = O(\alpha^2), \quad |F - 1| = O(\alpha), \tag{3b}$$

where λ is a typical wavelength, and a is a typical wave amplitude. $\epsilon \ll 1$ for all long waves, and $\alpha = O(\epsilon)$ holds for the even balance between the nonlinear and dispersive effects as for the gB model. The orders of magnitude of the forcing $(p_a + b)$ and of the detuning parameter $\delta_F = F - 1$ are consistent with the expansion criteria and the solvability conditions of the problem. The latter two conditions (3b) are more strict than that assumed for the gB model (in which they are of $O(\alpha)$ and $O(1)$ respectively), especially that imposed on the detuning parameter. Since the simple sum $(p_a + b)$ appears in (2), it implies that the surface pressure and the bottom topography of the same distribution are entirely equivalent for the fKdV model. We may therefore write $P(x,t) = p_a(x,t) + b(x,t)$ for the entirety of disturbances.

Numerical results were obtained from using (2) for the forcing distribution

$$P(x,t) = \frac{1}{2}P_m \left[1 + \cos\frac{2\pi x}{L}\right] \quad \left(-\frac{1}{2}L < x < \frac{1}{2}L, \ t > 0\right), \tag{4}$$

and the initial condition

$$\varsigma(x,0) = 0. \tag{5}$$

The numerical method originally developed by Wu & Wu[1] is found also very effective for the present task of calculating the fKdV equation (2).

As a typical example, we present here the result of a critical forcing prescribed by

$$P_m = 0.2, \ L = 2 \ (= 2h_o), \ F = 1.0. \tag{6}$$

The corresponding numerical results of the surface wave elevation ς, given at time intervals of $\Delta t = 2$, and the time-dependent wave resistance coefficient are shown in figure 1. A conspicuous feature of the result is that, after the surface pressure has been exerted and kept moving at the critical speed for a definite period of time, a solitary wave emerges just ahead of the disturbance, and eventually breaks away, at about $t = 30$, to propagate ahead as a free solitary wave, forming an entity which we call a 'precursor soliton', or a 'runaway soliton'. This is followed by another new solitary wave going through the same cycle, and this process seems to continue periodically and indefinitely.

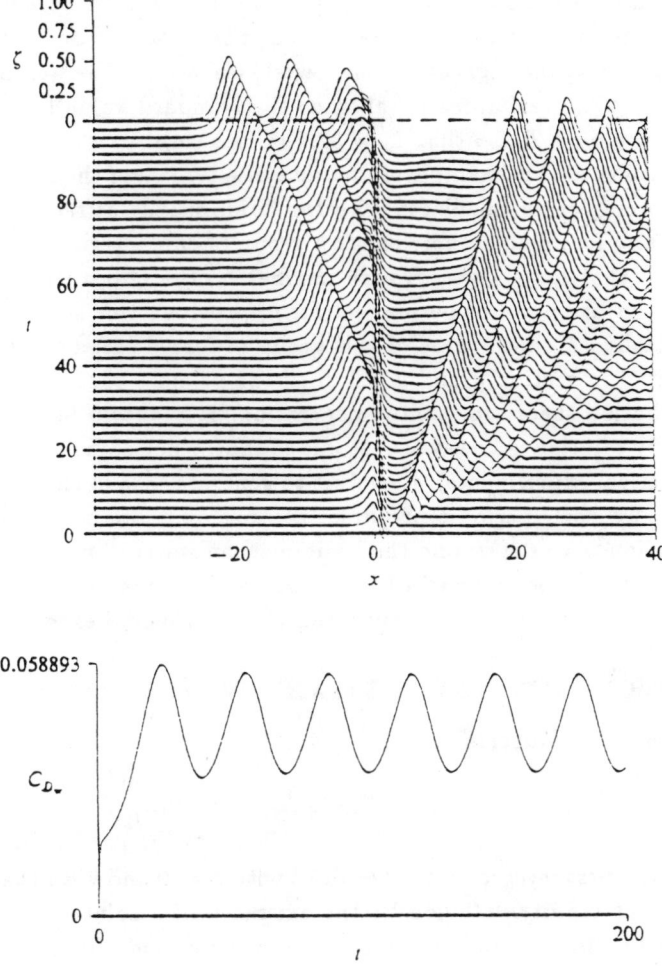

Figure 1. Numerical results of the fKdV model (2) for the surface wave elevation ζ (calculated with $\Delta x = 0.2$ and $\Delta t = 0.1$ and shown at time increment of 2.0) and the wave resistance coefficient C_{Dw} (with asymptotic mean of $\overline{C}_{Dw} = 0.044$) for the forcing distribution (4) with: $P_m = 0.2$, $L = 2.0$, and $F = 1.0$. The first three precursor solitons are generated at $t = 24.5$, 56.8, and 89.2 (the maxima of C_{Dw}), with a period of $T_s = 32.4$. — — —, initial undisturbed wave surface shown as reference at $t = 100$.

Immediately behind the moving topography there trails an ever-lengthening region of depressed water, of nearly uniform depth, which is in turn followed by a train of cnoidal-like waves oscillating about the originally undisturbed water surface, with the wave height decreasing with distance and with the train length increasing with time, and eventually emerging with the yet undisturbed water further downstream.

Also shown in figure 1 is the wave resistance curve, which oscillates nearly sinusoidally about the mean value of $\overline{C}_{Dw} = 0.044$, reaching a maximum (or minimum) when the free surface at the leading edge of the forcing disturbance is at the highest (or the lowest) elevation. The well defined period of the C_{Dw} variations is taken as the standard value for the soliton generation period T_s, so that $T_s = 32.4$ in this case.

The results of this example are found quite typical of the family of near-critical-forcing cases calculated with the fluid initially at rest. In addition, we have also found that there exists a group of characteristic forcing functions such that for a given Froude number F, the only response is but a local soliton-like wave of permanent form, provided this solitary wave existed initially. A simple case of this group of forced steady solitary waves is presented in section 3 and is taken as the primary motion to formulate a perturbation theory for soliton generation. It is shown in section 4 that this steady solution is however unstable so that, when sufficiently perturbed (by numerical experiments), it evolves to generate, like in the previous case shown in figure 1, periodic precursor solitons. The main objective of our investigation is to determine the basic mechanism that underlies the phenomenon in question, for which a preliminary attempt is made in section 5 by applying the method of truncated Galerkin modal expansion.

3. FORCED STEADY SOLITARY WAVES

Consider the equation

$$(F-1)\varsigma - \frac{3}{4}\varsigma^2 - \frac{1}{6}\varsigma_{xx} = \frac{1}{2}P(x), \qquad (7)$$

which is the first integral of (2) obtained with $\varsigma_t = 0$ and with the regularity condition at infinity. Obviously, the existence of a solution, $\varsigma_s(x)$ say, to equation (7), for some characteristic forcing function $P(x)$, is a necessary condition for (2) to have a steady solution of permanent form under forcing. It will also be sufficient for this to be a steady solution of (2) if it further satisfies the initial condition

$$\varsigma(x,0) = \varsigma_s(x). \qquad (8)$$

Then $\varsigma(x,t) = \varsigma_s(x)$ is a forced steady solution.

It is readily verified that

$$\varsigma_s = a \operatorname{sech}^2 kx, \quad a = \frac{4}{3}k^2 \qquad (9a)$$

$$P(x) = 2b \operatorname{sech}^2 kx, \quad b = F_1 a, \quad F_1 = F - 1 - \frac{2}{3}k^2, \qquad (9b)$$

is a solution of (7). It then follows that $\varsigma(x,t) = \varsigma_s(x)$ of (9) will be a forced steady solitary wave if it exists initially. The corresponding forcing (9b) is called the characteristic forcing for the forced steady soliton; it does no mechanical work since it makes no contribution to the wave resistance coefficient

$$C_{Dw} = -\int p_a(x,t) \frac{\partial \varsigma}{\partial x} dx, \qquad (10)$$

by virtue of p_a and ς_s being both even in x. Although there exists a family of an infinite number of solutions[3] of (7), we shall consider only (9) here as a simple example of the family.

Existence of these solutions is of significance in several ways. First, this class of solutions can be of value in application if it is found, because through further theoretical and experimental investigations, they can model certain types of forcing functions which yield only local effects, with no waves being radiated, thus broadening the family of solitary waves. Second, since (9) is an exact solution of (2) under the no-initial-perturbation condition (8), it provides a rigorous test for validating any approximation scheme that is supposed to hold for the Boussinesq-KdV class of theory. Furthermore, a rewarding result of using such a one parameter (the 'wave number' k) characteristic forcing is that when the forced steady waves are sufficiently perturbed to generate periodic runaway solitons by numerical calculation, the effects of the wavenumber k on the amplitude and period of generation is immediately deduced by the rule of similarity, as will be shown below. Finally, these exact solutions of the fKdV equation are indispensable in providing a primary wave state, possibly of finite amplitude, for perturbation studies of the stability criteria of these solutions.

4. PERTURBATION THEORY OF SOLITON GENERATION

Consider the solution of (2) in the form

$$\varsigma(x,t) = \varsigma_s(x) + \eta(x,t), \qquad (11)$$

where $\eta(x,t)$ is an arbitrary (but sufficiently smooth) perturbation of ς about the forced steady solution $\varsigma_s(x)$ as the primary state. Substituting (11) in (2) yields for η the following evolution equation:

$$\eta_t + \frac{\partial}{\partial x}\left[(F-1)\eta - \frac{3}{4}\eta^2 - \frac{1}{6}\eta_{xx} - \frac{3}{2}\varsigma_s(x)\eta\right] = 0, \qquad (12)$$

which is a nonlinear homogeneous equation with a variable coefficient. The problem then becomes one of calculating η from (12) under the initial condition

$$\eta(x,0) = \eta_0(x), \qquad (-\infty < x < \infty), \tag{13}$$

where $\eta_0(x)$ is an arbitrary function. In particular, if $\eta_0(x)$ assumes the form

$$\eta_0(x) \equiv \eta(x,0) = -\mu\,\varsigma_s(x), \tag{14}$$

then $\mu = 1$ restores the initial rest state of $\varsigma(x,0) \equiv 0$ $(-\infty < x < \infty)$ and the problem thereby becomes identical to the original problem of ς. However, with $\mu = 0$, the problem of η is nonlinear but homogeneous, which admits a trivial solution and is in a form convenient for investigating the stability of the solution for η, hence of $\varsigma(x,t)$ through (11). Although arbitrary perturbations can be applied to η in an infinitely many ways other than that specified by (14), e.g., by continuous perturbation in time, we shall confine ourselves to the initial perturbation (14) for definiteness.

At the inception of instability, if any, the amplitude of the unsteady flow would increase to a finite value, which is still small, so the problem then is closely related to the linear perturbation equation,

$$\eta_t + \frac{\partial}{\partial x}\left[(F-1)\eta - \frac{1}{6}\eta_{xx} - \frac{3}{2}\varsigma_s(x)\eta\right] = 0, \tag{15}$$

which is the linearized version of (12). By separation of variables,

$$\eta(x,t) = e^{\sigma t} f(x), \tag{16}$$

(15) becomes

$$\sigma f + \frac{\partial}{\partial x}\left[(F-1)f - \frac{1}{6}f_{xx} - \frac{3}{2}\varsigma_s(x)f\right] = 0, \tag{17}$$

where f is a complex-valued function of x, $\sigma = \gamma + i\omega$, γ and ω being real, and we reckon the real part of η for physical interpretation. Equation (17) is a linear ordinary differential equation of third order, with a non-selfadjoint operator, and together with the regularity conditions at infinity constitutes an eigenvalue problem to be solved. Here, we further note that under the regularity condition at infinity, (17) and (12) imply that

$$\int f(x)dx = 0, \tag{18}$$

and

$$\int \eta(x,t)dx = C, \tag{19}$$

where C is a constant which need not be zero and this may cause some difference between the two problems for f and η.

In parallel to the study of (12) by nonlinear stability theory, some of the main features of the solution can be exhibited by making resort to numerical experiments and to the approximate solution by modal expansion. Both approaches will be pursued below for comparison. To simplify the calculations involved, we apply the similarity transformation:

$$x' = kx, \quad t' = k^3 t, \quad \eta = k^2 \eta', \quad F - 1 = k^2(F' - 1), \tag{20}$$

so that (12) becomes

$$\eta_t + \frac{\partial}{\partial x}\left[(F' - 1)\eta - \frac{3}{4}\eta^2 - \frac{1}{6}\eta_{xx} - \frac{3}{2}\varsigma_s'\eta\right] = 0, \tag{21}$$

where the primes have been omitted for η', x' and t' (but not for F' and ς_s' for the ease of identification, and the primes may be restored when needed) and

$$\eta(x, 0) = -\mu \varsigma_s', \quad \varsigma_s' = \frac{4}{3}\operatorname{sech}^2 x. \tag{22}$$

Thus, all the parameter k's have been cancelled out in (21) and (22), and hence the solution η will be independent of k, which we shall call the normal form. In this form, the results corresponding to different values of k in the original problem can be deduced at once by the rule of similarity (20).

However, for convenience in the numerical computations, the k dependence was retained rather than using the normal form (corresponding to $k = 1$, $a = 4/3$, see (9a)), and Table 1 lists the main results for three chosen values of $a = 0.05$, 0.1 and 0.5, all for the critical case of $F = 1$ and with the rest state as the initial condition (i.e. $\mu = 1$, see (14) and (11)). With this initial disturbance, the three primary solutions (the forced steady solitary wave ς_s) are found to be unstable; they evolve to generate periodic precursor solitons, with period T_s given in Table 1. Also listed in Table 1 are the mean values of the drag coefficient $\overline{C}_{Dw}$.

Table 1

a	k	$P_m = -a^2$	T_s	$T_s' = T_s k^3$	α	$\overline{C}_{Dw}/\alpha^3$
0.05	0.1936	-0.0025	1,090	7.91	0.045	0.0395
0.10	0.2739	-0.01	387	7.94	0.092	0.0398
0.50	0.6124	-0.25	37	8.6	0.50	0.0404

Here, the α-value represents the numerical mean height of the first three free solitary waves generated, while a is the amplitude of a forced steady soliton. The detailed numerical results are exemplified in figure 2 for the case of $a = 0.10$.

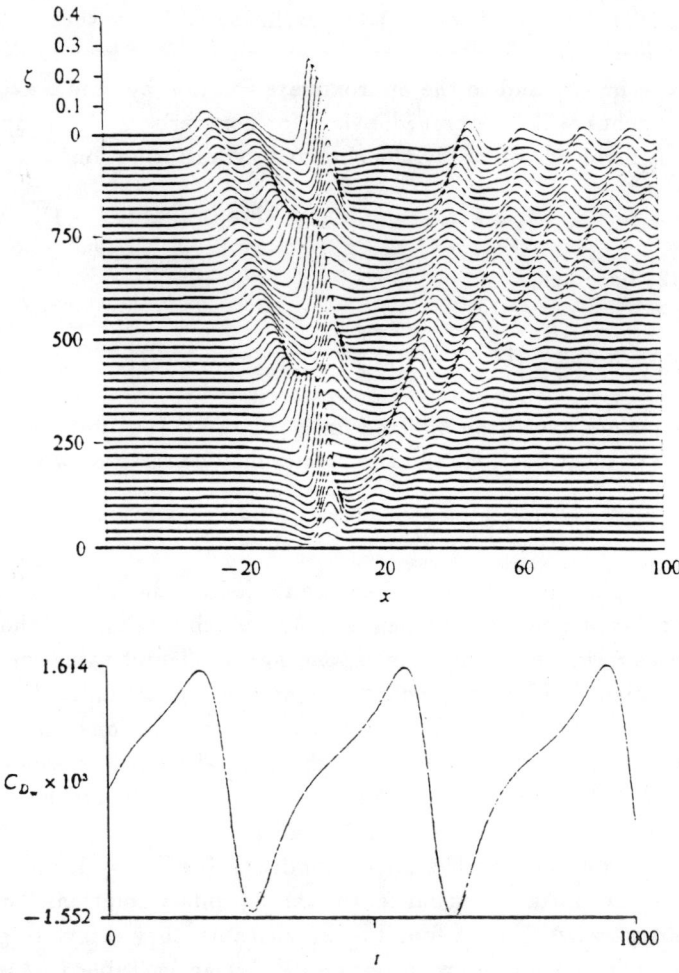

Figure 2. The fKdV - model results for waves generated from the initial rest state by stationary forcing (9b) at $F = 1$, with $k = 0.2739$ ($a = 0.10$) and amplitude $P_m = -0.01$. The wave resistance varies with period $T_s = 387$.

The slight discrepancy between the three values of T'_s is owing to the regularized fKdV equation which was used in the computation for numerical stability reasons, the difference between the original and the regularized equations vanish in the limit as $k \to 0$. From Table 1, we find by extrapolation to $k = 0$ that the period of generation of precursor solitons takes the value of $T'_s = 7.9$ (± 0.4) in the normal form. Hence, by restoring the original variable (20), we have

$$T_s = \frac{7.9}{k^3} , \qquad (23)$$

or in terms of the amplitude a ($= 4k^2/3$) of the forced steady soliton,

$$T_s = \frac{11.5}{a^{3/2}}. \tag{24}$$

This result is in good agreement with that derived from the mass- momentum-energy theorem, by Wu[3].

5. THE GALERKIN MODAL EXPANSION ANALYSIS

In an attempt to explore the basic mechanism underlying the phenomenon in question, we apply the method of modal expansion as follows. By making the change of variables:

$$z = \frac{1}{2}(1 + \tanh x), \qquad (0 \le z \le 1), \tag{25}$$

(21) and (22) become, with $\eta(x(z), t) = \hat{\eta}(z, t)$, $0 \le z \le 1$,

$$\hat{\eta}_t + 2B(z)\partial_z \left\{ \nu \hat{\eta} - \frac{3}{4}\hat{\eta}^2 - \frac{2}{3}B(z)\partial_z[B(z)\partial_z\hat{\eta}] - 8B(z)\hat{\eta} \right\} = 0, \tag{26}$$

$$\hat{\eta}(z, 0) = \hat{\eta}_0(z) = -\mu \frac{16}{3} B(z), \tag{27}$$

where

$$B(z) = z(1-z), \quad \nu = F' - 1 = (F-1)/k^2, \tag{28}$$

and $\partial_z \equiv \partial/\partial z$. The original regularity condition at $x = \pm\infty$ now becomes

$$\hat{\eta}(0, t) = \hat{\eta}(1, t) = 0, \quad (t \ge 0). \tag{29}$$

Let $\hat{\eta}$ be represented by the N-term truncated modal expansion:

$$\hat{\eta}(z, t) = \sum_{n=1}^{N} A_n(t) \sin n\pi z, \qquad (0 \le z \le 1). \tag{30}$$

For a two-term expansion, $N = 2$, projection of the differential equation (26) into the first and the second modes gives for $x \equiv A_1(t)$ and $y \equiv A_2(t)$ the following ordinary differential equations

$$\frac{dx}{dt} = -\omega_1 y - \beta xy, \quad \frac{dy}{dt} = \omega_2 x + \gamma_1 x^2 - \gamma_2 y^2, \tag{31}$$

where

$$\omega_1 = \omega_1^*(1 - \frac{\nu}{\nu_1}), \quad \omega_1^* = 0.8927, \quad \nu_1 = 0.5719,$$
$$\omega_2 = \omega_2^*(1 - \frac{\nu}{\nu_2}), \quad \omega_2^* = 1.512, \quad \nu_2 = 1.7987,$$
$$\beta = 1.561, \quad \gamma_1 = 0.8451, \quad \gamma_2 = 0.4244.$$

The two equations in (31) have four fixed points

$$
\begin{aligned}
P_1: & \quad (x_1, y_1) = (0, 0), \\
P_2: & \quad (x_2, y_2) = (-\omega_2/\gamma_1, 0), \\
P_3^{\pm}: & \quad (x_3, \pm y_3), \; x_3 = -\frac{\omega_1}{\beta}, \; y_3^2 = \frac{\gamma_1}{\gamma_2}\frac{\omega_1}{\beta}\left(\frac{\omega_1}{\beta} - \frac{\omega_2}{\gamma_1}\right).
\end{aligned}
\tag{32}
$$

The fixed point P_1 is always located at the origin; it is a center for $\nu < \nu_1$ and $\nu > \nu_2$, but is a saddle for $\nu_1 < \nu < \nu_2$. P_2 is located on the negative x-axis for $\nu_3 < \nu < \nu_2$ ($\nu_3 = -223$) in which region it is a center; it moves along the positive x-axis for $\nu > \nu_2$ and after crossing the y-axis at $\nu = \nu_2$ becomes a saddle. The fixed points $P_3^{\pm}$ degenerate into the line $x = x_3$ for $\nu_3 < \nu < \nu_1$ and for $\nu > \nu_1$, P_3^+ is a sink while P_3^- is a source.

Pertaining to the fixed point P_1, we see that for small x and y, the linearized equations of (31) are

$$
\begin{bmatrix} \dot{x} \\ \dot{y} \end{bmatrix} = A \begin{bmatrix} x \\ y \end{bmatrix}, \quad A = \begin{bmatrix} 0 & -\omega_1 \\ \omega_2 & 0 \end{bmatrix},
\tag{33}
$$

where $(\dot{\ }) = d/dt$. The eigenvalues of A, σ say, satisfy

$$\sigma^2 + \omega_1 \omega_2 = 0 \; ;$$

hence

$$
\sigma = \pm i\omega_o, \quad \omega_o = 1.162 \left[(1 - \frac{\nu}{\nu_1})(1 - \frac{\nu}{\nu_2})\right]^{1/2}, \quad (\nu < \nu_1, \nu > \nu_2),
\tag{34a}
$$

$$
\sigma = \pm \sigma_o, \quad \sigma_o = 1.162 \left[(\frac{\nu}{\nu_1} - 1)(1 - \frac{\nu}{\nu_2})\right]^{1/2}, \quad (\nu_1 < \nu < \nu_2).
\tag{34b}
$$

In particular, at $F = 1$, or $\nu = 0$, we have

$$\sigma = \pm i\omega_o, \quad \omega_o = 1.162.
\tag{35}$$

Here we note that when the parameter $\nu = (F-1)/k^2$ is either $< \nu_1$, or $> \nu_2$, the eigenvalues of the matrix A are purely imaginary, forming a conjugate pair. According to Camassa[28], this result is slightly improved to give $\sigma = \pm i\omega_o$, $\omega_o = 1.156$ for the $N = 4$ term approximation. At $F = 1$, the above values of $\omega_o = 1.162, 1.156$ obtained by the two- and four-mode expansions are in good agreement with Camassa's result, $\sigma = \pm i\omega_o$, $\omega_o = 1.143$, which was obtained numerically by truncating the series (30) at up to $N = 400$ terms.

Thus, the approximate solution of the two-mode expansion shows that the fixed point P_1 situated at the origin is a center within the range of the

ν-values specified by (34a), which includes the critical case of $F = 1$, corresponding to which the eigenvalue of the matrix A has the purely imaginary value of $\omega_o = 1.162$. However, this approximate solution is so crude in structure as to provide no indication of the possibility of bifurcation of a solution curve emanated from a point in the neighborhood of P_1 in the phase space into a new stable solution that describes a periodic generation of solitons. Nevertheless, it is of interest to note that the eigenvalue of $\sigma = \pm i\omega_o$, $\omega_o = 1.162$ at $F = 1$ for the two-mode approximation is in excellent accord with the more accurate solution of $\omega_o = 1.143$ obtained using the 400 modes expansion. It remains to be seen whether a bifurcation trajectory will emerge with more modes included in the analysis. In the event that a bifurcation of solution should arise in a higher order calculation, it still seems highly plausible that it would not be of the Hopf type in view of the fact that the eigenvalue σ is purely imaginary, with a zero real part.

To conclude our study at this stage, we can only say that the periodic process of generation of precursor solitary waves by a critical, steadily moving disturbance still seems to be of a new type which remains to be determined.

This work was jointly sponsored by ONR Contract N00014-85- K-0536, NR 655-005 and NSF Grant MSM-8118429. I am deeply indebted to George Yates and Roberto Camassa for helpful discussions, for their valuable assistance in obtaining the numerical results presented here and for letting me quote yet unpublished numerical results of the accurate eigenvalue calculations, which are effort-taking. The numerical calculations were done on the CRAY-X-MP/48 at San Diego Supercomputer Center (operated by the National Science Foundation).

References

1. Wu, T.Y. & Wu, D.M. 1982 In Proc. 14th Symp. on Naval Hydrodynamics, pp. 103-125. Washington, D.C.: National Academy of Sciences.
2. Lee, S.J. 1985 Generation of long water waves by moving disturbances. Ph.D. thesis, California Institute of Technology, Pasadena, CA.
3. Wu, T.Y. 1987 Generation of upstream advancing solitons by moving disturbances. To appear in J. Fluid Mech.
4. Huang, D.D., Sibul, O.J., Webster, W.C., Wehausen, J.V., Wu, D.M. & Wu, T.Y. 1982 In Proc. Conf. on Behavior of Ships in Restricted Waters, vol. II, pp. 26-1 to 26-10. Varna: Bulgarian Ship Hydrodynamics Centre.
5. Sun, M.-G. 1985 The evolution of waves created by a ship in a shallow canal. In The 60th Anniv. Volume-Zhongshan University, Mechanics Essays (in Chinese), pp. 17-25. China: Guangzhow.

6. Miura, R.M. 1968 J. Math. Phys. **9**, 1202-1204.
7. Whitham, G.B. 1974 Linear and Nonlinear Waves. Wiley.
8. Lin, C.C. & Clark, A. 1959 On the theory of shallow wake waves. Tsing Hua J. Chinese Studies, 1, 54-62.
9. Wu, T.Y. 1979 Tsunamis - Proc. National Science Foundation Workshop (May 7-9, 1979), pp. 110-149. Pasadena: Tetra Tech. Inc.
10. Wu, T.Y. 1981 J. Engng. Mech. Div. ASCE **107**, 501-522.
11. Schember, H.R. 1972 A new model for three-dimensional nonlinear dispersive long waves. Ph.D. thesis, California Institute of Technology, Pasadena, CA.
12. Lepelletier, T.G. 1981 Tsunamis-Harbor oscillations induced by nonlinear transient long waves. Ph.D. thesis, California Institute of Technology, Pasadena, CA.
13. Green, A.E. & Naghdi, P.M. 1976 J. Fluid Mech. **78**, 237-246.
14. Ertekin, R.C. 1984 Soliton generation by moving disturbances in shallow water: Theory computation and experiments. Ph.D. thesis, University of California, Berkeley.
15. Ertekin, R.C., Webster, W.C. & Wehausen, J.V. 1985 In Proc. 15th Symp. on Naval Hydrodynamics, pp. 347-364. Washington D.C.: National Academy Press.
16. Ertekin, R.C., & Wehausen, J.V. 1986 In Proc. 16th Symp. on Naval Hydrodynamics, pp. 167-185. Washington, D.C.: National Academy Press.
17. Ertekin, R.C., Webster, W.C. & Wehausen, J.V. 1986 J. Fluid Mech. **169**, 275-292.
18. Akylas, T.R. 1984 J. Fluid Mech. **141**, 455-466.
19. Cole, S.L. 1985 Wave Motion **7**, 579-587.
20. Lee, C.-Y. & Beardsley, R.C. 1974 J. Geophys. Res. **79**, 453- 462.
21. Grimshaw, R.H.J. & Smyth, N.F. 1986 J. Fluid Mech. **169**, 429- 464.
22. Smyth, N.F. 1986 Modulation theory solution for resonant flow over topography. Department of Math. Rep. 3. University of Melbourne, Victoria, Australia.
23. Zhu, J. 1986 Internal solitons generated by moving disturbances. Ph.D. thesis, California Institute of Technology, Pasadena, CA.
24. Zhu, J., Wu, T.Y. & Yates, G.T. 1986 Generation of internal runaway solitons by moving disturbances. 16th Symp. on Naval Hydrodynamics, July 14-18, 1986, University of California, Berkeley, CA.
25. Zhu, J., Wu, T.Y. & Yates, G.T. 1987 Internal solitary waves generated by moving disturbances. Third Intl. Symp. on Stratified Flows, February 3-5, 1987, California Institute of Technology, Pasadena, CA.
26. Zhu, J., Wu, T.Y. & Yates, G.T. 1987 Upstream internal solitons generated by moving disturbances. In Proc. Intern. Conf. on Fluid Mech. (ICFM'87), July 1-4, 1987, Beijing, China.
27. Lee, S.-J., Yates, G.T. & Wu, T.Y. 1987 In Proc. IUTAM Symp. on Nonlinear Water Waves. Aug. 25-28, 1987, Tokyo, Japan.
28. Camassa, R. 1987 Personal communication.

FINITE GROUPS OF GRAVITY WAVES

Chia-Shun Yih
The University of Michigan, Ann Arbor, Michigan 48109
and
The University of Florida, Gainesville, Florida 32611

Summary

Groups of gravity waves of finite length created in deep, originally quiescent water by an oscillating or moving surface pressure are constructed by superposition of the Cauchy-Poisson solution. This construction gives substance and reassurance to the concept of "wave packets" progressing with group velocity associated with the individual waves in the packets, a concept so important to water-wave research. The effects of viscosity are taken into account, thereby not only justifying the extensively used but completely artificial damping factor initiated by Lamb (1916, see Lamb 1945, p. 413), but also showing the hitherto little explored spatial damping of waves.

1. INTRODUCTION

The phenomenon of an isolated water-wave group and its velocity of advance were observed by Russel (1844, see Lamb 1945, p. 380). The first derivation of the group velocity of dispersive waves seems to have been given by Stokes (1876, see Lamb 1945, p. 381), although the significance of the group velocity as the velocity of propagation of the wave number was already at least implicit in the solution of Cauchy (1815, see Lamb 1945, p. 384, and p. 17 for reference to original work) and Poisson (1816, see Lamb 1945, p. 384) for waves created by an initial concentrated disturbance, long before Russel's observations and Stokes' work.

In Stokes' derivation, two wave trains of slightly different wave numbers and correspondingly slightly different frequencies are superposed, and the result is a train of waves of the mean wave number and the mean frequency bounded by an envelope with an amplitude sinusoidally and slowly varying with time and distance. The velocity of the envelope is the group velocity, which for gravity waves is less than the phase velocity of the individual waves contained in the envelope.

Stokes' derivation has the great merit of simplicity. Although the requirement of two wave trains of slightly different wave numbers and frequencies seems artificial at first sight, waves of neighboring wave numbers and frequencies do arise naturally in many problems to which the Fourier analysis can be applied, and these are essential for the formation of wave groups. Wave trains with two discrete wave numbers are merely an extreme idealization. However, these wave trains necessarily entail infinitely many wave groups, and this fact renders Stokes' construction inadequate for explaining Russel's observations or for supporting the many and frequent statements or implications in contemporary literature concerning isolated wave groups. It is thus very desirable to construct some examples of dispersive-wave groups of finite length.

As already mentioned in the fore-going, an interpretation of the Cauchy-Poisson solution, as given in pp. 384-398 of Lamb's book (1945), shows the significance of the group velocity as the velocity with which the wave number propagates – or a packet of waves of that wave number propagates. This strongly suggests that if the forcing at the free surface has a certain frequency, or moves with a certain velocity, a group of waves with that frequency, or moves with a phase velocity equal to that velocity, will be created. The present paper is the outcome of acting on that suggestion.

The deep mystery that when a wave maker oscillates $n(\gg 1)$ times in deep water (for instance) only $n/2$ waves are created can be dispelled satisfactorily only by considering the cancellation of waves, perhaps principally near the front of the group. The mechanics of that cancellation is already contained in the solution given in this paper, but has not been pursued in detail. However, the construction herein of wave groups of finite length, in showing the existence of such groups that propagate with their appropriate group velocities, and in being the *result* of that cancellation, provides an important step toward dispelling that mystery, and gives one reassurance when one talks about isolated wave packets.

In making the solutions determinate, the simple damping factor employed by Lamb (1916, see Lamb 1945, p. 413) will first be used, but will be justified later in this paper on the basis of the Navier-Stokes equations governing the dynamics of viscous fluids. This justification is a second purpose of this paper.

Stokes' construction of infinitely many wave groups and Lamb's artifice of the exponential damping factor have the merit of simplicity, and since they both contain a measure of what is needed, have been successful in explaining things and thus very useful to workers on water waves. The simplicity and the success have long been a blessing, but the very simplicity and success have in time become a curse, since they discourage the expenditure of arduous work to put something better in their place. The time for replacing them has arrived, and this work, motivated by the demand of reason, is a tribute to Professor C. C. Lin on the occasion of his retirement.

2. THE CAUCHY-POISSON SOLUTION

Let x and y be Cartesian coordinates, with x measured in a horizontal direction and y measured vertically upward from the free surface when the fluid (water) is at rest, and let t be the time. Consider irrotational gravity waves created in deep water by a distribution of the velocity potential ϕ applied instantaneously at the free surface at $t = 0$, and let this initial ϕ be denoted by ϕ_0 and given by

$$\phi_0 = aF(x) , \qquad (1)$$

where $F(x)$ is dimensionless and a has the dimension L^2/T, i.e., the dimension of the velocity potential.

The velocity potential ϕ has to satisfy the Laplace equation

$$\phi_{xx} + \phi_{yy} = 0 . \qquad (2)$$

At the free surface, where

$$y = \eta(x,t) , \qquad (3)$$

the kinematic condition is

$$\eta_t = \phi_y , \qquad (4)$$

with ϕ_y evaluated at $y = 0$, and the dynamic condition requiring constant pressure is

$$\phi_t + g\eta = 0 . \qquad (5)$$

where g is the gravitational acceleration. The ϕ_y and ϕ_t in (4) and (5) are evaluated at $y = 0$ in a linear theory. Another boundary condition is

$$\phi \to 0 \quad \text{as} \quad y \to -\infty . \qquad (6)$$

Let k denote the wave number and σ the corresponding frequency (defined as 2π divided by the period). Then the solution for ϕ satisfying (1), (2), (4), (5), and (6) is

$$\phi = \frac{a}{\pi} \int_0^\infty \cos \sigma t e^{ky} dk \int_{-\infty}^\infty F(\alpha) \cos k(x - \alpha) d\alpha , \qquad (7)$$

and the corresponding solution for η is

$$\eta = \frac{a}{\pi g} \int_0^\infty \sigma \sin \sigma t dk \int_{-\infty}^\infty F(\alpha) \cos k(x - \alpha) d\alpha , \qquad (8)$$

provided
$$\sigma^2 = gk. \qquad (9)$$

Equation (9) is the dispersion equation, which has a different form if surface-tension effects are included, or if the water depth is finite.

The famous Cauchy-Poisson solution is obtained if one lets $F(\alpha)$ in (7) and (8) be a Dirac distribution, i.e., if $F(\alpha)$ is zero everywhere except at $\alpha = 0$, in such a way that

$$\frac{1}{L}\int_{-\varepsilon}^{\varepsilon} F(\alpha)d\alpha = 1$$

for any ε however small, if L is the length scale. We can work with (7) and (8) without using the Cauchy-Poisson solution. But we note in passing that the Cauchy-Poisson solution demonstrates beautifully how wave packets of various wave numbers disperse, each packet of a given wave number propagating with the group velocity for that wave number. So this kinematic significance of the group velocity was already implied in the work of Cauchy in 1815 and Poisson in 1816 (see Lamb 1945, p. 17 and p. 384 for dates of the references.), long before this significance was reaffirmed and emphasized by others in the second half of the twentieth century.

Before we go on to use (7) and (8) to construct single wave groups, we note that (9) gives two values for σ for each real positive value of k, one positive and one negative. Taking the positive root gives the same ϕ and η as taking the negative root, as can be seen from (7) and (8). If we take both the positive and the negative roots of σ and add the results in each of (7) and (8), we merely get double the values of ϕ_0, ϕ, and η, with no other effects. Consequently we need not consider the negative root, and henceforth consider σ to be positive. Remembering this will remove a lot of ambiguities later.

The physical meaning of (1) has seldom been made clear. Lamb referred to it as an impulse. But since the pressure on the free surface is given by (for $y = 0$)

$$p = -\rho(\phi_t + g\eta), \qquad (10)$$

and the right-hand side is zero by virtue of (7) and (8), provided the integrals are convergent, the pressure on the free surface is always zero (or constant if we did not drop the constant in the Bernoulli equation above). So the use of the term impulse is rather confusing. The correct way of thinking is to regard (7) and (8) as valid only for $t \geq 0$, and to assume both ϕ and η to be zero for $t \leq 0$. Then ϕ_t is infinite at $t = 0$, and integration of ϕ_t between $t = -\varepsilon$ and $t = 0$ gives ϕ at $t = 0$ equal to

$$\phi = \frac{a}{\pi}\int_0^{\infty} e^{ky}dk \int_{-\infty}^{\infty} F(\alpha)\cos k(x-\alpha)d\alpha,$$

while integrating η in the same interval gives nothing. Hence the integration of p in the same interval at $y = 0$ gives $-\rho\phi_0$. This is what has been considered the impulse.

3. WAVE GROUPS PRODUCED BY AN OSCILLATING PRESSURE DISTRIBUTION

Using (7) and (8) as building blocks, one can obtain by timewise superposition a more general solution as follows:

$$\phi = \frac{a}{\pi} \int_0^\infty \left[\int_0^t \omega \cos\sigma(t-\tau) \sin\omega\tau e^{-\mu(t-\tau)} d\tau \right. $$
$$\left. \times \int_{-\infty}^\infty F(\alpha) \cos k(x-\alpha) d\alpha \right] e^{ky} dk , \qquad (11)$$

$$\eta = \frac{a}{\pi g} \int_0^\infty \left[\int_0^t \omega\sigma \sin\sigma(t-\tau) \sin\omega\tau e^{-\mu(t-\tau)} d\tau \right.$$
$$\left. \times \int_{-\infty}^\infty F(\alpha) \cos k(x-\alpha) d\alpha \right] dk . \qquad (12)$$

In (11) and (12), we have added the exponential damping factor $\exp[-\mu(t-\tau)]$, where μ is a damping coefficient much smaller than ω, and not the dynamic viscosity, to make the results determinate. This device may be called the device of fading memory, and was used by Lamb (1916, see Lamb 1945, p. 413). The only difference in its usage here is that we consider the damping to start from $t = \tau$, at which an element of ϕ_0 acts. Later in this paper, we shall provide a rational foundation for the use of some damping factor, though not exactly the same as the one above, by invoking the solution for water waves in a viscous fluid.

The solution for ϕ given by (11) certainly satisfies the Laplace equation and (6). Equations (11) and (12) also satisfy (4) at the free surface, provided (9) is satisfied.

A look at the remaining condition, the dynamic condition at the free surface, is most revealing. The linearized Bernoulli at the free surface is

$$p = -\rho(\phi_t + g\eta) . \qquad (13)$$

A calculation with (11) and (12) gives

$$p = -\frac{\rho a}{\pi} \left[\omega \sin\omega t \int_{-\infty}^\infty F(\alpha) \cos k(x-\alpha) d\alpha \right] dk , \qquad (14)$$

after y is equated to zero after the calculation. Note that differentiation of (11) with respect to t within the integral signs (for τ) contributes a term

that exactly cancels $g\eta$ on the right-hand side of (13), on account of (9). Since
$$\frac{1}{\pi}\int_0^\infty \left[\int_{-\infty}^\infty F(\alpha)\cos k(x-\alpha)d\alpha\right]dk = F(x),$$
equation (14) is
$$p = -\rho a\omega \sin\omega t F(x). \tag{15}$$

Hence (11) and (12) are solutions for wave motion created by the pressure distribution (15), from $t = 0$ up to time t, at the free surface, in water otherwise at rest.

It is important to note that (14) results from the variable upper limit of the τ-integral in (11). This explains why, in spite of (9), which is derived from the condition of constant or zero pressure at the free surface, an x-dependent pressure distribution is nonetheless obtained. This fact is related to the interpretation of (7) as giving an impulse at $t = 0$. But interpreting (11) as corresponding to a pressure distribution (15) makes things much easier to grasp, since pressure is far more familiar a quantity than impulse. Without (11) and (12), the formulation of the problem of wave generation by a pressure distribution is cumbersome at best.

Now let
$$F(x) = 1 \text{ in } -b \le x \le b, \tag{16}$$
and zero elsewhere. The visible quantity η then can be calculated from (12), and is
$$\eta = \frac{a\omega}{\pi g}\int_0^\infty \sigma I_1 I_2 dk, \tag{17}$$
where
$$I_1 = \int_0^t \sin\sigma(t-\tau)\sin\omega\tau e^{-\mu(t-\tau)}d\tau$$
$$= \frac{1}{2}e^{-\mu t}\int_0^t \{\cos[\sigma t - (\sigma+\omega)\tau] - \cos[\sigma t - (\sigma-\omega)\tau]\}e^{\mu\tau}d\tau, \tag{18}$$
and
$$I_2 = \int_{-b}^b \cos k(x-\alpha)d\alpha = -\frac{1}{k}[\sin k(x-b) - \sin k(x+b)]. \tag{19}$$

Multiplying I_1 to I_2, expressing the product in terms of sine functions, and treating these as the imaginary parts of exponential functions, one can carry out the integration with respect to τ, and obtain
$$\eta = -\frac{a\omega e^{-\mu t}}{4\pi g}\int_0^\infty \frac{\sigma I}{k}dk, \tag{20}$$

where
$$I = f(x_1) - f(x_2), \qquad (21)$$

$$\begin{aligned}f(x_1) =\operatorname{Im} \Bigl[&\frac{1}{i(\sigma+\omega)+\mu}\{\exp i(kx_1+\omega t-i\mu t)-\exp i(kx_1-\sigma t)\} \\ +&\frac{1}{-i(\sigma+\omega)+\mu}\{\exp i(kx_1-\omega t-i\mu t)-\exp i(kx_1+\sigma t)\} \\ -&\frac{1}{i(\sigma-\omega)+\mu}\{\exp i(kx_1-\omega t-i\mu t)-\exp i(kx_1-\sigma t)\} \\ -&\frac{1}{-i(\sigma-\omega)+\mu}\{\exp i(kx_1+\omega t-i\mu t)-\exp i(kx_1+\sigma t)\}\Bigr], \\ & \qquad (22)\end{aligned}$$

and
$$x_1 = x - b, \quad x_2 = x + b. \qquad (23)$$

In (22), Im means "the coefficient of i" in the expression that follows it.

Recalling that we need only consider positive values of σ, we see that the first two members within the brackets of (22) will make no contributions to waves in the flow, that the third member will give waves propagating to the right, and the fourth member will give waves propagating to the left. Similarly the term containing x_2 in (20) will contain a term giving waves going to the left and one going to the right. Obviously the solution for η will be symmetric with respect to $x = 0$. We need therefore consider only waves propagating to the right. Doing that, and writing

$$dk = 2\sigma d\sigma/g,$$

we see from (18) and (19) that the part of η corresponding to waves propagating to the right is contained in

$$-\frac{a\omega}{2\pi g}\operatorname{Im}(J_1 + J_2), \qquad (24)$$

where

$$J_1 = \int_0^\infty \frac{1}{\sigma-(\omega+i\mu)}[\exp i(kx_1-\omega t)-\exp i(kx_2-\omega t)]d\sigma \qquad (25)$$

and

$$J_2 = -\int_0^\infty \frac{e^{-\mu t}}{\sigma-(\omega+i\mu)}[\exp i(kx_1-\sigma t)-\exp i(kx_2-\sigma t)]d\sigma. \qquad (26)$$

The integral J_1 is obtained upon taking the contour in Figure 1, for positive x_1 or x_2, and the contour in Figure 2, for negative x_1 or x_2. In Figures 1 and 2, $\sigma = \sigma_r + i\sigma_i$, and the radius of the circular portion is very large. The angle of inclination of the slanted portion of the contour in these figures can have any value between zero and $\pi/2$. It has been chosen to be $\pi/8$ because when viscous effects are taken into account later, it will be seen that the angle must not exceed $\pi/6$. The contribution to J_1 from the circular part is zero. The contribution from the slanted lines can be shown not to contain any discrete wave component in the following way. Let J_{1s} be the integral J_1, but with lower and upper limits changed to

$$\sigma_\infty = \lim_{|\sigma| \to \infty} |\sigma| e^{i\pi/8} \quad \text{and zero},$$

respectively. Multiply J_{1s} by

$$\exp i(k'x - \sigma't), \quad \text{with} \quad k'g = \sigma'^2,$$

and integrate with respect to x over the entire x-axis, from minus to plus infinity. The integrals with respect to x and σ are both convergent along the slanted line in Figure 1, and the result is not singular in any way. If J_{1s} contained a discrete Fourier component, the result would be infinite for some value of k'. A similar argument applies to the slanted line in Figure 2. Thus, the wavy part of J_1 is, with $k_e = \omega^2/g$,

$$2\pi i [\exp\{i(k_e x_1 - \omega t) - 2\mu\omega g^{-1} x_1\} - \exp\{i(k_e x_2 - \omega t) - 2\mu\omega g^{-1} x_2\}] \quad (27)$$

for positive x_1, zero for negative x_2, and

$$-2\pi i \exp\{i(k_e x_2 - \omega t) - 2\mu\omega g^{-1} x_2\} \quad (28)$$

for

$$-b < x < b.$$

The integral J_2 requires more care. Since $k = \sigma^2/g$,

$$\exp i(kx_1 - \sigma t) = \exp[i(\sigma_r^2 - \sigma_i^2)g^{-1} x_1 - i\sigma_r t - \sigma_i(2\sigma_r g^{-1} x_1 - t)], \quad (29)$$

and similarly when x_2 replaces x_1. For a given x_1 and a given t, and for the first term in J_2, if

$$x_1 - \frac{gt}{2\sigma_r} \quad (30)$$

is positive we use a circular contour above the σ_r - axis, followed by a slanted line, as shown in Figure 3. At the value of σ_r, denoted by $\hat{\sigma}_r$ that makes (30) vanish, the contour follows a vertical path from P to its image point Q below the σ_r - axis. Then the lower slanted line is followed all the way to the origin, whereby the circuit is completed. Use Figure 4 if $\hat{\sigma}_r$ is reached before $\pi/8$.

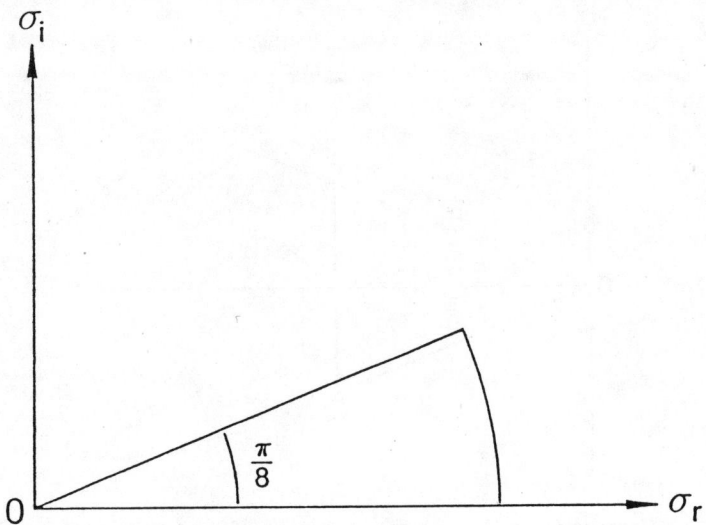

Figure 1. Contour for evaluating the integral in (25), for positive z_1 or z_2. $\sigma = \sigma_r + i\sigma_i$.

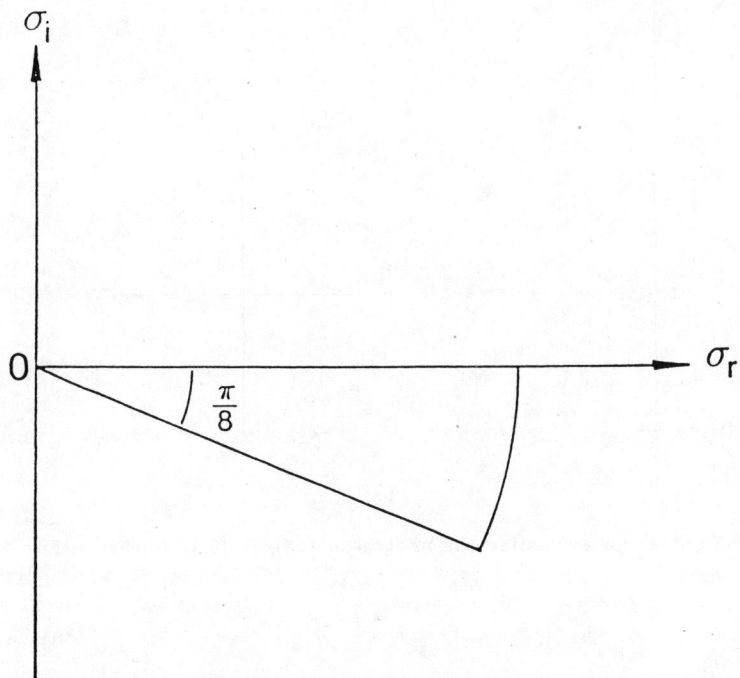

Figure 2. Contour for evaluating the integral in (25), for negative z_1 or z_2. $\sigma = \sigma_r + i\sigma_i$.

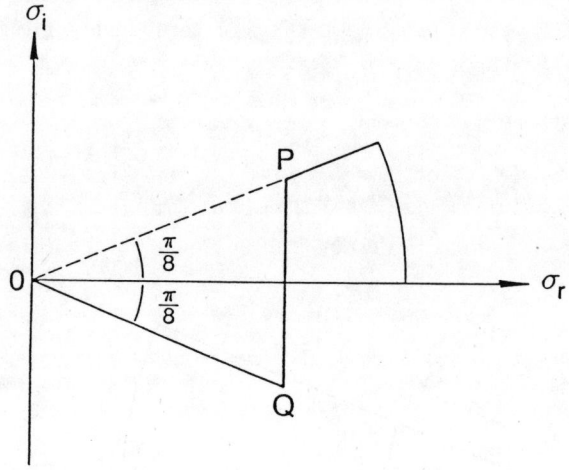

Figure 3. Contour for evaluating the integral in (28), if $\pi/8$ is reached before $\hat{\sigma}_r$. $\sigma = \sigma_r + i\sigma_i$.

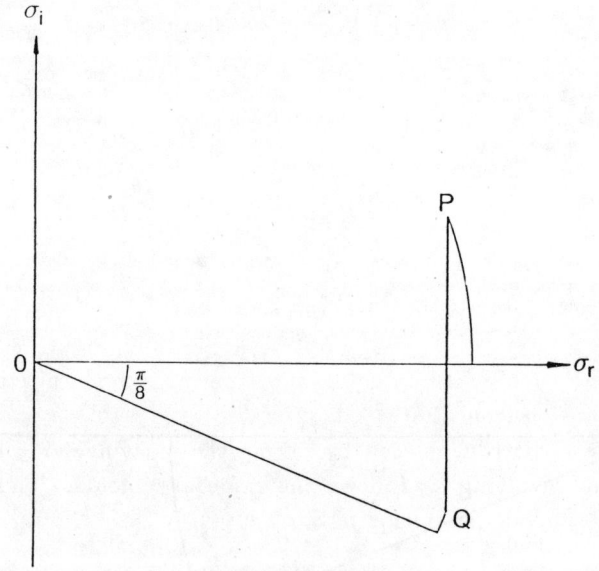

Figure 4. Contour for evaluating the integral in (28), if $\hat{\sigma}_r$ is reached before $\pi/8$. $\sigma = \sigma_r + i\sigma_i$.

Hence, if $\hat{\sigma}_r > \omega$, the pole $\omega + i\mu$ is not within the circuit, and the residue of the contour integral of the first term in the integral of (26) is zero. Otherwise it is

$$2\pi i E(x_1), \tag{31}$$

where

$$E(x) = \exp[i(k_e x - \omega t) - 2\mu\omega g^{-1}x]. \tag{32}$$

Similarly for the second integral in (26). Hence we have the wavy part of J_2 equal to

$$-2\pi i[E(x_1) - E(x_2)] \tag{33}$$

if

$$x_1 - \tilde{c}_g t > 0, \tag{34}$$

where

$$\tilde{c}_g = \frac{\omega}{2k_e} = \frac{g}{2\omega} \tag{35}$$

is the group velocity for gravity waves in deep water with frequency ω. On the other hand, the wavy part of J_2 is zero if

$$x_2 - \tilde{c}_g t < 0, \tag{36}$$

and is

$$2\pi i E(x_2) \tag{37}$$

if

$$-b + \tilde{c}_g t < x < b + \tilde{c}_g t. \tag{38}$$

In the evaluation of J_2 by using the contour in Figure 3, the integral over the slanted lines again contributes nothing to the wavy part of J_2. The contribution from the vertical leg in Figure 3 to the wavy part of J_2 is also zero. This can be seen by taking the part involving x_1, since the development involving x_2 follows the same arguments. That part is, in view of (29) and the definition of $\hat{\sigma}_r$,

$$J_3 = -\exp(i\hat{\sigma}_r t/2 - \mu t) \int_{\sigma_i P}^{\sigma_i Q} \frac{i}{\sigma - (\omega + i\mu)} \exp(-i\sigma_i^2 g^{-1} x_1) d\sigma_i, \tag{39}$$

where $\hat{\sigma}_r \neq \omega$. For a fixed x_1 and large t, it can be shown from (40) that $J_3 \sim t^{-1}$, since $\hat{\sigma}_r$ is proportional to t for fixed x_1, and $\sigma = \hat{\sigma}_r + i\sigma_i$. The same conclusion holds for the part of J_2 containing x_2. Hence the contribution of the vertical leg in the contour in Figure 3 to J_2 is only transient for any x_1.

Let the part of η that corresponds to right-going waves be denoted by η_{wr}. Going back to (24), and using the results obtained for the wavy parts of J_1 and J_2, we can evaluate η_{wr}. Let

$$G(x) = \sin(k_e x - \omega t) \exp(-2\mu\omega g^{-1} x) \ .$$

Then the final results are:

(i) If $x < -b$, $\eta_{wr} = 0$,
(ii) If $x_2 - \tilde{c}_g t < 0$ and $x_1 < 0 < x_2$, $\eta_{wr} = -\frac{a\omega}{g} G(x_2)$,
(iii) If $x_2 - \tilde{c}_g t < 0$ and $x_1 > 0$, $\eta_{wr} = \frac{a\omega}{g}[G(x_1) - G(x_2)]$,
(iv) If $b < -b + \tilde{c}_g t < x < b + \tilde{c}_g t$, $\eta_{wr} = \frac{a\omega}{g} G(x_1)$,
(v) If $x_1 > \tilde{c}_g t$, $\eta_{wr} = 0$.

There is a slight reduction of k_e, of $O(\mu^2)$, which has been neglected.

These results may seem complicated. They become immediately easy to grasp if one restates them as follows:

Two wave trains both with wave number k_e and frequency ω, one starting from $x = -b$ and the other starting from $x = b$, progress to the right with group velocity $\tilde{c}_g$ into otherwise quiet water.

Since the flow is symmetric with respect to $x = 0$, there is a set of results for left-going waves which can be obtained by symmetry arguments from results (i) to (v). These can be restated as follows:

Two other wave trains, both with wave number k_e and frequency ω, one starting from $x = b$ and the other starting from $x = -b$, progress to the left with group velocity $\tilde{c}_g$ into otherwise quiet water.

Note that where the two wave trains both exist [see (iii).], they reinforce each other if $k_e b = \pi/2$ or $(2n + 1)\pi/2$, but tend to cancel each other if $k_e b = n\pi$. (The reinforcement or cancellation would be complete if $\mu = 0$.)

The investigation for Section 3 is now finished, and we call attention to the fact that the radiation condition of Sommerfeld has not been applied because it is not at all needed, that the wave trains are exponentially damped with respect to x_1 and x_2, that they may reinforce or cancel each other where they both exist, and that they progress into wave-free water. The artificial factor $\exp[-\mu(t - \tau)]$ will be discussed in terms of the true viscosity of the fluid in a later section.

4. WAVE TRAINS OF FINITE LENGTH

If, in the problem treated in Section 3, the oscillating pressure is removed at $t = T$, the integral I_1 given by (18) is now replaced by

$$I_1 = \int_0^T \sin \sigma(t - \tau) \sin \omega\tau e^{-\mu(t-\tau)} d\tau$$

$$= \frac{1}{2} e^{-\mu t} \int_0^T \{\cos[\sigma t - (\sigma + \omega)\tau] - \cos[\sigma t - (\sigma - \omega)\tau]\} e^{\mu\tau} d\tau \ . \tag{40}$$

The development in Section 3 can be repeated, and one obtains the result that two wave trains with wave number k_e, frequency ω, and length $\tilde{c}_g T$, progressing to the right. One of these terminate at a point which is at a distance $\tilde{c}_g(t-T)$ from $x = b$, and the other terminates at the same distance from $x = -b$. Similarly there are two left-going wave trains. The flow is symmetric with respect to $x = 0$. Again where the right-going wave trains co-exist, they may reinforce or (partially) cancel each other. Similarly for the left-going wave trains.

Thus we have constructed single wave groups of finite length. Each would progress into wave-free water and leave the water behind wave-free, except for the waves of the other trains.

For inviscid fluids, the μ in the exponential factors in (40), (41), and (42) can be put to zero, since it has served the purpose of making the flow determinate. The factors can then be dropped.

5. WAVE GROUPS PRODUCED BY A MOVING PRESSURE DISTRIBUTION

Consider the waves created by a pressure distribution moving to the left with speed c:

$$p = \beta \rho c^2 \quad \text{in} \ -b < x + ct < b . \tag{43}$$

Then (11) and (12) are replaced by

$$\phi = \frac{\beta c^2}{\pi} \int_0^\infty \left[\int_0^t \cos \sigma(t-\tau) e^{-\mu(t-\tau)} d\tau \int_{-b-c\tau}^{b-c\tau} \cos k(x-\alpha) d\alpha \right] e^{ky} dk , \tag{44}$$

$$\eta = \frac{\beta c^2}{\pi g} \int_0^\infty \left[\sigma \int_0^t \sin \sigma(t-\tau) e^{-\mu(t-\tau)} d\tau \int_{-b-c\tau}^{b-c\tau} \cos k(x-\alpha) d\alpha \right] dk . \tag{45}$$

Proceeding as in Section 3, we have

$$\eta = -\frac{\beta c^2}{\pi} \int_0^\infty \frac{\sigma}{k} I \, dk \tag{46}$$

where

$$I = \int_0^t \sin \sigma(t-\tau)[\sin k(x_1 + c\tau) - \sin k(x_2 + c\tau)] e^{-\mu(t-\tau)} d\tau$$

$$= \frac{1}{2} RP[H(x_1) - H(x_2)] , \tag{47}$$

$$H(x_1) = \frac{1}{i(kc+\sigma)+\mu}[e^{ik(x_1+ct)} - e^{i(kx_1+\sigma t)-\mu t}]$$

$$- \frac{1}{i(kc-\sigma)+\mu}[e^{ik(x_1+ct)} - e^{i(kx_1+\sigma t)-\mu t}] , \tag{48}$$

where x_1 and x_2 are given by (23). The dependence of $H(x_1)$ on the other variables has not been exhibited, for brevity.

The roots of
$$kc - \sigma - i\mu = 0 \tag{49}$$
are, since $kg = \sigma^2$,
$$\sigma = \frac{g}{2c} \pm \left(\frac{g^2}{4c^2} + \frac{ig\mu}{c}\right)^{1/2}. \tag{50}$$

One of them is in the first quadrant in the complex-σ plane, and the other in the fourth. For small μ the roots can be approximated by
$$\sigma = -i\mu - \frac{c}{g}\mu^2, \quad \sigma = \frac{g}{c}(1+\mu^2) + i\mu. \tag{51}$$

The roots of
$$kc + \sigma - i\mu = 0 \tag{52}$$
are obviously the negatives of those of (49). So the roots of (49) and (52) are in the first or fourth quadrant only. Furthermore, the root of (52) in the first quadrant is outside of the contour in Figure 1, since μ is assumed much smaller than g/c. This can be seen from the first root given by (50) after the signs have been changed. Thus, in evaluating I in (47) by using the contours shown in the figures, as the situation demands, it is only the root given by (50) with the positive sign that is significant in determining the wavy part of η.

The terms in (48) corresponding to waves are, after the relevant part is extracted from the second bracket,
$$\frac{ig}{(g+i2\mu c)(\sigma - gc^{-1} - i\mu)}[\exp ik(x_1+ct) - \exp\{i(kx_1+\sigma t) - \mu t\}],$$

if terms of $0(\mu^2)$ are neglected. Similarly for the terms in $H(x_2)$ corresponding to waves. The rest of the development follows closely the steps described in Section 3, and is omitted here. The final results are as follows:

(i) There are no waves ahead of the moving disturbance.
(ii) Behind the disturbance there are two overlapping wave trains, both of wave velocity c and wave number $k_e = g/c^2$, and both of length $(c_g)_e t$. where $(c_g)e$ is the group velocity of the waves, and is equal to $c/2$. One of the trains starts at $x_2 = 0$ and the other at $x_1 = 0$. Depending on the length over which the disturbance acts, the two trains may reinforce or partially cancel each other where they overlap.

(iii) The damping factor for the train starting at $x_1 = 0$ is

$$\exp[-2\mu(x_1 + ct)/c],$$

and the damping factor for the other train is the same factor with x_2 replacing x_1.

(iv) There is a slight increase of k_e of $0(\mu^2)$ over g/c^2, as a result of the second equation in (51). This is neglected.

If the moving disturbance is a moving body, floating or submerged, the development and the results are similar. The creation of an ever-lengthening gravity-wave group (which is the sum of the two trains, with the parts outside of their common interval neglected) behind the body allows one to calculate the wave drag from the rate of increase of the wave energy behind the body, upon letting μ be zero. This is a much more direct way of seeing things than calculating the wave drag from an infinite wave train behind the body. In that case, as is well known, one has to calculate the energy flux (or rate of work done) at a section behind the body.

6. GRAVITY-WAVE TRAINS OF FINITE LENGTH CREATED BY A MOVING DISTURBANCE IN DEEP WATER

If a surface pressure is applied at $t = 0$ and moves to the left with speed c, and is then removed at time T, a group of waves of length $cT/2$ will be formed, and will move to the left with the group velocity $c/2$ (if the effect of the spread of the disturbance is neglected). Only gravity waves have been considered here. Had surface-tension effects been included, one would expect two wave groups, one of the gravity type and the other of the capillary type. When the disturbance is removed, the two groups will separate, since the train of the capillary type has a greater group velocity, even though the individual waves in either group still moves with the same phase velocity c.

7. EFFECTS OF VISCOSITY

When viscous effects are taken into account, but the Reynolds number ($g^2/\omega^3\nu$ or $c^3/g\nu$, as the case may be, ν being the kinematic viscosity) is large, Lamb's solution (1945, pp. 625-627) applies, and in the solution for ϕ or η instead of the factor $\exp i(kx - \sigma t)$ one now has, with σ^2 still equal to gk,

$$\exp[i(kx - \sigma t) - 2\nu k^2 t].$$

Thus one may consider the factor $\exp(-\mu t)$ as a useful but empirical representation of the true factor $\exp(-2\nu k^2 t)$. Replacing the former by the latter, one can carry out the calculation in Section 3 or 5 as before, and the results are the same. Of course, since the new factor involves k, one has to go through the calculation to see that it will cause no new difficulties. But the contours in the figures have been chosen with the factor $\exp(-2\nu k^2 t)$ in mind, and in the following we shall show that indeed no new difficulties arise.

The factor in Section 3
$$(\sigma - \omega - i\mu)^{-1}$$
is now replaced by
$$(\sigma - \omega - i2\nu k^2)^{-1}, \tag{53}$$
and the factors in Section 5
$$(kc - \sigma - i\mu)^{-1} \quad \text{and} \quad (kc + \sigma - i\mu)^{-1}$$
are now replaced by
$$(kc - \sigma - i2\nu k^2)^{-1} \quad \text{and} \quad (kc + \sigma - i2\nu k^2)^{-1}. \tag{54}$$

We have to determine the poles of (53) and (54). Aside from the important one which is, for small ν, at
$$\sigma = \omega + i2\nu\omega^4 g^{-2} \tag{55}$$
approximately, the other three of (53) are at large values of $|\sigma|$, given approximately for small ν by
$$1 - i2\nu g^{-1}\sigma^3 = 0. \tag{56}$$

If we write
$$\sigma = |\sigma|e^{i\theta},$$
then (56) gives
$$\theta = -\frac{\pi}{6}, \frac{\pi}{2}, \frac{7\pi}{6}. \tag{57}$$

The contours in the figures avoid all three poles with these values of θ. Indeed, they were chosen with this avoidance in mind in the first place.

As to the poles of (54), one is at $\sigma = 0$. Examination of (47) and (48) with μ replaced by $2\nu k^2$ reveals that this is not really a pole, since the numerators of (48) also vanish at $\sigma = 0$. Thus $H(x_1)$ and $H(x_2)$ do not become infinite at $\sigma = 0$.

The important pole of the first factor in (54) is at

$$\sigma = \frac{g}{c} + \frac{i2\nu g^2}{c^4} \tag{58}$$

approximately, for small ν. The other two poles of that factor are at

$$\sigma = \left(\frac{gc}{2\nu}\right)^{1/2} \left[\exp\left(\frac{-i\pi}{4}\right), \exp\left(\frac{i3\pi}{4}\right)\right] \tag{59}$$

approximately, for small ν. The poles of the second factor in (54) are at values of σ which are the negatives of those given by (58) and (59). All the poles except the one given by (58) are outside of the contours chosen in the figures. Thus for Section 5 new difficulties do not arise either when the new damping factor $\exp(-2\nu k^2 t)$ is used.

Note that in Lamb's solution the stress layer at the free surface has been ignored, since the Reynolds number is assumed high, and therefore the normal stress at the free surface is simply represented by $-p$.

When μ is replaced by $2\nu k^2$ to begin with in the development in Section 3 and 5, all the results remain valid after μ in the results is replaced by $2\nu k_e^2$, as the mathematics requires. Thus the damping factor has been replaced by one involving the wave number, as required by the Navier-Stokes equations, and the artificiality of a frequently invoked device in wave dynamics has been removed. This and the construction of gravity-wave groups of finite length constitute the dual purpose of this paper.

ACKNOWLEDGMENT

This work has been supported by the Fluid-dynamics Division of the Office of Naval Research through the grant N00014-87-C-0194, for which the author wishes to express his appreciation.

Reference
1. Lamb, H. 1945, *Hydrodynamics*, Dover, New York.

SOLITONS AND OTHER MACROSTRUCTURES IN FLUID DYNAMICS

Ru Ling Chou[†], C. K. Chu[‡]

1. INTRODUCTION

In fluid dynamics, there are many large-scale structures which exhibit remarkable stability and permanency. Most noticeable among these are vortices, solitons and coherent structures in turbulent flow. In this paper, we shall review some of these features, which have been studied by various colleagues and students at Columbia in recent years.

The problems can be studied from linear stability analysis, or from a variational principle, in which under reasonable constraints, some suitably defined entropy or energy is maximized or minimized. Also, the formation dynamics of these objects—numerical solution of the nonlinear equations for their formation from reasonable initial conditions—will insure not only their attainability but their stability, with respect to perturbations included in the calculations. The last viewpoint is the main viewpoint adopted by us, except for the shear layer coherent structures, where we also include a variational principle as formulated by Cousins.

We dedicate this paper to Prof. C. C. Lin on his retirement as a token of respect and gratitude. While neither of us has ever taken a course from Prof. Lin, he has given each of us valuable advice at various junctures of our careers. The senior author particularly benefited from much encouragement from Prof. Lin, prior to returning to graduate school, at the start of his dissertation, and at many later points in life. We thank him and wish him many more happy and productive years.

[†] Postdoct. Res. Sci., Lamont-Doherty Geol. Observ., Columbia Univ., and NASA/Goddard Inst. for Space Studies

[‡] Prof., Dept. of Applied Phys. & Nuc. Eng., Columbia Univ.

2. SOLITONS

Since the work of Zabusky and Kruskal [1], we recognize that solitary waves described by the Korteweg-deVries equation (KdV equation) undergo nonlinear interactions with each other without loss of form or identity—hence the name solitons. Most of the studies on soliton formation and dynamics, including the ingenious inverse-scattering transform, have dealt with solitons generated from initial conditions. More recently, Chu, Xiang and Baransky[2] studied solitons generated from boundary conditions.

They found that for the KdV equation solved in the $x > 0$, $T > 0$ quarterplane, for a constant boundary value prescribed at $x = 0$, i.e.,

$$\zeta_t + \zeta\zeta_x + \epsilon\zeta_{xxx} = 0 \tag{1.1}$$

$$\zeta(x,0) = 0, \quad \text{and} \quad \zeta(0,t) = \zeta_b = \text{constant}$$

an unending series of identical solitons emerge from the boundary. The amplitude of the solitons is exactly double the boundary value, and the speed of each soliton is as predicted by classical theory, 1/3 of the amplitude. Fig. 1, taken from Ref. 2, shows this case. The zero undisturbed state, $\zeta(x,0) = 0$, actually corresponds to a zero limit water level in a channel.

This problem was recently generalized by R. L. Chou in her doctoral thesis at CUNY[3]. In particular, she extended the previous work[2] to include: various different initial water depths, treatment of solitons by Boussinesq equation, soliton head-on as well as overtaking collisions, wall reflections, and excitation by random boundary values. The use of the Boussinesq equations permits the study of waves in both directions, hence head-on collisions and wall reflections, hitherto not treatable with KdV equation, can be studied. The results from the Boussinesq equations, whenever the same initial and boundary conditions also apply for the KdV equation, indeed agree with the KdV results very well.

Two most interesting and unexpected phenomena were uncovered in this series of studies. The first is the solitons excited by a constant boundary conditions ζ_b for various initial depths ζ_0. In contrast to the case of $\zeta_0 = 0$ of Fig. 1, where solitons mature soon after they were born, greater values of ζ_0 result in much slower maturing of the emitted solitons. Such a case is shown in Fig. 2. This wave pattern is not to be confused with a collisionless shock or a weak bore, which is a fixed train of fixed length, resulting from having dissipation added into the KdV or Boussinesq equations. The present wave train is time-dependent, and the leading waves indeed saturate or mature at the soliton amplitude of twice ζ_b, but does so more slowly than in the case of Fig. 1. In fact, the soliton maturing time is directly dependent on the ratio ζ_b/ζ_0 : the greater this ratio, the longer the soliton maturing time.

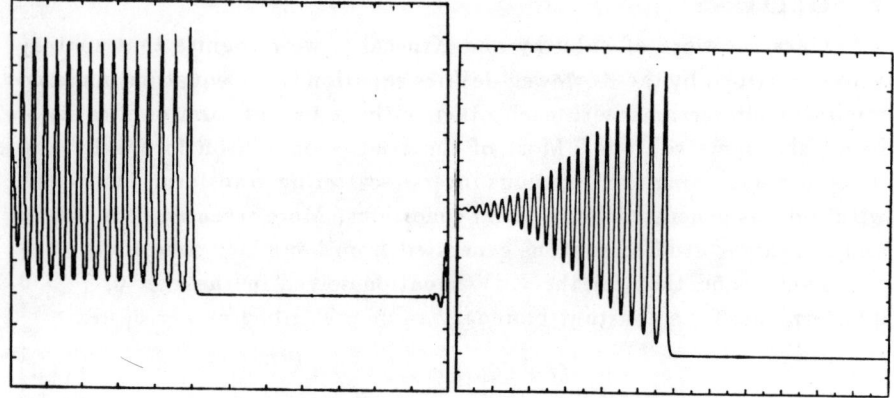

Fig. 1 KdV Solitons: $\zeta_0 = 0$, $\zeta_b =$ constant. (Ref. 2)

Fig. 2 KdV Solitons: $\zeta_0 \neq 0$, $\zeta_b =$ constant. (Ref. 3)

The other striking feature uncovered is the solitons generated by random boundary conditions. This study was initially undertaken to study the possibility of generating "soliton turbulence", suggested to us by M. Tabor. The calculations were performed with ζ_b obtained from a random number generator. The resulting solitons are shown in Fig. 3. In fact, no soliton turbulence showed up at all: the solitons produced are virtually identical to those formed from a constant wall boundary condition, equal to the expected value of the random numbers. The calculation was then repeated with each random boundary value held fixed for 10 time steps, and still the same conclusion obtained. The wave patterns are modulated by the boundary conditions only for slowly varying boundary values, i.e., when the characteristic time of the boundary variations is a significant fraction of the soliton formation time.

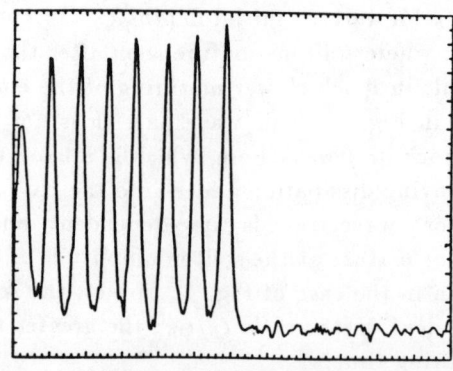

Fig. 3 Kdv Solitons: $\zeta_0 = 0$, $\zeta_b =$ random (Ref. 3)

These results further strengthen the well-recognized fact that solitons are very stable objects, and maintain their own identity and macrostructure under all kinds of perturbations and interactions.

3. COHERENT STRUCTURES IN TURBULENT FLOW.

Coherent structures in turbulent shear layers, which have been known for about a decade, are large vortical patterns, essentially independent of Reynolds number and with a mean flow nearly two-dimensional. In his recent Ph.D. thesis, A. Cousins[5] addressed this problem both variationally (as a two-dimensional inviscid problem), and numerically (as formation dynamics).

Consider a plane shear flow, of periodicity L in the x-direction, and velocity $+U$ and $-U$ at large $+y$ and $-y$ respectively. In terms of the stream function ψ and the vorticity ω, with

$$\omega = -\triangle\psi$$

Cousins introduced an entropy

$$S \equiv -\int \omega \ln \omega \, dx \, dy$$

where the integral is taken over $|x| < L$, $|y| < \infty$. This entropy is maximized subject to the constraints that the circulation Γ and the kinetic energy E remain constant:

$$\Gamma = 2UL$$

$$E = \int |\nabla\psi|^2 \, dx \, dy + \text{constant}$$

The resulting Euler-Lagrange equation is

$$\triangle\psi + C\exp(\lambda\psi) = 0$$

where C is a constant parametrizing the thickness of the unperturbed shear layer, and λ is the Lagrange multiplier. The solution of this equation is the Stuart stream function, obtained previously in other context:

$$\psi = -\frac{LU}{2\pi} \ln\left[\cosh(2\pi\frac{y}{L}) - \rho\cos(2\pi\frac{y}{L})\right]$$

where the Lagvrange multiplier λ has been determined, and the constant ρ is related to the parameter C thus:

$$\lambda = \frac{\pi}{LU}$$

$$\rho = \sqrt{1 + \frac{C}{2\lambda U^2}}$$

$\rho = 0$ corresponds to a zero thickness shear layer or inviscid slipstream, $\rho = 1$ corresponds to concentrated vortices, while intermediate values of ρ corresponds to various distributed vorticity and shear layer thickness. Fig. 4 shows the streamlines of such a flow with $\rho = 1/4$. As this flow has a greater "entropy" than the unperturbed shear flow, it is the physically realized case, corresponding to the occurrence of the vortical structures.

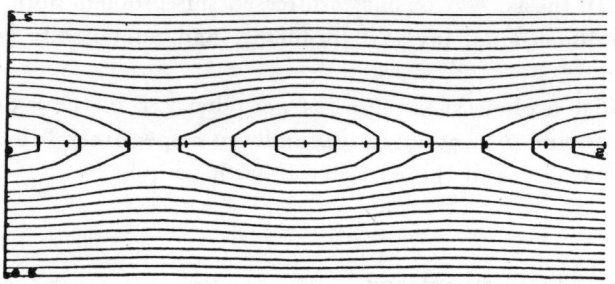

Fig. 4 Streamlines for Shear Layer, $\rho = 1/4$. (Ref. 4)

In his thesis[4], Cousins also used a vortex-in-cell method to simulate the formation of such flows from a periodic finite-thickness vortex sheet. In a subsequent work,[5], he made a more realistic calculation by injecting vorticity from a flat plate upstream, and solving a mixed initial-boundary value problem. Fig. 5 is the result of one such calculation, which is in remarkable agreement with the observed flow patterns.

Fig. 5 Vortex Structure in Shear Layer from Simulation. (Ref. 5)

4. SPHERICAL VORTICES.

Spherical vortices have recently regained interest partly because of their similarity to an experiment in plasma physics. In the spheromak experiment, the confined plasma has the idealized shape of a sphere. The theory of this configuration was given by, among others, Rosenbluth and Bussac[6], using a variational principle popularized by Taylor[7]. The total energy of the plasma (sum of internal energy and magnetic energy) is minimized subject to the constraint that the "helicity" be constant. The helicity is defined as $K \equiv \int \mathbf{A} \cdot \mathbf{B} \, dV$, $\mathbf{B}$ and $\mathbf{A}$ being respectively the magnetic field and its vector potential. K is constant on every flux surface in ideal MHD, but Taylor[7] hypothesized that in a real plasma, there is enough dissipation for this invariant to change on flux surfaces but to remain constnat over the entire volume. The resulting equilibria have the property that pressure is constant everywhere, and the current is everywhere parallel to the magnetic field (force-free). One of these equilibria is exactly spherical[6], and because of the energy principle, stability is automatically assured. The formation dynamics of such a configuration has been studied by Lui et al[8], and a complicated set of currents and fields are programmed in sequence to achieve this configuration.

Since the pressure is uniform, the internal energy plays no role in the energy being minimized. Thus, making the correspondence between velocity and magnetic field, kinetic energy and magnetic field energy, the same analysis and reasoning leads to a spherical vortex. The flow has vorticity everywhere parallel to velocity (Beltrami flow). Stability follows from the energy principle, but the formation dynamics of such a vortex is not obvious, and to our knowledge, not worked out thus far.

Another kind of spherical vortex is the Hill's vortex, whose plasma counterpart is called a field-reversed configuration. In cylindrical coordinates, the flow is in the r-z plane, and the vorticity in θ direction. We do not have a variational principle for such vortices, but the formation dynamics has been worked out some time ago by Lui[9].

These few examples, hopefully, illustrate the remarkable identity and stability of fluid dynamical macrostructures— solitons, vortices, coherent structures— that occur in nature.

References

1. Zabusky, N. J. & Kruskal, M. D., *Phys.Rev.Letts.* **15**, 240 (1965)
2. Chu, C. K., Xiang, L. W. & Baransky,Y., *Comm. Pure Appl. Math.* **36**,495 (1983)
3. Zhou, R. L., "Solitons Induced by Boundary Motion", Ph.D thesis, City University of New York 1987
4. Cousins, A. & Chevray,R., "Large Scale Order in Turbulent Shear Flows", Columbia Univ. Turb. Res. Lab. Report 101 1986
5. Cousins, A. and Chu, C. K. , "A Vortex-in-Cell Algorithm for the Spatially Developing Turbulent Mixing Layer", submitted *Phys. Fluids*
6. Rosenbluth, M. N. & Bussac, M. N. *Nuc. Fusion,* **24**,489 (1979)
7. Taylor, J. B. *Phys. Rev. Letts.,* **33**,1139 (1974)
8. Lui, H. C. , Chu, C. K. & Aydemir, A., *Phys Fuilds*, **24**, 673 (1981)
9. Lui, H. C., unpublished; details given in Chu, C. K. , "Computational Fluid Dynamics", chapter in Parter, S.V. (ed.), "Numerical methods for Partial Differential Equations", Acad. Press 1979

WAVE PROPAGATION ON AN ICE-COVERED OCEAN

Erik Mollo-Christensen and Antony K. Liu

NASA/Goddard Space Flight Center
Greenbelt, MD 20771

Motivated by a report of surface waves of one meter amplitude being observed 700 km inside the Antarctic ice pack in the Weddell Sea, we have analyzed the dynamics of wave propagation in sea ice and found that the mean compressive stress in the ice has a strong influence. Sufficiently high mean compressive stress can reduce the group velocity to zero and still not exceed the buckling stress. By examining the momentum exchange between waves and pack stress, one can suggest a sequence of events that describe the processes that may lead to rafting and breakup in the ice pack, and to pressure ridge formation.

INTRODUCTION

The thickness of the ocean ice cover is variable from place to place, depending on past history of air and sea temperature and even more on rafting, where one ice floe slides partly over another and then freezes in place; this process can repeat and results in variations in thickness from less than a meter to tens of meters at pressure ridges.

It is of interest to explore the processes that lead to rafting of ice floes and the formation of pressure ridges. The interactions between surface waves and pack compression play a significant role, as we show in our analysis that we summarize here. We consider surface gravity waves that on an ice-covered ocean are modified by the effects of flexural stiffness of the ice, compression in the ice field and the changes of inertia due to the ice. We have represented the ice cover as a slender beam under horizontal compression, floating on the water.

FORMULATION

Ignoring viscosity, we take the water velocity field to be derivable from a potential, $\phi(x, z, t)$, satisfying

$$\nabla^2 \phi(x, z, t) = 0 \tag{1}$$

The dynamic boundary condition[1] is taken as

$$B\frac{\partial^2}{\partial X^2}\left(\frac{1}{R}\right) + \frac{Q}{R} + M\frac{\partial \eta^2}{\partial \tau^2} = -\frac{\partial \phi}{\partial \tau} - g\eta - \frac{(\nabla \phi)^2}{2} \quad \text{at} \quad z = \eta, \quad (2)$$

with $B = Eh^3/[12(1-s^2)\rho_w]$, $Q = Kh/\rho_w$, and $M = \rho_i h/\rho_w$.

E is Young's modulus for ice, h the ice thickness, ρ_w is the water density, ρ_i ice density, K is the horizontal compressive stress in the ice, s is Poisson's ratio, and g is the acceleration of gravity. R is the radius of curvature.

The kinematic boundary condition is

$$W = \frac{d\eta}{dt} = u\frac{\partial \eta}{\partial x}; \quad \text{at} \quad z = \eta. \quad (3)$$

For the linear problem, the dispersion relation for waves of frequency and wavenumber k is:

$$\sigma^2 = (gk + Bk^5 - Qk^3)/(1 + kM). \quad (4)$$

From the dispersion relation, it is apparent that buckling divergence is possible. There is a minimum critical stress for buckling failure and there exists a corresponding wavenumber for buckling at the critical stress.

Also of interest is the fact that the group velocity

$$C_o = \frac{\partial \sigma}{\partial k} = [g + (5 + 4kM)Bk^4 - (3 + 2kM)Qk^2]/[2\sigma(1 + kM)^2] \quad (5)$$

can be zero for sufficiently high compressive stress, and that this stress can be less than the critical stress for buckling or compressive failure of the ice.

A wave that encounters a region with a very low value for group velocity will attain higher amplitude to maintain energy conservation. Where the group velocity decreases to zero along the propagation direction, a caustic will occur, and the wave amplitude will be singular, were it not for the fact that linear theory would lose its validity.

DISCUSSION

We also analyzed the weakly nonlinear modulation of the wave field[2], taking accounts of the higher order effects of curvature upon flexure and the effects of finite amplitude in the hydrodynamic boundary conditions. We found that the effects of mean compressive stress in the ice cover can significantly change the time scale of wave packet development, and even change the modulation dynamics as compressive stress increases from unstable (as in the free surface wave case) to stable and then back again[3]. We have not looked into the further significance in this stability change, and do not claim to be able to interpret this last result. The change in modulational stability due to changing compressive stress may lead to wave packet formation inside the ice pack, and cause large amplitudes to occur. These, in turn, may lead to failure and rafting of the floes formed.

The waves that occur inside the ice pack were generated by wind in the open ocean, and they impinged on the pack edge. We have analyzed the reflection and refraction of waves at a straight pack edge and found that, for a given angle of impingement, short waves are totally reflected while longer waves penetrate. The critical wave number for total reflection is dependent on angle of impingement.

As the longer waves propagate inside the ice pack, they will be damped, and we have also analyzed their damping rate due to viscous effects. The waves also contain and carry momentum, and as they are damped, amplified or destroyed, there has to be exchange of momentum between the waves and the mean stress field in the ice and the mean water velocity field. We have also calculated the relation between wave momentum changes and compressive stress in the ice pack.

The parameter space for wave propagation in sea ice has the coordinate axes corresponding to frequency, wavenumber, ice thickness, and compressive stress; it is difficult to describe in detail, and one must look for particular cases, observation and events to apply the results of our analysis. Such a case has fortunately presented itself, in August 1986, the Research Vessel "Polarstern" of the Alfred Wegener Institut fur Polar und Meeresforschung was located in the Weddell Sea ice pack, approximately 700 km from the ice edge as a series of waves of approximately 1 m amplitude caused the ship to roll and heave, and caused failure and rafting in the ice pack. The ice pack thickness had been measured as part of the research program, and the wavelength and wave period were estimated, aided by the use of radar.

The ice thickness was 2 m, the observed wave period 18 seconds and the wavelength was 250 m. The only remaining unknown in the dispersion relations Eq. (2) is the compressive stress K.

Figure 1 shows how the compressive stress varies with wavelength for waves of 18 second period in ice of 2 m thickness. Also shown is the corresponding group velocity, and we note that the group velocity is very close to zero. We conclude that the wave field had a caustic at the ship, and the large amplitude is due to local energy accumulation. Calculating the time constant of wave packet formation from the nonlinear modulation equation, we found a time constant of 8 minutes, which at the zero group velocity occurs at a point, but which at a very low group velocity would occur over a small region. The packet of waves could thus also have been formed close to the ship from a long train of impinging waves.

We also calculated the wave damping rate for 18 second period waves and found the $1/e$-folding distance to be approximately 750 km for waves of group velocity of 5 m/s. It therefore appears possible for the waves to have propagated from the ice edge. There were gale force winds and corresponding large waves off the ice edge in the open ocean, and it is likely that the needed wave energy was available.

Our analysis has illustrated several of the elements of the dynamics that participate in wave propagation in an ice pack, and which contribute to rafting of ice floes, local failures in the pack, and the maintenance of ice pack non-uniformity.

We thank the chief scientist on board the "Polarstern", Dr. Ernst Augstein, for furnishing us with the data from the observations.

References
1. Mollo-Christensen, E., 1983, "Interaction between waves and mean drift in an ice pack", *J. Geophys. Res.*, **88**, 2971-2972.
2. Benjamin, T.B. and J.E. Feir, 1967, "The disintegration of wave trains on deep water", *J. Fluid Mech.* **27**, 417-430.
3. Liu, A.K. and Benney, D.J., 1981, "The evolution of nonlinear wave trains in stratified shear flows", *Studies in Applied Math.*, **64**, 247-269.

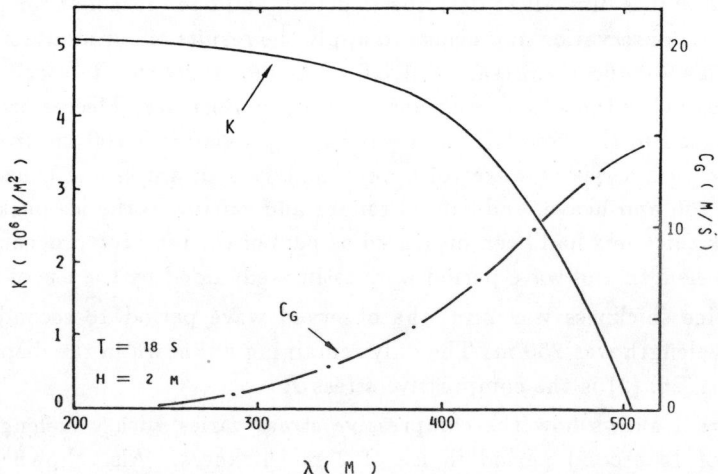

Figure 1. Ice compression stress effects on wave length (with ice thickness $h = 2$ m and wave period $T=18$ sec) and group velocity.

Diffusion-Controlled Reaction in a Vortex Field

Ronald G. Rehm,[*] Howard R. Baum[†] and Daniel W. Lozier[‡]
National Bureau of Standards
Gaithersburg, MD 20899

June 25, 1987

Abstract

A two-dimensional model of a constant-density diffusion-controlled reaction between unmixed species initially occupying adjacent half-spaces is formulated and analyzed. An axisymmetric viscous vortex field satisfying the Navier-Stokes equations winds up the interface between the species as they diffuse together and react. A flame-sheet approximation of the rapid reaction is made using Shvab-Zeldovich dependent variables. The model was originally proposed by F. Marble, who performed a local analysis and determined the total consumption rate along the flame sheet. The present paper describes a global similarity solution to the problem which is Fourier analyzed in a Lagrangian coordinate system. An asymptotic analysis of the Fourier amplitudes, valid for large Schmidt numbers is presented. The solution is evaluated numerically in Lagrangian and Eulerian coordinate systems.

1 Introduction

It is pleasure to participate in this symposium to honor Professor C.C. Lin. His dedication to the profession of applied mathematics and his high standards of scholarship are greatly appreciated. The variety of contributions at this symposium demonstrate the breadth of his technical interests.

The theoretical study of chemical reactions in complex flow fields has received increased attention lately; see references given in [1],[2]. This attention is warranted because a much clearer physical picture of turbulent flows and their

---

[*]Center for Applied Mathematics

[†]Center for Fire Research

[‡]Center for Applied Mathematics

coupling to combustion processes has been achieved through experiments over the past fifteen years. Turbulent combustion is difficult to analyze because it is highly nonlinear, transient and involves a wide range of length and time scales. When the Reynolds number is large, experiments indicate that the length and time scales can be separated into large and small scales. The large, or geometrical, scale is essentially nondissipative and is related to the geometry defining the flow configuration and the fuel and oxidizer distributions. Combustion, on the other hand, takes place on the small scale associated with the diffusion and establishes the rate at which the reactants disappear and heat is released.

An outline of a general approach for studying the problem of turbulent reacting flows, based on these observations is given in a recent paper by Baum, Corley and Rehm [2]. The approach is to analyze large-scale flow fields separately from the small-scale mixing and reaction but in a way that will allow the phenomena to be coupled through analytical and computational techniques. Also presented in [2] is a model for three-dimensional small-scale mixing and reaction in a stretched vortex flow field.

In [3], Marble originally posed a two-dimensional model of small-scale mixing and reaction and studied it analytically. The Marble problem has several important features; it is a diffusion-controlled reaction in a viscously spreading vortex which stretches the flame sheet. The model was analyzed locally along the flame front to provide global species consumption rates.

Two-dimensional numerical computations of the flow properties in reacting mixing layers have recently been carried out. Experiments indicate that during the early stages of development of these mixing layers, the flow fields remain primarily two dimensional. Then, before vortex pairing begins, the Marble problem can be regarded as an analytical approximation to combustion in one of the vortices. Also, recently, a direct numerical computation of the Marble problem has been undertaken by Laverdant and Candel. The computations demonstrate that, for the small-scale problem of mixing, diffusion and reaction, the computations are large, complex and difficult.

The work presented in this paper is essentially analytical, and, within the context of the mathematical model, permits one to calculate combustion properties as accurately as one likes. Therefore, it could be used, for example, to test the accuracy of the numerical computations cited above. The novel features of our work compared with that of Marble [3] are (i) it is observed that the convection-diffusion equation for the Shvab-Zeldovich variable permits a similarity solution, reducing the number of independent variables from three (radius, angle and time) to two (angle and similarity variable); (ii) as in [2], a Lagrangian coordinate system is used to eliminate flame-sheet resolution problems induced by vortex winding; (iii) Fourier analysis in angle and a combined numerical and analytical treatment in the similarity variable allow one to solve the global problem essentially exactly. More details of the analysis outlined here and references are presented in [1].

2 Formulation of the Problem

Consider the model in which initially there is fuel in the left half-plane and oxidizer in the right half-plane in arbitrary proportions. These half-spaces are brought into contact and simultaneously a line vortex with axis at the origin is imposed (see Figure 1). The vortex induces a convective mixing of the interface between the two species, increasing the area of the separating surface in the neighborhood of the origin and enhancing the diffusion of the species into each other. It is assumed that the reaction rate is rapid and a flame-sheet approximation is made. The chemical reaction is assumed to take place at constant density and all diffusion coefficients (kinematic viscosity, thermal and concentration coefficients) are assumed to be constant. The tangential velocity v_θ imposed is

$$v_\theta(r,t) = r\frac{d\theta}{dt} = \frac{\Gamma}{2\pi r}[1 - \exp(-\eta)] \qquad (1)$$

where Γ is the circulation of the vortex, ν is the kinematic viscosity and $\eta = r^2/4\nu t$ is a similarity variable.

With the assumptions described above, the equations for species and energy are decoupled from the momentum and continuity equations; they are equations representing a balance between convection, diffusion and reaction. The reaction rate, which is very nonlinear in concentrations and temperature, can be eliminated by taking a linear combination of the dependent variables when the additional assumption is made that the thermal and species diffusion coefficients are equal. The linear combinations are often called coupling functions or Shvab-Zeldovich variables, and in this case all satisfy the same equation.

The Shvab-Zeldovich variable Z satisfies the convection-diffusion equation

$$\frac{\partial Z}{\partial t} + \frac{v_\theta}{r}\frac{\partial Z}{\partial \theta} = D\left(\frac{\partial^2 Z}{\partial r^2} + \frac{1}{r}\frac{\partial Z}{\partial r} + \frac{1}{r^2}\frac{\partial^2 Z}{\partial \theta^2}\right) \qquad (2)$$

where D is the species diffusion coefficient.

Integrating the tangential velocity gives the angle θ at time t for any fluid element initially located at r, θ_0. A change of variables to the Lagrangian coordinates, ρ, θ_0, τ,

$$\begin{aligned} r &= \rho \\ \theta &= \theta_0 + \frac{\Gamma}{8\pi\nu}\frac{1 - E_2(\eta)}{\eta} \\ t &= \tau \end{aligned} \qquad (3)$$

where $E_j(z) = \int_1^\infty t^{-j}\exp(-zt)dt$, can then be made in the equation for Z. Finally, assuming that Z is only a function of the similarity variable η and the angle θ_0, and performing a Fourier decomposition in the angle,

$$Z(\eta, \theta_0) = \sum_{n=-\infty}^{\infty} Z_n(\eta)\exp(in\theta_0) \qquad (4)$$

a system of equations for each of the Fourier mode amplitudes Z_n is obtained:

$$\frac{d^2 Z_n}{d\eta^2} + f_n(\eta) \frac{dZ_n}{d\eta} + g_n(\eta) Z_n = 0 \qquad (5)$$

where

$$f_n(\eta) = \text{Sc} + \frac{1}{\eta} + in \frac{1 - \exp(-\eta)}{\eta^2} \text{Re}$$

$$g_n(\eta) = -\left(\left[\frac{1 - \exp(-\eta)}{\eta} \text{Re}\right]^2 + 1\right) \frac{n^2}{4\eta^2}$$

$$+ \frac{in}{2} \frac{\exp(-\eta) - [(1 - \exp(-\eta))/\eta]}{\eta^2} \text{Re}$$

and where $\text{Sc} = \nu/D$ is the Schmidt number and $\text{Re} = \Gamma/4\pi\nu$ is the Reynolds number based upon the circulation Γ.

Boundary conditions are that the solution remain bounded as η goes to zero and that the initial conditions are recovered as η goes to infinity. From the symmetry of the problem, for even values of n, $Z_n(\eta \to \infty) = 0$, and for odd values of $n = 2m + 1$,

$$Z_{2m+1}(\eta \to \infty) = \frac{(-1)^{m+1}}{\pi} \frac{1}{2m + 1} \qquad (6)$$

For computation, the infinite endpoint is replaced by a finite parameter.

The general solution for $Z_n(\eta)$ is a complex-valued function of the real variable η. From the governing equation, it is seen that $Z_{-n}(\eta)$ is the complex conjugate of $Z_n(\eta)$. The modes Z_n are synthesized using an FFT routine to determine $Z(\eta, \theta_0)$, and information about the location of the flame-sheet and the rate at which fuel is consumed can be determined.

3 Solution

3.1 Large Schmidt Number Analysis

When the Schmidt number is large, asymptotic methods allow one to determine an analytical solution to Eq.(5) from which the character of the solution to Eq.(2) for large Reynolds number can be determined. For large Sc

$$\tilde{Z}(\eta, \theta_0) = 1/2 + 2 \sum_{m=0}^{\infty} \frac{\exp[-\tilde{A}_m(\eta)]}{\pi(2m + 1)} \sin[(2m + 1)\Phi_0] \qquad (7)$$

where

$$\tilde{A}_m(\eta; \text{Re}, \text{Sc}) = (2m + 1)^2 \frac{1 + \text{Re}^2 \tilde{f}_1(\eta)}{4 \, \text{Sc} \, \eta}$$

$$\Phi_0 = \theta_0 - \pi \pm \frac{\pi}{2} - \frac{\text{Re}}{2\,\text{Sc}} \tilde{f}_2(\eta) \qquad (8)$$

and where

$$\tilde{f}_1(\eta) = [1/3 - 2E_4(\eta) + E_4(2\eta)]/\eta^2$$
$$\tilde{f}_2(\eta) = [(E_3(\eta) - 1/2)/\eta + E_2(\eta)]/\eta$$

Some important observations can be made from the asymptotic solution Eq.(7). All terms of the Fourier sine series are exactly zero when $\Phi_0 = 0$, and this condition determines the equation for the flame sheet for a stoichiometric mixture. However, the terms in the series will also be small when the argument of the exponential is large. The physical interpretation of these statements is as follows. For a stoichiometric mixture, there are two reaction regions. In the outer region there is a flame sheet, which remains close to the convectively mixed interface in the absence of diffusion. This interface is determined by the equation $\Phi_0 = 0$, and for moderately large η, is very close to the initial interface in the Lagrangian coordinate system. In the inner region, there is a burnt core in which both fuel and oxidizer are depleted, and the growth of this core is determined by the condition that the arguments of the exponentials are large enough that each of the terms in the Fourier series is negligible. For a large Reynolds number (and Schmidt number) this condition determines a value of the similarity variable, η^* say, and the growth of the burnt core is determined then by the equation $r^2/4\nu t = \eta^*$. The observations are consistent with those made by Marble[3].

3.2 Results

Numerical solution of Eq.(5) has been performed for many values of the parameters Re and Sc using a central difference approximation to the differential operators. The interval of integration for the differential equation is truncated to carry out the numerical integration and asymptotic values for the Fourier amplitudes are applied at the truncated location. A finite number of Fourier modes has been computed, and, as noted above, only odd modes of integer greater than zero have been computed. The analytical behavior of the solution for small values of the independent variable η must be handled carefully for lower order modes; therefore, a different dependent variable is computed near the origin. The numerical computations become more difficult as the Reynolds number is increased, and, therefore, the large Schmidt number analysis is very useful. It is also quite realistic for diffusion-controlled reactions in miscible liquids where the Schmidt number really is large.

In Figure 2 is shown plots of the interface shape in Eulerian coordinates for three values of Reynolds number 1, 10 and 100, and for a Schmidt number of 10. These plots are drawn approximately to scale. In each plot the area has been blackened within the burnt core to indicate no reaction activity. The burnt core grows with increasing Reynolds number. It should be emphasized that these plots do not represent time evolution, but rather different parametric configurations. The amount of convective mixing increases dramatically with Reynolds number.

Important quantities of interest from this analysis are the global rates of species consumption and of heat release; it is desired to calculate these quantities as functions of the Reynolds number, Schmidt number and initial concentrations of fuel and oxidizer. The consumption rates under general conditions can be calculated from the analysis presented earlier but require computation of Fourier amplitudes using an ODE solver, synthesis of the solution using FFT routines, location of the flame surface using a root finder and integration over the whole sheet to obtain the global rates. When the Schmidt number is large and the initial mixture of fuel and oxidizer is stoichoimetric, the analytical results presented earlier can be used to obtain simple analytical expressions for the rates.

We determine the enhancement of the consumption rate C caused by the imposition of the vortex. If we denote by C_0 the consumption rate in the pure diffusion case ($\text{Re} = 0$), then we desire $C - C_0$. We further assume that the Reynolds number is large enough that the Fourier series can be approximated by only one term. A Reynolds number of 100 or more is adequate for example. Integration of the approximate expression for the local enhancement of the consumption rate over the flame sheet, then gives the global enhanced species consumption rate:

$$C - C_0 \approx \frac{\hat{\Gamma}(2/3)}{\pi} D \left(\frac{2\,\text{Re}^2\,\text{Sc}^2}{3} \right)^{1/3} \tag{9}$$

The parametric dependence of this expression agrees with that reported by Marble [3].

4 Conclusions

In this paper we analyzed the Marble problem by methods that accurately solve the equations globally. The Marble problem has no natural length or time scale, and, therefore, allows a global similarity solution; we have exploited this observation. Perhaps the most important analytical result is that which expresses the dependence of the species consumption rate on the Reynolds and Schmidt numbers for large values of these parameters, and this result corroborates that found by Marble [3].

The solution presented here elucidates, we believe, the nature of the problem, and the method of solution provides all of the machinery necessary to calculate any results accurately. The Marble problem is inherently interesting because it addresses the question of enhancement of species consumption and heat release rates by flame stretching in a simple geometry. It is also of interest because, as discussed earlier, it simulates experiments in a two dimensional shear layer.

Finally, the work presented here can be viewed in a larger context, where it, or its generalization, is regarded as a submodel of a more general model for turbulent combustion; see Figure 3. The large scale fluid motion must be computed in a large Reynolds number or essentially inviscid approximation. We have developed such a large scale model for buoyant convection induced by a room fire [1],[2]. However, the large scale flow could as well be calculated for other configurations, and then the small scale combustion submodel would be embedded into this flow field.

We wish to thank Drs. G.B. McFadden and C. Fenimore for useful comments. This research was partially supported by the Air Force Office of Scientific Research under contract AFOSR-ISSA-87-0018.

References

[1] Rehm, R.G., Baum, H.R. and Lozier, D.W., "Diffusion- Controlled Reaction in a Vortex Field", National Bureau of Standards Report NBSIR 87-3572, June, 1987.

[2] Baum, H.R., Corley, D.M. and Rehm, R.G. "Time-Dependent Simulation of Small-Scale Turbulent Mixing and Reaction", Paper presented at the Twenty-First International Symposium on Combustion, Munich, West Germany, August 3-8, 1986.

[3] Marble, F.E. "Growth of a Diffusion Flame in the Field of a Vortex", *Recent Advances in Aerospace Sciences* (C. Cassci, Ed.) 1985, p.315.

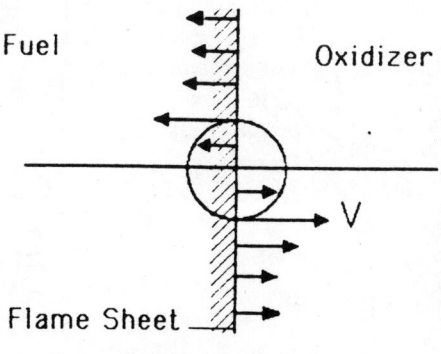

Figure 1

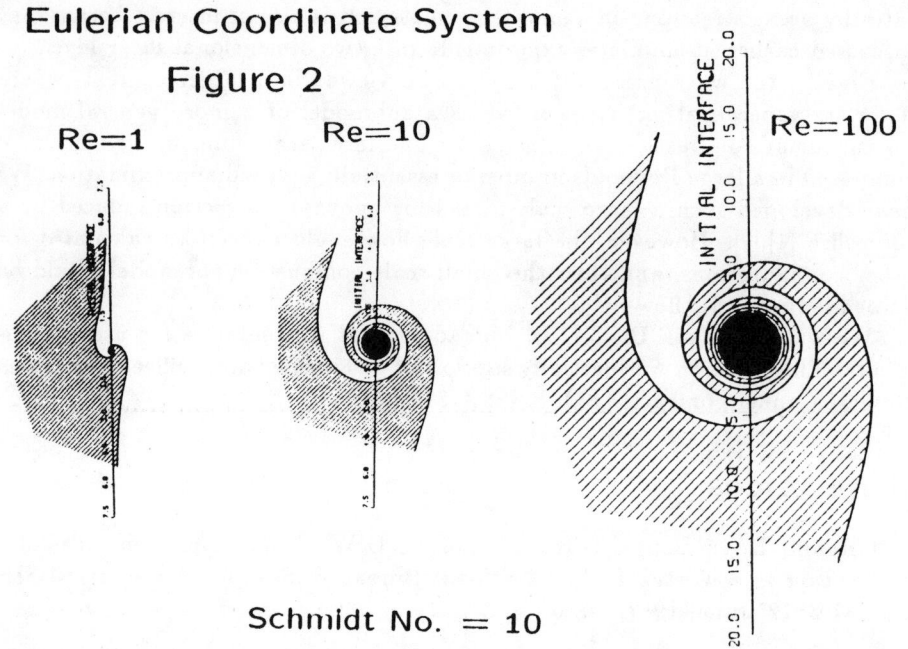

Eulerian Coordinate System
Figure 2

Re=1 Re=10 Re=100

Schmidt No. = 10

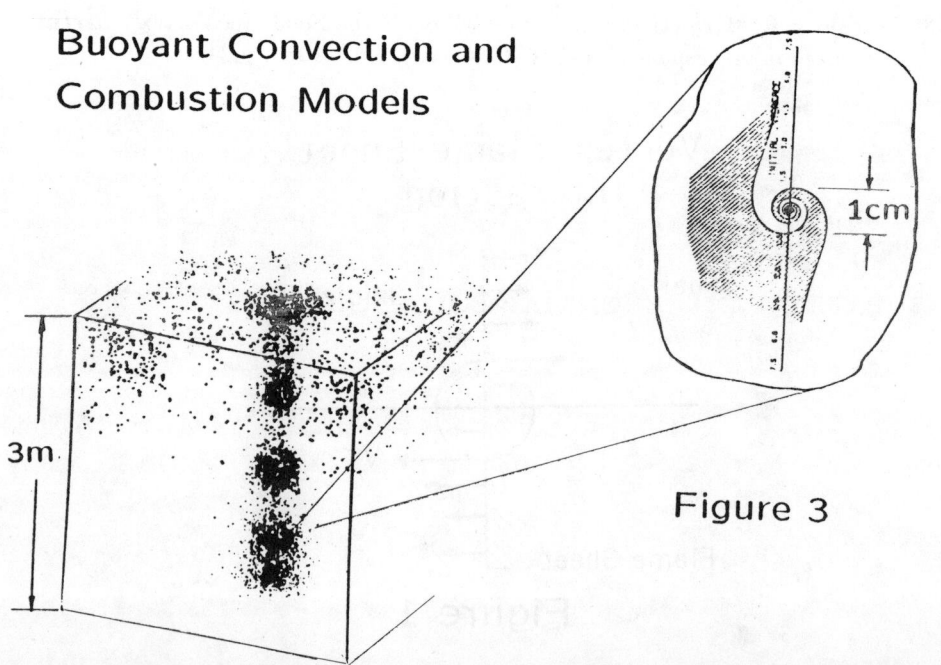

Buoyant Convection and Combustion Models

Figure 3

AMUNDSON'S MODEL OF THE COMBUSTION OF A CARBON BALL

S. I. ROSENCRANS

Mathematics Department
Tulane University
New Orleans, Louisiana 70118 USA

DEDICATED TO PROFESSOR C. C. LIN
with affection and respect

1. INTRODUCTION.

N. Amundson[1] has formulated the partial differential equations and boundary conditions which govern the combustion of an impermeable carbon ball. On the burning surface itself, the reaction

$$C + \tfrac{1}{2}O_2 \to CO$$

occurs, and the carbon monoxide thus formed may diffuse out of the ambient boundary (a concentric sphere of radius r_a) or else react with the oxygen in the stagnant layer D between the two spheres

$$CO + \tfrac{1}{2}O_2 \to CO_2;$$

the carbon dioxide thus formed can diffuse out of the ambient boundary or else react on the burning surface to produce more carbon monoxide. In the paper referred to Amundson has considered a hierarchy of models, one of which ignores the reaction producing carbon dioxide. In that case all reactions take place on the carbon surface, and a significant simplification occurs: boundary values of the unknown quantities uncouple from interior values and we can by the use of Green's functions find separate integral equations for each. In this way we can find the solution without resorting to the "quasi-steady" approximation often used in the chemical engineering literature.

2. PARTIAL DIFFERENTIAL EQUATIONS.

Unknowns in this problem are the radius of the ball, and the temperature and oxygen density on the burning ball and in the stagnant layer D between the burning ball and the ambient sphere. Let

$$w = \text{oxygen density}$$

z = temperature

$r(t)$ = radius of the ball.

There are two heat equations for the diffusion of oxygen and temperature,

$$\frac{\partial w}{\partial t} = D_1 \left(\frac{\partial^2 w}{\partial r^2} + \frac{2}{r} \frac{\partial w}{\partial r} \right)$$

$$w(0, x) = w_0$$

$$\frac{\partial z}{\partial t} = D_1 \left(\frac{\partial^2 z}{\partial r^2} + \frac{2}{r} \frac{\partial w}{\partial r} \right)$$

$$z(0, x) = z_0,$$

two conditions for the fluxes of oxygen and temperature on the carbon surface (these are derived from the reaction rate of $C + \frac{1}{2}O_2 \to CO$),

$$z_r = -c_2 w e^{-c_4/z}$$

$$w_r = -c_1 w e^{-c_4/z},$$

and an equation for the rate of change of the moving boundary $r(t)$:

$$r_t = -c_3 w e^{-c_4/z}.$$

We assume constant initial conditions $w = w_o$ and $z = z_o$, and constant ambient conditions at $r = r_a$: $w = w_a$, $z = z_a$.

3. INTEGRAL EQUATIONS.

To convert the above system of equations into integral equations we need the Green's function for the heat equation with zero boundary value on the ambient sphere $r = r_a$ and singularity at $r = r_0$, $t = t_0$. It is

$$g_i = \frac{r_0}{r} \left\{ \frac{e^{-\frac{(r-r_0)^2}{4D_i(t_0-t)}} - e^{-\frac{(r+r_0-2r_a)^2}{4D_i(t_0-t)}}}{\sqrt{4\pi D_i(t_0-t)}} \right\}.$$

Routine application of Green's identities to the region D with due attention paid to the moving boundary yields integral equations for $w(t)$ and $z(t)$, the values of oxygen density and temperature on the burning surface. These, and the equation for $r(t)$, are as follows. First we define five expressions which occur in the integrands.

$$e_i^- = e^{\frac{-(r(s)-r(t))^2}{4D_i(t-s)}}$$

$$e_i^+ = e^{\frac{-(r(s)+r(t)-2r_a)^2}{4D_i(t-s)}}$$

$$G_i = \frac{r(t)}{r(s)} \frac{e_i^- - e_i^+}{\sqrt{4\pi D_i(t-s)}}$$

$$H_i = -\frac{G_i}{r(s)} + \frac{r(t)}{r(s)} \frac{1}{4D_i^{\frac{3}{2}}\pi^{\frac{1}{2}}(t-s)^{\frac{1}{2}}}$$
$$\times \left(\frac{r(t)-r(s)}{(t-s)} e^- + \frac{r(t)+r(s)-2r_a}{(t-s)} e^+ \right)$$

$$S_i(u_o, u_a) = 2u_0 \sqrt{\frac{D_i t}{\pi}} \frac{1}{r(t)} \left\{ e^{\frac{-(r_0-r(t))^2}{4D_i t}} - e^{\frac{-(r_0-2r_a+r(t))^2}{4D_i t}} \right\}$$
$$- u_0 erf\left(\frac{r_0-r(t)}{\sqrt{4D_i t}}\right) + u_0 erf\left(\frac{r_0-2r_a+r(t)}{\sqrt{4D_i t}}\right)\left(\frac{2r_a-r(t)}{r(t)}\right)$$
$$+ (u_0-u_a)erf\left(\frac{r_a-r(t)}{\sqrt{4D_i t}}\right)\left(\frac{2r_a}{r(t)}\right) + \frac{2r_a u_a}{r(t)}.$$

In terms of these expressions the equations for w and z are

$$w(t) = S_1(w_0, w_a) - 2\int_0^t \frac{r(s)^2}{r(t)^2}\left\{-D_1 w(s)H_1 + c_1 D_1 w(s)e^{-c_4/z(s)}G_1\right.$$
$$\left. + c_3 w(s)^2 e^{-c_4/z(s)}G_1\right\}ds$$

$$z(t) = S_2(z_0, z_a) - 2\int_0^t \frac{r(s)^2}{r(t)^2}\left\{-D_2 z(s)H_2 - c_2 D_2 w(s)e^{-c_4/z(s)}G_2\right.$$
$$\left. + c_3 w(s)z(s)e^{-c_4/z(s)}G_2\right\}ds.$$

4. METHOD OF SOLUTION.

Questions of existence and uniqueness of solutions, *a priori* estimates, etc., will be dealt with in a forthcoming paper[2], dealing with one-dimensional, i.e., "slab" geometry. As far as those issues are concerned, the difference between the two geometries is not very significant. (That paper also contains some asymptotic results as $t \to \infty$.) Here we shall explain how we solve the system of three integral equations numerically. Note the mild square-root singularity at an endpoint. For such problems Kershaw[3] has suggested using the trapezoid rule obtained by linear interpolation of f in $\int f(s)/\sqrt{t-s}\, ds$. This results in three simultaneous nonlinear equations to be solved at each time step. The method is rapid and accurate. Kershaw's paper contains a proof of the convergence of this scheme.

5. SOME NUMERICAL RESULTS.

We have not yet fully explored the variety of results that can be found by varying the nine parameters, nor is there room here to discuss all those results which we have found. A fairly typical course of combustion is shown in the two figures. Three quite distinct stages are evident from the graphs. It is planned to explore these results further in a sequel to this paper in which we also hope to address the question of estimation of the extinction time and other characteristic parameters.

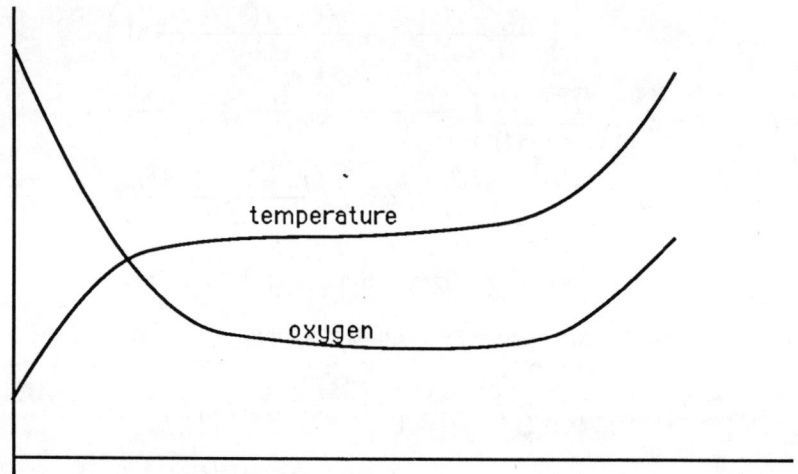

Figure 1. Temperature and oxygen density versus time.

References

1. Amundson, N., in *Partial Differential Equations and Dynamical Systems*, ed. W. E. Fitzgibbon, Pitman, 1984.
2. Rosencrans, S.I., to appear in *Quart. Appl. Math.*
3. Kershaw, D., in *Treatment of Integral Equations by Numerical Methods*, ed. C. T. H. Baker, Academic Press, 1982.

INERTIA EFFECTS OF A LOCALIZED FORCE NEAR A WALL IN A SLOW SHEAR FLOW

M. Shibata and C. C. Mei
Massachusetts Institute of Technology
Cambridge, Ma. USA

Abstract

By oscillating horizontally a water tank with a sand layer of uniform depth on the bottom, Bagnold[1] found that two principal types of ripples can be generated from the surface of the sand layer. If the oscillation amplitude is large, then any initial protrusion from the sand bed may initiate flow separation. In each half period vortices are generated which tend to excavate sand particles on both sides of the protrusion, and to bring them in suspension. Some of the excavated particles are deposited over the protrusion. When the excavations are large enough, their edges becomes new sources of vortices, which in turn form new troughs and crests further away from the initial protrusion. Thus sand ripples spread outwards from the original protrusion. Bagnold calls these the *vortex ripples*. On the other hand, for small amplitude oscillations over a flat bed, only a few of the particles are moved initially. Their rolling motion mobilizes other particles which gradually form line ripples. These ripples have heights of just one or two grain diameters. Bagnold calls these the *rolling grain ripples*. In later experiments by Carter, Liu and Mei[2], a monolayer of sand is sparsely spread on the aluminium bed of a wave tank. Progressive surface waves produced by a piston at one end are observed to form ripples of nearly uniform wave length. Turbulence is absent. The grains may either roll or slide on the smooth bed. Because of the low density of particles per unit bed area, particle collision is not prevalent.

More extensive experiments have been performed by Kaneko and Honji[3] in a closed water tunnel. With glass beads observations of two types of bed were made: i) a monolayer of beads distributed sparsely and randomly over a rigid plane floor, and ii) a thick layer of the same beads. In case i) the oscillating beads quickly form particle waves along lines transverse to the flow. The distance between these lines increases with time until an equilibrium value is reached. No particle waves are found if the average distance between particles is initially greater than 20 radii. They further examined more closely a single row of glass beads initially in line with the flow. As soon as the oscillation begins, two neighbouring particles first approach each other to form an aggregate. Later the axis of the aggregate tends to rotate so as to become perpendicular to the flow. Gradually two neighboring aggregates attract each other to form new aggregates involving more beads. Eventually short lines are formed in the transverse direction. In case ii) Kaneko and Honji first placed some colored beads on the top of a layer of transparent beads. Under oscillating water, the colored beads first form particle waves similar to those found on a smooth and rigid bottom. As time goes on some transparent beads are also mobilized and take part in forming ripples. The wavelength of ripples are of the same order of magnitude as that of the corresponding particle waves in a monolayer. Thus the initial stage of ripple formation in the two types of bed appear to be governed by a similar mechanism. In both experiments, the oscillatory boundary layer is laminar and the Stokes layer thickness is comparable to the bead diameter. From these observations, Kaneko and Honji conclude that the particle waves are originated by hydrodynamical interaction at a distance. They have also performed computer simulation[4] for a single layer of particles on a smooth bed. Starting from a rectangular array of identical particles, they include, in the equation of motion for each particle, the inertia of each particle, friction on the bottom and a hypothesized interaction force between pairs of particles. More specifically the assumed force is nonzero only when the interparticle distance is less than a specified value L_1, attractive if the distance is greater than $L_2(< L_1)$ and repulsive if it is less than L_2. With either a single row or a rectangular array of particles, they found particle waves resembling the experiments qualitatively.

Because of the strong similarity between a monolayer and a thick layer of particles in the process of ripple formation, it is worthwhile to examine more closely the mechanism in the former case which should be the simpler. The phenomenon must still involve a myriad of physical elements. For example, the motion of particles must be affected by the random distribution of shapes and sizes of, and friction among, the particles. As an initial step, it is useful (i) to calculate, rather than to hypothesize, the hydrodynamical interaction among identical particles, in order (ii) to provide the basis for examining the initial instability of a particle array when they are slightly out of alignment. This paper aims at providing partial answer to step (i). In order to facilitate calculations we shall consider particles which are fine enough so that the Reynolds number is small. Thus our results are expected to be at best of qualitative relevance to the experiments of Kaneko and Honji, but are more relevant to finer particles such as silt, smaller ambient velocity or higher fluid viscosity. Specifically we assume the particle radius to be much smaller than the spacing l which is in turn much smaller than the thickness of the Stokes boundary layer δ, i.e., $a/l, l/\delta \ll 1$. This assumption enables us to use the Stokes approximation in the near field of each particle. But for the interaction among particles spaced far apart, the far field is the most important. To this end we circumvent the difficulties of the boundary condition on the surface of a particle and seek an equivalent representation by a localized force. Known theories on the total force on a sphere near a wall is invoked as a guide. Now in the far field, convective inertia is important so that Oseen approximation accounting for the ambient shear is needed. Integral solutions of the Oseen equations is first obtained and then calculated numerically. Numerical results are used to infer that two particles at the bottom of an oscillatory boundary layer will attract each other if in tandem, and repel each other if side by side. Further work on the instability of a rectangular array of particles is being continued.

References
1. Bagnold, R. A. *Proc. Roy. Soc. London*, **187**, 1 (1946).
2. Carter, T. G., Liu, P. L. F. & Mei, C. C., *J. Waterways, Harbors, Coastal Eng. Div. ASCE*, **99**, 165 (1976).
3. Kaneko, A. & Honji, H. *Sedimentology*, **26**, 101 (1979).
4. Kaneko, A. & Honji, H. *Report, Res. Inst. Appl. Mech. Kyushu Univ.* **24**, 241 (1982).

SOME MODEL EQUATIONS FOR SURFACE WAVES ON A FLUID

Dedicated to Professor C. C. Lin

M. C. Shen

Department of Mathematics, University of Wisconsin

Madison, WI 53706

This paper deals with some approximate equations derived for nonlinear surface waves on a fluid. An inhomogeneous Burgers equation is presented for surface waves on a viscous fluid flow down an inclined plane with a prescribed moving disturbance on the free surface. It is found that a soliton-like wave with decaying amplitude can appear if the deviation of the speed of the moving disturbance from a critical speed exceeds some limit. An existence theorem for the Korteweg-de Vries equation with variable coefficients is used to justify an approximate method for the development of a solitary wave in a channel of variable cross section. The method applied to elliptical channels yields results in contrast to previous findings.

1. INTRODUCTION

We first consider the problem of surface waves on a viscous flow down an inclined plane [1-8]. It is known that within the framework of long wave approximation, Burgers equation is obtained as an approximate equation for the evolution of surface waves on the fluid when nonlinearity and dissipation reach a balance [3,5]. If a moving surface disturbance is prescribed on the free surface, an inhomogeneous Burgers equation appears and can also be reduced to a linear equation by the Cole-Hopf transformation [7]. It is found that the moving disturbance may generate various types of waves, especially a solition-like wave with decaying amplitude [7,8].

Another problem considered here concerns the development of a solitary wave over an uneven bottom [9-14]. An equation of the $K - dV$ type with variable coefficients has been used for the investigation of the fission of solitons in a channel of variable cross section. Suppose that a solitary wave moves from a uniform section of a channel to another through a transition section. If the effect of the transition section is neglected, we may use the solitary wave as the initial condition of the $K-dV$ equation for the second uniform section, and solve the initial value problem by the inverse scattering method. Then a criterion can be derived for the number of solitons to be generated in the second section. Here we shall justify this approximation and show by an example that in contrast to previous results a solitary wave may still split into a sequence of smaller solitons when it moves from a section of an elliptical channel of constant depth and width to one of larger depth but the same width [14].

2. INHOMOGENEOUS BURGERS EQUATION

We choose a coordinate system moving with the prescribed disturbance at the speed λ in the $x-direction$, and the $x-axis$ coincides with the inclined plane. The approximate equation for the free surface is given by [7]

$$\eta_t - \lambda_1 \eta_x + 2R\sin\theta\ \eta\eta_x + R[-(\frac{1}{3})\cos\theta$$

$$+ (\frac{2}{15})R^2\sin^2\theta]\eta_{xx} + RP_x/2 = 0, \qquad (1)$$

where R is the Reynolds number, θ is the angle between the incline and the horizontal, $0 < \theta < \frac{\pi}{2}$, $\lambda = R\sin\theta + \epsilon\lambda_1$, ϵ is a small positive parameter, and P is the prescribed shear stress on the free surface. Let

$$\nu = R[(\frac{1}{3})\cos\theta - (\frac{2}{15})R^2\sin^2\theta] > 0,$$

and

$$\eta = -[\frac{\nu}{(R\sin\theta)}]\frac{\varphi_x}{\varphi}. \tag{2}$$

Substitution of (2) in (1) yields

$$\varphi_t - \lambda_1\varphi_x - \nu\varphi_{xx} = cP\varphi, \tag{3}$$

where $c = \frac{R^2 \sin\theta}{2\nu}$. We prescribe $\eta = 0$ initially and choose $\varphi = 1$ at $t = 0$. First consider a uniformly distributed shear stress

$$P_h(x) = a \text{ for } -h/2 \le x \le h/2$$
$$= 0 \text{ elsewhere,}$$

where $a > 0$, $h > 0$. The solution of (3) subject to $\varphi = 1$ at $t = 0$ is obtained by Laplace transform [7]. Then we consider the limiting solution for $ah = 1$ as $h \to 0$. The types of waves generated depend upon $|\lambda_1| > c$, $0 < |\lambda_1| < c$ and $|\lambda_1| = c$. In particular, if $\lambda_1 < -c$, the dominant term in (2) for η, is

$$f(x,t) = (2R\sin\theta)^{-1}c\lambda_1(\lambda_1 - c)^{-1}exp(-\lambda_1\frac{x}{\nu})$$

$$\times erfc[(xt^{-1/2} - t^{1/2})/(2\nu^{1/2})].$$

For a fixed t, $(2R\sin\theta)f(x,t)$ has a maximum at $x = -\lambda_1 t$ with an amplitude $-\nu^{1/2}/[\lambda_1(\pi t)^{1/2}]$, and represents a soliton-like wave moving with speed $-\lambda_1$ toward $x = \infty$. For $x < 0$ and a fixed t, η attenuates exponentially. For other cases, we only have shock-like fronts. For a general moving disturbance we have the following result [8]. Let $P(x)$ is a smooth function with support in $(-\alpha, \alpha)$, $\alpha > h$, and denote by η_h, φ_h and $\eta_\delta, \varphi_\delta$ respectively the solution η, φ corresponding to the uniformly distributed shear stress and its limiting case as $h \to 0$. Assume that $\int_{-\alpha}^{\alpha} |P(x)|dx \le (\lambda_1^2 + 4\nu\ell)^{1/2}/2c$ where ℓ is a positive constant, which can be chosen arbitrarily small.

THEOREM 1. If $\int_{-\alpha}^{\alpha} |P_h(x) - P(x)|dx \le \epsilon$ and $\sqrt{h} \le \epsilon$, then $\sup_{|x|\le 2\alpha} |\eta - \eta_\delta| \le K\epsilon exp(\ell t)$ for any $t > 0$, where K is a positive constant. The proof is omitted here. We only note that η can be approximated by η_δ up to $t = O(\ell^{-1}\log\epsilon^{-1})$.

3. K-dV EQUATION WITH VARIABLE COEFFICIENTS

Consider the motion of an irrotational, inviscid fluid in a channel of variable cross section. To be definite, assume that the channel cross section is convex and changes slowly in the axial direction. By the long wave approximation again, we obtain [12]

$$m_1 \eta_{1x} + m_2 \eta_1 + m_3 \eta_1 \eta_{1\xi} + m_4 \varsigma_{1\xi\xi\xi} = 0, \qquad (4)$$

where m_1 to m_4 are functions of x, $\xi = \beta S(x,t), \beta$ is a large parameter and $S(x,t)$ satisfies

$$(S_t)^2 = (S_x)^2 A(x)/w(x),$$

where $A(x)$ is the area, and $w(x)$ the width, of a cross section of the channel. Let $G(x) = [A(x)/w(x)]^{1/2}$ and choose S so that $S_t = -1$ and $S_x = G(x)$. It can be shown that, if $A(x)$, $w(x) > 0$ and the slope of the boundary curve in a cross section are infinite at the two endpoints of the width line, m_1 to m_4 are positive. Assume that they are all continuous. We introduce new variables

$$\tau = \int_0^x m_4(x') m_1^{-1}(x') dx',$$

$$\varsigma = m_3(x) m_4^{-1}(x) \eta,$$

and (4), in terms of τ and ς, becomes

$$\varsigma_\tau + \varsigma \varsigma_\xi + \varsigma_{\xi\xi\xi} = H(\tau)\varsigma, \quad \tau > 0, \quad -\infty < \xi < \infty, \qquad (5)$$

subject to

$$\varsigma(0, \xi) = \varsigma_0(\xi), \quad -\infty < \xi < \infty, \qquad (6)$$

where

$$H(\tau) = -m_2 m_4^{-1} - m_1 m_4^{-2} d(m_4 m_3^{-1})/dx.$$

Let $H^S(-\infty,\infty)$ denote the Sobolev space of order s of the L^2-type. We have

THEOREM 2. *Equation* (5) *subject to* (6) *with* $\varsigma_0 \in H^S, s \geq 2$, *has a unique solution* ς *for any* $T > 0$. $\varsigma \in C[0,T;H^S] \cap C'[0,T;H^{S-3}]$, *and depends upon* ς_0 *continuously in* $H^S -$ *norm*.

In the study of the development of a solitary wave in a channel with two uniform sections, we made the assumption that the distance x_1 to cover the transition section is small so that its effect on the solitary wave coming from the first uniform cross section may be neglected. Indeed this assumption can be readily justified by Theorem 2. Let $\eta(x,\xi), \eta^*(x,\xi)$ be respectively the solutions of (x) with the initial conditions $\eta(0,\xi)$ at $x = 0$ and $x = x_1$.

THEOREM 3. *For any given* $X > 0$ *and* $\epsilon > 0$, *there exists* $\delta > 0$ *such that*

$$\|\eta^*(x,\xi) - \eta(x,\xi)\|_s < \epsilon \;\; for \;\; x_1 < \delta, x \in [0,X]$$

where $\|\cdot\|_s$ is the H^S-norm and $S \geq 2$. As an example consider an elliptical channel with cross sections defined by $y^2/a^2(x) + z^2/b^2(x) - 1 = 0$ where $2a(x)$ and $b(x)$ are respectively the variable width and depth of the channel. The criterion for a solitary wave to split into n solitons assumes the form [14]

$$(\gamma_1/\gamma_2)^{1/2}(b_1/b_2)^{1/4}[E(\gamma_1)/E(\gamma_2)] = n(n+1)/2, \quad (7)$$

where $\gamma_i = a_i/b_i$, $i = 1,2$, are respectively the constant ratios of a/b for the first and second uniform sections, and

$$E(\gamma) = 3(13-\pi^2) - \gamma^2$$

$$+24\gamma \sum_{n=1}^{\infty}[n(1-4n^2)]^{-1}[1+(\gamma-1)^{2n}/(\gamma+1)^{2n}]\times[1-(\gamma-1)^{2n}/(\gamma+1)^{2n}]^{-1}.$$

If $a_1 = a_2 = 1$, $b_1 = 1/3$ say, (7) yields

$$b_2^{9/4}E(b_2^{-1}) = 2(3)^{-9/4}E(3)/n(n+1),$$

which possesses solutions $b_2 > 1/3$ for $n > 1$. We note that in the case of rectangular and triangular channels fission of solitons can only take place when the depth of the second uniform section is smaller [10,11,13].

Acknowledgements. The research reported here was partly supported by the National Science Foundation under Grant NO. MCS-521-5064.

REFERENCES

1. Benjamin, T. B., "Wave Formation in Laminar Flow Down an Inclined Plane", *J. Fluid Mech.* 2, 554-574 (1957).

2. Yih, C. S., "Stability of Liquid Flow Down an Inclined Plane", *Phys. Fluids* 6, 321-334 (1963).

3. Benney, D. J., "Long Waves on Liquid Films", *J. Math. and Phys.* 45, 150-155 (1966).

4. Mei, C. C., "Nonlinear Gravity Waves in a Thin Sheet of Viscous Fluid", *J. Math. and Phys.* 45, 266-288 (1966).

5. Shen, M. C. and Shih, S. M., "Asymptotic Theory of Nonlinear Surface Waves on a Viscous Fluid in an Inclined Channel of Arbitrary Cross Section", *Phys. Fluids*, 17, 280-286 (1974).

6. Shih, S. M. and Shen, M. C., "Uniform Asymptotic Approximation for Viscous Fluid Flow Down an Inclined Plane", *SIAM J. Appl. Math.* 6, 560-582 (1975).

7. Shen, M. C. and S. M. Sun, "Critical Viscous Surface Waves over an Incline", *Wave Motion*, to appear.

8. Sun, S. M. and Shen, M. C., "An Inhomogeneous Burgers Equation", to appear.

9. Madsen, O. S. and C. C. Mei, "The Transformation of a Solitary Wave over an Uneven Bottom", *J. Fluid Mech.* 39, 781-791 (1969).

10. Tappert, F. D. and Zabusky, N. J., "Gradient-induced Fission of Solitons", *Phys. Rev. Lett.* 27, 1774-1776 (1971).

11. Johnson, R. S., "On the Development of a Solitary Wave Moving over an Uneven Bottom", *Proc. Camb. Phil. Soc.*, 73, 183-203 (1973).

12. Shen, M. C. "Nonlinear Waves in a Channel", *Nonlinear Waves*, Ed. L. Debnoth, Cambridge University Press, 69-83 (1983).

13. Zhong, X. C. and Shen, M. C., "Fission of Solitons in a Symmetric Triangular Channel with Variable Cross Section", *Wave Motion*, 5, 167-176 (1983).

14. Cai, Y. H. and Shen, M. C., "The Development of a Solitary Wave in a Channel of Variable Elliptical Cross Section", *Phys. Fluids* 28, 2352-2356 (1985).

IV. PLASMA PHYSICS

PHYSICS OF SPACE AND LABORATORY PLASMAS

Bruno Coppi
Massachusetts Institute of Technology

The aspect of the scientific career of Professor C. C. Lin that I have always found most attractive is his ability to move through different fields (engineering, applied mathematics and astrophysics) with a uniquely open mind and preserving his depth of thinking for all the problems he is dealing with. In other words, one does not find in Lin any trace of scientific snobbishness, that is the attitude of considering one's own field of interest nobler than another's, nor the weakness to give up on scientific standards when moving in a new, less explored domain of research such as galactic dynamics was when he first became involved in it. Therefore, his presence at MIT as a colleague and friend has been a constant source of inspiration and moral support.

Here is a summary of my presentation in Professor Lin's honor:

Recent experimental developments such as the attainment of well-confined plasmas by the Alcator machine at MIT, where the peak particle density has reached nearly $2 \times 10^{15} cm^{-3}$, and the plasma measurements preceeding the Uranus bowshock, where the density is about $2.5 \times 10^{-2} cm^{-3}$ has allowed us to test the most recent developments of plasma theory over nearly 17 orders of magnitude of density variation. The most important aspect of these plasmas is that their dynamics are controlled by collective modes, which in their linearized mathematical descriptions are called microinstabilities, and not by the known collisional (so-called classical) transport theory. Space and astrophysical plasmas have provided the initial incentive to undertake the investigation of basic processes such as those that go under the generic name of magnetic reconnection and that have to be considered, in order to describe the interaction of the magnetic field "carried" by the Solar Wind with that of the Earth, or the nature of solar flares. Similarly the interest in processes that produce an effective viscosity, that is transfer of angular momentum, has been particularly stimulated by the key role that they play in the dynamics of accretion disks around compact stars.

On the other hand, laboratory experiments have produced "high energy" plasmas where the characteristic mean free paths are very long relative to all their dimensions and where the onset and the effects of collective modes can be studied and compared with available theories. Of particular importance is the evidence of macroscopic phenomena that involve a decoupling of the plasma motion from the magnetic field (i.e., magnetic reconnection) and manifest themselves as major disruptions of the plasma column or as large scale oscillation-relaxation processes (so-called "sawtooth oscillations") in the center of the plasma column. The decay of the rate of rotation induced in a toroidal plasma column by the (tangential) injection of energetic neutrals has been observed to be higher than that predicted by the relevant collisional transport theory. Thus the excitation of so-called ion-mixing modes, that can transport angular momentum and thermal energy without producing a net particle flow can be proposed as a possible explanation for this direct evidence of effective viscosity.

Finally, a phenomenon whose analysis has yet to find a significant application in astrophysics, is that of the so-called anomalous (that is, higher than that predicted by the collisional transport theory) electron thermal energy conductivity. In fact, on the basis of experimental observations as well as of theoretical considerations, we have pointed out that this cannot be formulated in terms of a linear thermal conduction equation. Rather, the observations can be interpreted following the so-called "principle of profile consistency"[1] that is in direct violation of an equation of this type and does not rely on transport coefficients that involve functions of local plasma parameters. Specifically, the electron temperature profiles that have been measured in a rather large variety of regimes and with different spatial distributions of the electron thermal energy source has been shown to remain nearly unchanged as implied by this "principle".

References
1. B. Coppi, *Comments Pl. Phys. Cont. Fus.* **5**, 261 (1980).

STABILITY OF ROTATING ELECTRON BEAMS*

Y.Y. Lau

Naval Research Laboratory
Washington, DC 20375-5000

It is shown that the longitudinal stability of a thin rotating electron beam depends sensitively on the manner in which the equilibrium rotation is supported. Maximization or elimination of small signal growth can be achieved by adjusting the radial electric field and the axial magnetic field which are needed to support the equilibrium rotation. A simple dispersion relation is given for general combinations of electric and magnetic fields, and for arbitrary electron energy and beam current. Potential applications and proof-of-principle experiments are suggested.

1. INTRODUCTION

Recent developments in high power microwave electronics and high current cyclic accelerators have led to renewed interest in the stability of rotating electron beams. Depending on the device, this rotation is supported either by an axial magnetic field, or by a radial electric field, or by a combination of both. Even for the case of a thin beam, the crucial dependence of the beam stability on the equilibrium type was noted only in the last few years. Here, we summarize some of these recent findings.* We shall show that, for a given geometry and a given kinetic energy of the beam, highest small signal growth is obtained if the rotation is supported by a radial electric field alone. We shall also give the condition under which the dynamical instabilities and the resistive wall instabilities are minimized. As we shall see, these are the basic properties of space charge waves on a rotating electron beam. It is appropriate to remark here that the analysis of space charge wave in a rotating electron beam is analogous to that of the density waves in galaxies.[1]

---

*This work is an outgrowth of a collaboration with David Chernin (Refs. 3,4), whom I thank. It is supported by the Office of Naval Research.

2. EQUILIBRIUM

For simplicity consider a thin cylindrical layer of electrons which rotates concentrically about the z axis at velocity $\bar{v}_o = \hat{\theta}\, v_o(r) = \hat{\theta}\, r\, \omega_o(r)$ between two coaxial metallic pipes. Whatever axial magnetic field B_o and radial electric field E_o which are required to support the equilibrium rotation, v_o must satisfy the radial force balance:

$$\gamma_o \frac{v_o^2}{r} = - \frac{e}{m_o}(E_o + v_o B_o) \qquad (1)$$

Here, $e < 0$ is the electron charge, m_o is the electron rest mass, $\gamma_o = (1 - v_o^2/c^2)^{-1/2}$ is the relativistic mass factor, and c is the speed of light. It is important to include the relativistic effects once the electron kinetic energy exceeds 5keV^2.

To label various types of equilibrium corresponding to different devices, we introduce a dimensionless quantity h, defined by

$$h = \frac{-er\, E_o}{m_o\, \gamma_o^3\, v_o^2} \qquad (2)$$

Physically, $\gamma_o^2 h$ is the ratio of the electric force to the centripedal force in equilibrium. From Eqs. (1) and (2) we may characterize various electron devices according to the value of h [Fig. 1]:

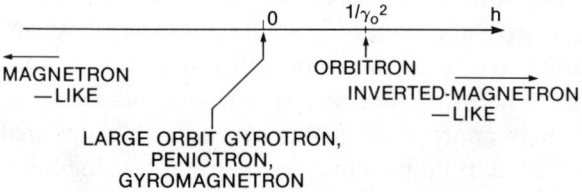

Fig. 1. Correspondence between various electron devices and the values of h. Discussion of the individual devices may be found in Refs. (4),(5).

(i) $h \rightarrow 0$ corresponds to the large orbit gyrotron, peniotron, gyromagnetron, Astron, betatron, etc., where the equilibrium rotation is supported by an axial magnetic field alone, and the only electric field is due to the beam's own charge ($E_0 \approx 0$).

(ii) $h = 1/\gamma_o^2$ corresponds to the orbitron model, in which the rotation is supported solely by a radial electric field [$B_o = 0$].

(iii) $h >> 1/\gamma_o^2$ corresponds to an *inverted* "smooth-bore" magnetron, with the cathode at the outer conductor and the anode at the inner conductor. The rotation is approximately given by the $E \times B$ drift (centrifugal force is small).

(iv) $h << -1/\gamma_o^2$ corresponds to a conventional "smooth-bore" magnetron, with the cathode at the inner conductor and the anode at the outer conductor. Again, the rotation is approximately given by the $E \times B$ drift and the centrifugal force is small compared with either the electric or Lorentz force in equilibrium.

(v) The planar limit is recovered formally as $r \to \infty$ (fixing E_o, v_o). That is, $|h| \to \infty$ corresponds to the planar limit.

3. DISPERSION RELATIONSHIP

The self-excited modes in the system described in Sec. 2 are governed by a rather complicated second order ordinary differential equation subject to the appropriate boundary conditions at the metallic walls.[3,4] For a thin beam, the growth rates have been obtained analytically to two orders in τ/R, where τ is the beam thickness and R is the mean radius of the beam. The dispersion relation takes into account of the effects of the DC self field of the beam, and has passed various tests, including comparisons with a direct numerical integration of the governing equation.[3,4] Given below is a heuristic derivation of just the leading term (in τ/R) of the dispersion relation, intending to illustrate the dominant physical process and the salient features. A more detailed discussion of various issues is given elsewhere.[5]

As in the density wave theory,[1] we shall first calculate the density response of the beam to some imposed electric field. A dispersion relation is obtained when this electric field is required to be excited by the density perturbation.

Ignoring axial motion and axial variation, one anticipates that the rotational motion of the *thin* beam interacts most strongly with the azimuthal component of the perturbed electric field ($E_{1\theta}$) which the beam experiences. This interaction would be very strong if this field corotates with the beam. Conservation of energy gives

$$e\, v_o\, E_{1\theta} \cong d\,\epsilon/dt \tag{3}$$

where ϵ is the total energy (kinetic and potential) of the electron beam. Upon using the chain rule, we express

$$d\epsilon/dt = (d\dot\theta/dt)/(d\,\dot\theta/d\epsilon) = (\ddot\eta/R)/(d\dot\theta/d\epsilon) \tag{4}$$

in terms of the linear azimuthal displacement η of an electron from its unperturbed position. Thus, (3) becomes, upon linearization,

$$\ddot\eta = e\, R\, v_o\, E_{1\theta}\,(d\omega_o/d\epsilon) = e\, E_{1\theta}/M_{\text{eff}} \tag{5}$$

where the equilibrium value $\dot\theta = \omega_o$ is expressed as a function of the particle energy. In analogy with the force law "$F = ma$", we define in (5) an effective mass $M_{\text{eff}} = (R\, v_o\, d\omega_o/d\epsilon)^{-1}$. It is not difficult to show from (1) and (2) that, with $\beta_o = v_o/c$,

$$M_{\text{eff}} = -m_o\,\gamma_o \left[\frac{1 + \gamma_o^2\, h^2}{\beta_o^2 + 2h}\right]. \tag{6}$$

which can either be positive or negative [Fig. 2].

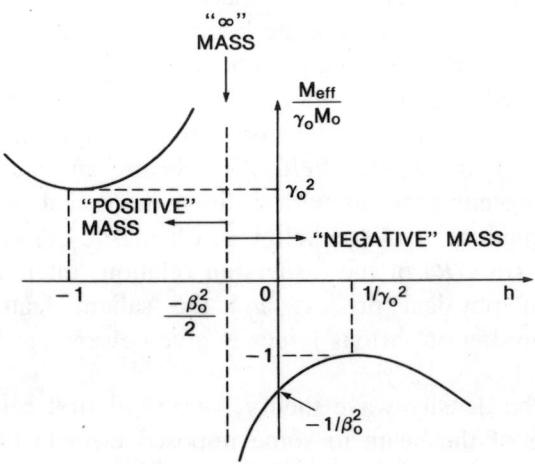

Fig. 2. The normalized effective mass ($M_{\text{eff}}/\gamma_o\, m_o$) as a function of h.

For a perturbation proportional to $\exp[i\omega t - il\theta]$, d/dt stands for $i(\omega - l\omega_o)$ where l is the azimuthal mode number. Associated with the azimuthal displacement η is a surface charge density perturbation σ_1, given by

$$\sigma_1 = -\frac{\sigma_o}{R}\frac{\partial \eta}{\partial \theta} = \frac{-il\sigma_o}{R}\frac{e\,E_{1\theta}}{(\omega - l\omega_o)^2\,M_{\text{eff}}}, \tag{7}$$

in which σ_0 is the unperburted surface charge density and $E_{1\theta}$ is evaluated at the beam radius $r = R$. In writing the last expression, we have used (5). Equation (7) expresses the surface charge density perturbation in the beam in response to some imposed azimuthal electric field. Note that the various equilibrium types and the dynamics enter in M_{eff} (Fig. 2) and that the electromagnetic properties of the structure has thus far not entered into consideration.

To complete the analysis, we need to determine what kind of perturbed electric field would be excited if there is a charge perturbation σ_1 on the beam. Since Maxwell equations are linear the field $E_{1\theta}$ at the beam radius is proportional to σ_1; the proportionality constant is ig:

$$E_{1\theta} = ig\,\sigma_1 \tag{8}$$

where g is proportional to the "impedance" which depends only on ω, l and on the surrounding structure.[6] Inserting (8) into (7), we obtain the dispersion relationship for density waves on a rotating electron beam:

$$(\omega - l\,\omega_o)^2 = -\left[\frac{lg}{R}\right]\left[\frac{|e\,\sigma_o|}{\gamma_o\,m_o}\right]\left[\frac{\beta_o^2 + 2h}{1 + \gamma_o^2\,h^2}\right] \tag{9}$$

This is the dispersion relationship to the lowest order in τ/R, if we represent the surface density $\sigma_o = \rho_o\,\tau$, ρ_o being the volume density of the beam.

4. BEAM STABILITY

The dispersion relation (9) is valid[3,4] for general container geometry and for arbitrary energy γ_o and h, as long as the beam is sufficiently thin. For convenience of discussion of the beam dynamics, we shall assume g to be real and positive, (except if otherwise stated). We may draw the following conclusions regarding the stability of the electron beam:

(a) When $h = 0$, Eq. (9) indicates that an instability exists as a result of the relativistic effect. Physically, the rotation frequency

$|e|B_o/\gamma_o m_o$ is a decreasing function of energy (γ_0). A test electron in front of some charge condensation on the beam is accelerated; its mass increases but its rotational frequency decreases. Thus, the test particle in effect falls back to the condensation and the condensation grows. This "negative mass" instability [c.f. Eqs. (5), (6)] places a limit on the beam current in cyclic accelerators.[7] It turns out to be identical to the cyclotron maser instability, which is responsible for the radiation generated in gyrotrons.

(b) Instability arises as long as $h > -\beta_o^2/2$. It is maximized with respect to h when $h = 1/\gamma_o^2$, as is readily demonstrated from Eq. (9). This case corresponds to the orbitron configuration [Figs. 1,2]. In other words, for a given rotational energy and a given geometry, the instability is most pronounced when the equilibrium is supported by a radial electric field alone as in the orbitron model, regardless of the beam energy (γ_o).

(c) The dispersion relation (9) suggests that the negative mass instability is suppressed if [3,4]

$$h < -\beta_o^2/2. \qquad (10)$$

That is, negative mass instability may be stabilized by a negative radial DC electric field of a suitable magnitude. In terms of an external potential V imposed between the inner conductor at $r = a$ and the outer conductor at $r = b$, the stability condition (10) reads

$$|eV| > (m_o c^2/2) \beta_o^4 \gamma_o^3 \ln(b/a). \qquad (11)$$

Note that this stabilization mechanism is independent of the beam velocity spread, and is insensitive to the beam current or container geometry, or mode number. Although it is impractical to stabilize a high energy electron beam against the negative mass instability by this method due to the γ_o^3 dependence in (11), it becomes attractive, however, if this method is applied to cyclic acceleration of high energy ions ($\sim$ 500 MeV) of intermediate atomic mass (atomic number of order twenty).

(d) If the wall is lossy, g becomes complex and the resistive instabilities would result whether the effective mass is positive or negative. However, even this resistive instability can be stabilized if $h = -\beta_o^2/2$, as is evident in the dispersion relation (9). Physically, when $h = -\beta_o^2/2$, the effective mass of a rotating electron is infinite [cf. Fig. 2]. The beam is very rigid azimuthally and is reluctant to transfer its rotational energy to the resistive wall, which is the physical mechanism for the excitation of the resistive instability.

(e) The negative mass instability should disappear in the planar geometry limit. This intuition is also reflected in the dispersion relation (9). In the planar limit, $h \to \infty$ by (2) and the right hand member of (9) tends to zero. What remains is then the diocotron instability (i.e., Kelvin-Helmholtz instability) which arises from the velocity shear in the equilibrium $E_o \times B_o$ drift. This shear is due to the DC self electric field of the beam and its effect enters only in the higher order term (in τ/R) not displayed in the dispersion relation (9). Thus, the diocotron instability is the residual instability when the curvature effect is absent.[3,4]

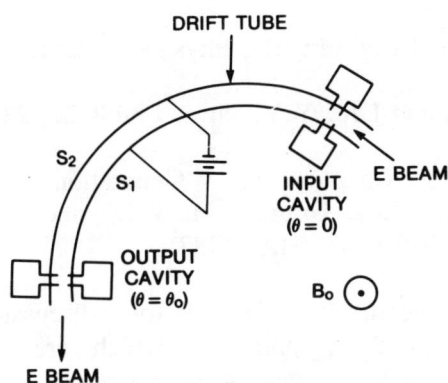

Fig. 3. Schematic drawing of an amplifier configuration to test the response of an electron beam to an input signal.

All of above predictions regarding the dynamical dependence of the equilibrium type could be tested in a controlled experiment such as the one proposed in Fig. 3. The response of the electron beam may be monitored at the output cavity, after an external radio frequency (rf) signal is impressed upon the beam at the input cavity. The orbit of the beam is bent either by a magnetic field B_o or by an electrostatic field, or by both. The polarity and the magnitude of the externally imposed voltage (V) and the external magnetic field B_o may be adjusted to correspond to various values of h [Figs. 1,3]. Such an experiment may be carried out with an electron beam of energy $\leq$ 10 KeV, $B_o <$ 100G, $|E_o| <$ 5 KeV/cm, and beam current $<$ 0.1A.

Note that under suitable conditions, the configuration in Fig. 3 serves as a power amplifier[8]: The input *rf* signal modulates the beam. The space charge density wave grows as the beam propagates along the circular drift tube. Accompanying a propagating density perturbation is a strong *rf* current which excites the output cavity, where the amplified signal is extracted.

REFERENCES

1. Lin, C.C. and Shu, F.H., Astrophys. J. **164**, 646 (1964); also, Proc. Natl. Acad. Sci. U.S.A. **55**, 229 (1966).

2. Hirshfield, J.L. and Wachtel, J.M., Phys. Rev. Lett. **12**, 533 (1964).

3. Lau, Y.Y. and Chernin, D., Phys. Rev. Lett. **52**, 1425 (1984).

4. Chernin, D. and Lau, Y.Y. Phys. Fluids **27**, 2319 (1984).

5. Lau, Y.Y., Chapter in "Generation of High Power Microwaves," Eds. Granatstein, V.L. and Alexeff, I., Artech House Inc., Norwood, MA (1987).

6. See, e.g., Appendix B of Ref. (4) for a discussion of the normalized admittances b_+ and b_-, which are related to g by $\epsilon_o (b_+ + b_-) = 1/g$. Here, ϵ_o is the free space permittivity.

7. Kapetanakos, C. and Sprangle, P., Physics Today, Feb. 1985 issue.

8. Lau, Y.Y., Phys. Rev. Lett. **53**, 395 (1984); Also U.S. Patent 4,617,493 (issued Oct. 14, 1986).

ION BEAM INERTIAL FUSION: SOME ISSUES OF INTEREST FOR APPLIED MATHEMATICS[*]

James W-K. Mark
Lawrence Livermore National Laboratory
Livermore, California 94550

The goal of ion beam inertial fusion is to accelerate and focus 2-15 megajoules of particle beams at 150-1500 terawatts of peak power and to use them to implode and ignite pellets of DT fusion fuel for energy generation. For Heavy Ion Fusion, the requirements include 15-450 kiloamperes of say thallium ions in low charge state at about 10 GeV of kinetic energy; or for Light Ion Fusion, many megamperes of Lithium ions at some tens of MeV kinetic energy. In this paper we point out from this field of research a few topics which, in the author's opinion, could be of interest for applied mathematics, where further development would also be useful. One of these involves numerical simulations of the "statistical-fluids" relevant to the field of highly inhomogeneous non-neutral plasmas (similar to galactic dynamics). Another topic involves issues of geometrical symmetry of axially-placed beam illumination versus spherical implosions of gases in targets; judicious analytical developments could be used to guide simulations of directly driven ion beam targets.

1. INTRODUCTION

One method for obtaining electrical power from fusion energy is to develop high current particle beams. These could be used to illuminate, heat and implode spherical target pellets involving deuterium and tritium (DT) to temperatures over one hundred million degrees and close to a thousand times liquid densities. Under these conditions, the targets produce net energy gain from DT fusion reactions. With the target energy gain over input energy $G \gtrsim 40 - 70$, we can drive fusion power plants providing electricity at economically competitive rates.

There is a close mathematical analogy between some problems of high current particle beams and galactic dynamics (which latter subject is discussed elsewhere in this conference). The governing kinetic equations are most similar for the case of nonneutral plasmas in the main accelerator

---

[*] Work performed under the auspices of the U.S. Department of Energy by the Lawrence Livermore National Laboratory under contract number W-7405-ENG-48.

of Heavy Ion Fusion (HIF). Here the repulsive interparticle space-charge forces require external focusing forces to generate beam "equilibria" or near steady beam transport, whereas for galaxies the self-gravity of stars and gas maintains force balance against their rotation and effective pressure. For particle beams propagating in fusion chamber gases, additional processes occur such as stripping of electrons from beam ions or ionization of background gases due to collisional processes. For example the propagation of self-pinched particle beams of small beam radius is facilitated by the neutralization of the repulsive space charge by electrons but incomplete neutralization of the self-current pinch effect of the highly stripped beam ions. There are on occasion effects of collective density waves, particle resonances, instabilities and hydrodynamic limits as in galactic dynamics. Although physical details differ, similar analytical tools apply. Numerical simulations are being carried out with $2D$ and now also $3D$ particle in cell codes. We briefly outline one such class of nonneutral plasma problems in Section 2, mentioning some helpful additional developments in Section 4.

Another topic relates geometrical symmetry considerations to the gas-dynamical implosion of spherical targets. In Section 3, we briefly describe investigations into several examples of ion beam targets which utilize the energy efficiency of direct drive while at the same time optimizing on the symmetry requirements. Heavy ion beams of charge state $Z \geq 3$ at 5-10 GeV have $\lesssim 15-20$ m bending radii with 3.5 Telsa fields. Beams like them could be used with targets involving direct drive. Control of asymmetries in direct-drive ion beam targets depends on control of the effects of residual target asymmetries after an appropriate illumination scheme has been adopted. As part of this article, we outline results on two ion beam target concepts in which the effects of residual asymmetries are ameliorated. Thirty-two beams are placed according to our axially symmetric Gaussian-quadrature illumination scheme.[1] The targets survive the effects of residual asymmetries in our recent $2D$ hydrodynamic simulations.

Axially symmetric illumination is more convenient for fusion chamber design (see e.g. Ref. 2). Our illumination scheme[1] represents an initial step in the direction of providing adequate spherically symmetric implosions with an axially symmetric illumination scheme. Further developments in this direction would allow the use of the efficiency of direct drive with an illumination geometry most convenient for reactor design.

2. BEAM COMPRESSION: A PROBLEM OF NONNEUTRAL PLASMAS

The ion pulse emerging from a heavy-ion reactor driver will have the energy required to drive fusion targets, but the pulse will require longitudinal compression to give it enough peak power or current.[3] We are studying techniques for this power multiplication process as part of the U.S. induction accelerator development effort for heavy ion fusion (HIF), funded by the Office of Energy Research of the U.S. Department of Energy.

The hardware that initiates compression or current amplification will lie between the main accelerator and final focus at the fusion chamber (see Fig. 1). There are important interactions between pulse compression and the designs of main accelerator, final focus, and targets: it turns out that compression can strongly affect the peak power, the focal spot size, and even the pulse shape demanded by the target.

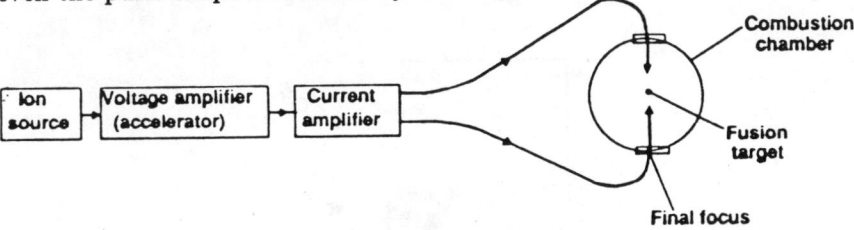

Fig. 1. Conceptual diagram illustrating position of current amplifier or beam compression relative to main accelerator and target in HIF.

The governing equations include the Vlasov equation describing the motion of beam ions as a collisionless fluid in the $6D$ phase space of particle position and momenta, just as the Boltzmann-Jeans equation applies to the statistical description of stars in galactic dynamics. There is also Poisson's equation of the beam self-space charge in addition to equations specifying the periodically occuring transverse (x, y) focusing forces on the beam which is propagating in the longitudinal (or z) direction. These focusing forces are usually provided by external magnetic or electrostatic quadrupoles.

Because of the intrinsically $3D$ geometry of this beam-focusing force system, analytical descriptions require use of circularly averaged focusing forces in the transverse direction or more usually use of moment or envelope equations which describe the rms transverse radius or outer envelope of the beam as a function of longitudinal distance. Numerical simulations are being increasingly used by means of particle in cell codes.

2.1 Beam Compression and Simulations with Particle in Cell Codes

An accelerated beam pulse has finite extent in space and time. Beam compression is initiated by pulsing the accelerator modules so that the tail of the beam moves faster than the head. Especially when the velocity varies linearly with distance (measured from the center of the pulse), this

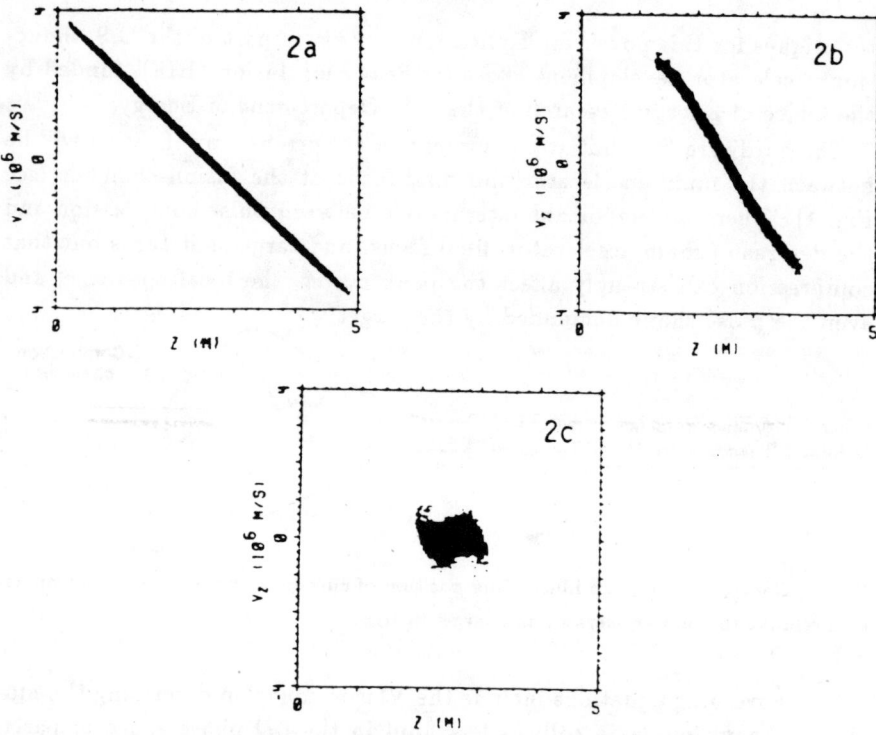

Fig. 2. (z, v_z) phase space picture from a 2.5-D particle-code simulation for $Z = 3$ and $(\delta v_z)_0 = 3 \times 10^4$ m/s: (a) at $t = 0$ (initial velocity tilt), (b) $t = 0.36 \mu s$, (c) $t = 0.72 \mu s$ (final compressed state). Δv_z is the systematic velocity tilt given by the full height of frame (a); δv_z refers to rms values such as those of frame (c).

differential velocity can be viewed as a velocity "tilt" in longitudinal (z, v_z) phase space [see Fig. 2(a)].

In our studies of beam compression we used our 2.5-dimensional particle in cell code CONDOR. Simulated particles are followed in 3-D while fields are evaluated in 2-$D(r, z)$ cylindrical geometry. We assumed typical numbers, for example 18Z-μC beams of lead ions, with $Z = 1, 3$, and 5. Figures 2 and 3 show a typical example. The beamlet initial rms random velocity dispersion $(\delta v_z)_0$ were approximately $(1, 3,$ and $10) \times 10^4$ m/s (values change slightly after particle loading, because of initial internal adjustments in beam parameters in the simulations). We use an electrostatic field solver in a frame defined by the average velocity of the beam, so we ignore relativistic and self-current effects (a good first approximation). If the beamlet ions were at ~ 10 GeV kinetic energy, the $(\delta v_z)_0$ values used correspond roughly to $(\delta p_z/p_z)_0 = (1, 3,$ and $10) \times 10^{-4}$. The beam radius is $a \simeq 3$ cm, and the beam is assumed to be centered in a perfectly conducting cylinder with radius $b = 6$ cm. The boundary conditions at $z \to \pm\infty$ are approximated accurately by doubling the periodicity length of the simulation region in z, because the Green's function falls off exponentially.

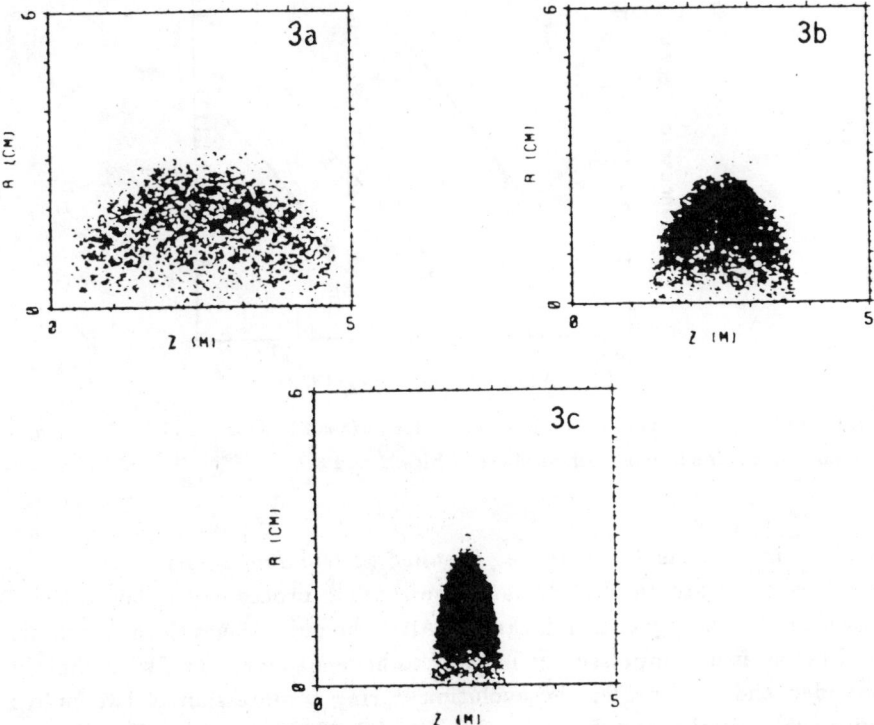

Fig. 3. Positions of a representative sample of the particles of Fig. 2 at corresponding times. Horizontal and vertical length scales differ.

The transverse quadrupole focusing forces on an isolated drifting beam bunch are approximated by an axially symmetric radial electric field with the simple time history

$$E_r = E_0(1 + \alpha t)\frac{r}{b}, \qquad (1)$$

where E_0 and α are chosen for each simulation so that the beam radius approximates half the cylinder radius (i.e., 3 cm) throughout the simulation. Typically 4×10^4 to 6×10^4 particles are used on a 30×512 grid, which represents a region 6cm×5m in the r and z directions, respectively. We made the initial line charge density $\lambda(z, 0)$ nearly parabolic, because the pulse shape then remains relatively invariant despite the high space charge of the beams for $Z = 3$ and 5. We discuss pulse-shape modifications below.

Figure 4 summarizes some of our results for various Z and for $(\delta v_z)_0 = 3 \times 10^4$ m/s, obtained from simulations such as those of Figs. 2 and 3. It shows the achievable compression as functions of Δv_z and Z.

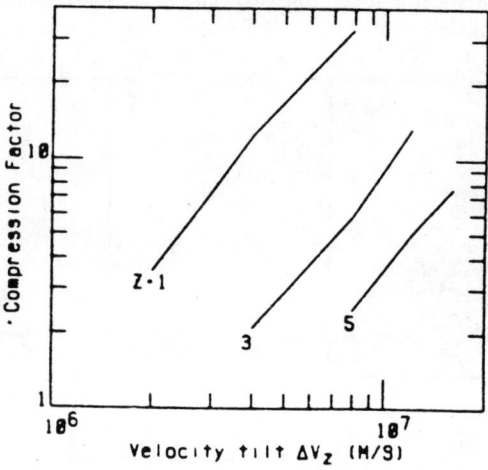

Fig. 4. Pulse compression vs Δv_z (full height) for various Z and for $(\delta v_z)_0 = 3 \times 10^4$ m/s, obtained from simulations such as those of Figs. 2 and 3.

2.2 Obtaining the Pulse Shapes Required by Our Ion Targets

We concentrate on $Z = 3$, but pulse shaping processes similar to those discussed here apply to all relevant Z. After the pulse leaves the accelerator, and before it is compressed, it is likely to have a square $\lambda(z)$, with slightly rounded ends. Consider the evolution during compression of two pulses with such initial shape, but with different initial velocity tilts. Recall from Fig. 4 that the initial tilt (or dv_z/dz) is directly related to the longitudinal steepness of the final pulse shape. This steepness (measured, for example, by $d\lambda/dz$) is a direct consequence of the higher final peak $\lambda(z)$ attained for greater initial dv_z/dz.

Our discussion of final $d\lambda/dz$ vs initial dv_z/dz motivates our pulse-shaping idea, which is to give the beam head a tilt that is less steep than that for the tail. We do this, as shown in Fig. 5(a), by imposing an initial tilt with two different slopes but with continuous v_z. The initially square pulse evolves to the shape shown in Fig. 5(b) by the time $\lambda(z)$ (or the power) has increased by a factor of 7.

The pulse shape of Fig. 5(b) is one that could drive high-gain targets. With the addition of an initial long, low-power prepulse or "front porch" this pulse shape is like that required by the direct-drive target[4] of Fig. 7. (The target sees the pulse reversed: the beam head reaches the target before the beam tail.) The front porch of the pulse of Fig. 7 does not

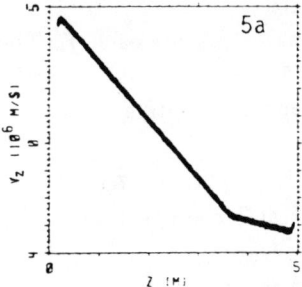
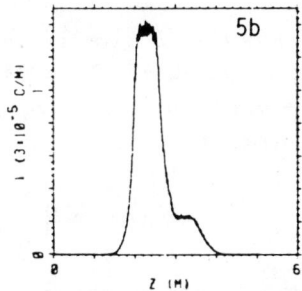

Fig. 5(a) Two tilts that are combined, resulting (b) in a pulse with different ramps in the front and rear. This pulse shape is like that required by the target of Ref. 4

require compression. Its effect could be included by the simple device of creating an initial velocity tilt that approximates three connected straight lines in (z, p_z) phase space; the front porch would require near-zero velocity tilt.

Pulse shapes of great interest in connection with our radiation-driven targets could also be obtained. We can produce[5] different rates of rise to the front porch of the pulse seen by the target. More complicated shapes would require more linear segments in $v_z(z)$. Similar techniques might reproduce two ramped pulses; we might even approach the pulse shapes shown in Ref. 6 if we used multiple pulses of this type.

3. ACHIEVING SPHERICAL IMPLOSION IN ION BEAM TARGETS WHILE DEVIATING FROM SPHERICAL ILLUMINATION

As we mentioned in the introduction, to use the efficiency of direct drive we need an appropriate illumination scheme as well as methods to reduce residual asymmetries. We are developing an axially symmetric illumination scheme[1] in which beams are situated on the rims of cones whose angles are the zeroes of Legendre polynomials. This Gaussian-quadrature scheme achieves symmetry comparable to that achievable with the same number of beams placed uniformly over the surface of a sphere. The scheme improves on uniform beam placement in that it initiates the desirable trend towards allowing axial symmetry and the limiting of the number of azimuthal planes occupied by beamlets

3.1 Gaussian-Quadrature Beam Illumination Scheme

The incident directions of the beamlets lie in several cones about the z axis. The cone angles θ_j are determined by the Legendre-Gauss quadrature points $x_j = \cos\theta_j, j = 1, 2, \ldots, L_0$; the azimuthal angles are given by $\phi_{jk} = 2\pi k/M_j, k = 1, 2, \ldots, M_j$. The constant L_0 gives the number of θ_j

angles, and the M_j give the numbers of angles ϕ_{jk} for each j. The total number of beamlets is $N_b = \Sigma M_j$.

If the nth beamlet is at (ϕ_j, ϕ_{jk}), we assume that its deposition profile in a spherical target is

$$E_n(r,\theta,\phi,t) = \frac{w_j}{M_j} \sum_{l=0}^{\infty} E_l(r,t) P_l(\cos\gamma_n) , \qquad (2)$$

where the angle γ_n is defined in Fig. 6, $P_l(x)$ is a Legendre polynomial, and w_j is the Legendre-Gauss quadrature weight corresponding to the quadrature point x_j. The result of summing E_n over all beamlets (i.e., over j and k) is[1]

$$E(r,\theta,\phi,t) = A_{0,0} E_0(r,t) + \sum_{l=2L_0}^{\infty} A_{l,0} E_l(r,t) P_l(\cos\theta)$$
$$+ 2 \sum_{m=M_0}^{\infty} \sum_{l=m}^{\infty} \mathrm{Re}(A_{lm} e^{im\phi}) E_l(r,t) P_l^m(\cos\theta) , \qquad (3)$$

where the coefficients of the Legendre functions cancel exactly for $m = 0, 1 \le l < 2L_0$ and for $1 \le m < M_0, l \ge m$, where $M_0 = \min(M_j)$. The coefficients $A_{l,m}$ of Eq. (3) are determined by the geometry of the beam placement scheme and are given in Ref. 1.

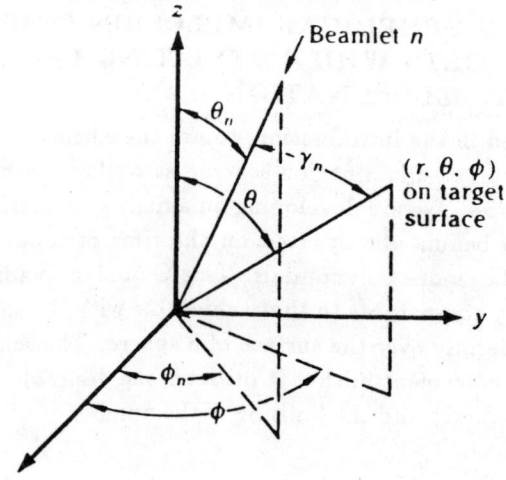

Fig. 6. Geometry used in deriving Eq. (3). The angle γ_n is measured between the axis of beamlet n and an arbitrary point on the target (r, θ, ϕ).

3.2 Target Concepts to Ameliorate Residual Asymmetries.

A number of authors have studied direct-drive ion beam targets for ICF (early examples include Refs. 7 to 10). At least in the initial stages of implosion of most direct-drive targets, ion energy is deposited deep in the target, essentially right on top of the payload mass. There is therefore less smoothing of nonuniformities in deposition in this case than in indirect (radiation) drive (e.g. Ref. 11-12 or see Ref. 13 for references to other early results). For these targets to work, it is necessary to develop illumination schemes with greater intrinsic uniformity than radiation drive.

For a given illumination geometry, deposition uniformity can typically be optimized only in the neighborhood of a single target radius, which we expect to be determined by and close to the radius R_{beam} of the beamlet focal spot. As the target implodes, this optimal target radius for symmetry eventually deviates significantly from the radius R_{dep} of peak ion energy deposition, so that symmetry becomes significantly degraded after some characteristic time t_s for detuning of optimal symmetry. As a consequence, an initially satisfactory symmetry will be degraded during the implosion.

This degradation (for fixed number of beamlets and fixed R_{beam}) occurs because for most target designs, the symmetry detuning time t_s is less than the total duration t_d of the direct-drive pulse required to drive the target to ignition (i.e., $t_s < t_d$). However, residual asymmetry of direct drive can be ameliorated in two ways. Firstly, in Ref. 4 we showed that for ion targets (such as those shown in Fig. 7) with an external high-Z tamper, we can limit the motion of the deposition region to less than $\pm 10\%$ of the initial radius, by varying the ratio of tamper and ablator inertias.[4] This is much less than the typical motion of a factor of 2 in radius that takes place without a tamper. It is one method of ensuring that $t_s > t_d$ because judicious tamping sufficiently lengthens t_s.

The target of Fig. 7 uses this method, it gives in spherically symmetric $1D$ simulations an energy gain of 160 over the required 1.8 MJ of input (beam) energy with focal radius $R_{beam} = 2.6$ mm. The drive pulse of Fig. 7 can be generated in the final parts of the induction heavy-ion accelerator as the pulse is compressed in length to reach peak power. (In Section 2.2, for example, we show that such a pulse shape could be obtained by relatively simple manipulation.)

On the other hand, a second method to reduce residual asymmetry is to omit modifying the existing symmetry detuning time t_s but use sufficiently short bursts of direct drive pulses t_d so that again we satisfy $t_s > t_d$. One example is the hybrid-drive implosion concept, which uses short length t_D of direct drive pulse. It separates the normal drive pulse into two phases, and thus allows for example the combined use of two drivers to provide the pulses at reduced total energy and peak power (but is not limited to this application). We divide the implosion of a high-gain ICF target into two phases (see Fig. 8): (1) At $t = 0$, short-wavelength laser or ion beams compress the ablator to higher density; (2) At some later $t \simeq t_h$, an ion driver delivers energy into the dense ablator to provide the required ignition velocity at low convergence ratio.

With the appropriate ion species and kinetic energy, we can also obtain very good coupling of the ion beam energy to the target ablator and thereby maximize the hydrodynamic efficiency. We can further improve the hydrodynamic efficiency by tamping the implosion (see Ref. 14, also for comparisons with other targets, comments on stability issues, etc.). In hybrid drive, the thin, dense, compressed ablator does not move appreciably during the short heavy-ion pulse. For example, in one of our simulations the radius of peak ion deposition R_{dep} varies from its average value by only 13% during the ion pulse. Our two-dimensional simulations indicate that this motion is small enough that adequate illumination symmetry is maintained.

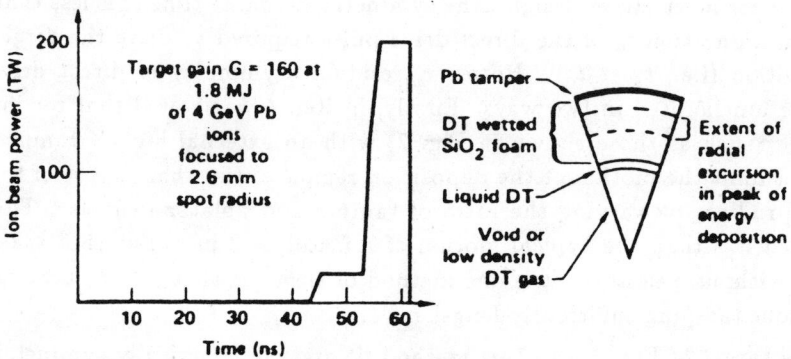

Fig. 7. Direct-drive ion beam target with deposition radius motion controlled for symmetry considerations; and drive pulse used.

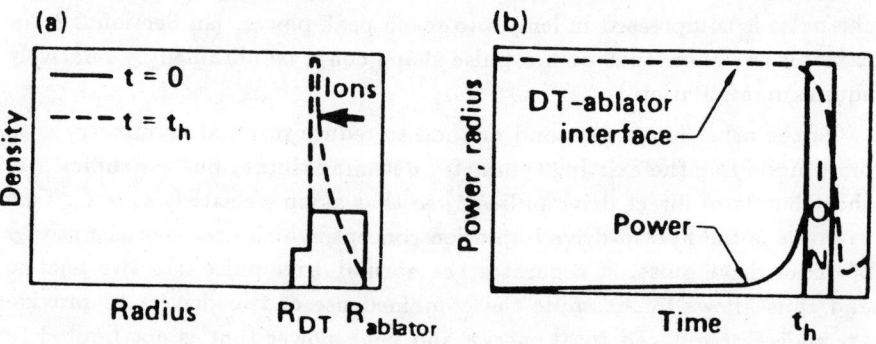

Fig. 8. Two-step hybrid-drive implosion. (a) Density vs radius. Ions deposit energy in compressed ablator. (b) Power and ablator radius vs time.

3.3 Results of a 32-Beam Axially-Symmetric Illumination

With the above illumination scheme and a thin-layer model of ion energy deposition, we estimate that we need 32 beams to attain 1% to 2% average deposition asymmetry, which should be adequate for direct-drive pulses with convergence ratios of 20 to 50. To the extent that our present pair of two-dimensional LASNEX simulations are representative, we have confirmed that we can symmetrically implode and ignite targets under these conditions.

Details similar to Eqs. (2) and (3) suggest that the residual asymmetries in such an illumination scheme have an angular structure related to the Legendre polynomial $P_8(\cos\theta)$, where θ is the angle between the beamlet and the symmetry axis of the illumination scheme. Clearly, even with 50 angular zones and 16 beamlets in a two-dimensional half-sphere simulation, we have only provided 10 to 15 angular zones per full cycle in $P_8(\cos\theta)$. Moreover, to obtain statistics yielding errors of no more than 1 to 10%, we need 10^5 to 10^6 rays with three-dimensional ray tracing, and an estimate of up to an hour of Cray time per time step.

However, Eqs. (2) and (3) indicate that we need only obtain $E_8(r,t)$ from the effective deposition of a single beam. We do this by converting[15] the spherically symmetric one-dimensional problem of Figs. 7-8 into a two-dimensional problem with zones in the r and θ directions. The energy of that beam is then deposited in these zones with the beam axis along the axis of symmetry (Fig. 9).

We now use our new code capability called GOAL to link our target hydrodynamics code LASNEX at suitably chosen time steps to sum the single-beam result into that of a 32-beam array in a full sphere. This equivalent 32-beam result is used as individual zone-by-zone energy sources to a full two-dimensional LASNEX run, as shown in Fig. 9. We thus reduce our problem into one involving only 1 to 2×10^4 ion rays at a select number of time steps, and treating one beam directed onto the symmetry axis, for which two-dimensional ray tracing suffices.[15] Since even this problem requires 1 to 2 min. of Cray time for ray tracing per time step, it is convenient that we can use the ray tracing with one beam to provide energy sources for the second two-dimensional simulation of Fig. 9 at a selected subset of time steps, guided by the one-dimensional problem.

Figure 10 shows density and ion-temperature contours of the fuel zones involved in DT burn for a hybrid-drive target near the time of peak energy production (after the drive ions have been turned off). Sixty-five percent of one-dimensional energy yield is obtained. In a similar manner, for the tamped target of Section 3.1, we obtained 45% of full yield in our two-dimensional hydrodynamic simulation. The DT burn progressed well

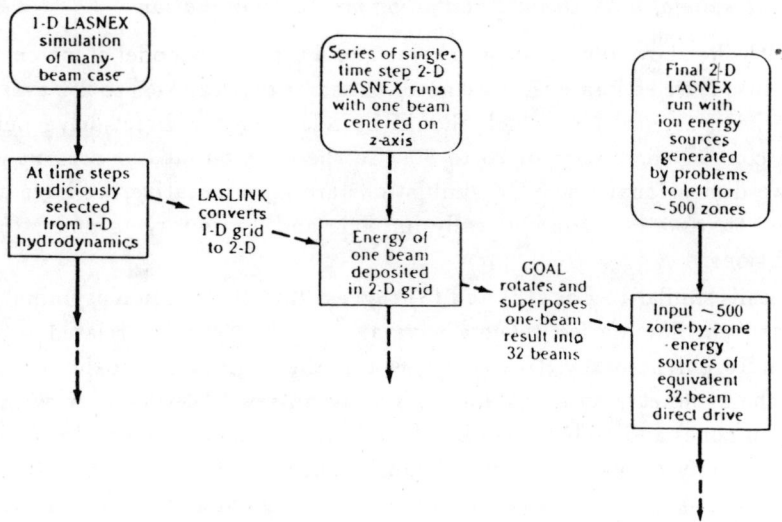

Fig. 9. Scheme for simultaneous running of 1-D and 2-D LASNEX jobs linked by GOAL and LASLINK.

and is relatively uniform. We obtain a reasonably self-consistent computer simulation if the target remains nearly spherical during beam energy deposition. Figure 8 of Ref. 15 indicates the self-consistency of the method for obtaining the energy deposition of 32 beams. The effects of asymmetries of higher-order than P_8 have yet to be accounted for, and we have not evaluated the full effects of random beam displacements from the intended axes. However, we have also not yet included positive effects, such as tuning the target pulse timing in two dimensions, choosing an optimal tamper thickness for two-dimensional effects, and choosing a beam transverse profile to optimize symmetry.

4. SUMMARY AND DISCUSSION

Ion beam targets involving direct drive are efficient but satisfying symmetry requirements have been considered difficult. From our recent work, initial indications are that some ion targets driven directly by only 32 beams might survive the full effects of beam deposition asymmetries.

We can expect at least two times reduction of driver output energy or peak power the same target energy yield. Alternately, we may expect similarly larger energy gain at fixed target energy input. This expectation is based on an analysis that includes judicious choice of beam placement and/or energy. Our beam placement scheme of Ref. 1 represents the desirable direction towards providing spherically converging shocks in direct drive to implode targets, but using an axially symmetric illumination

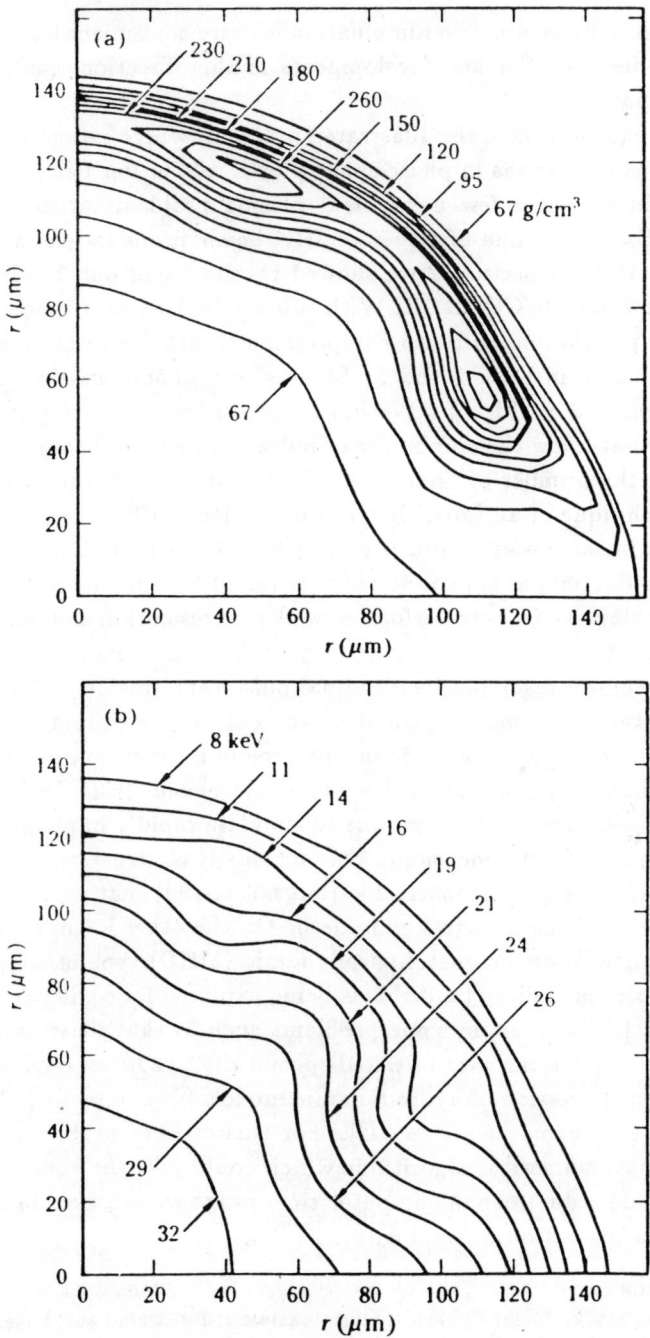

Fig. 10. Fuel region of a hybrid-drive target at 49 ns, near the time of peak DT burn: (a) density; (b) ion temperature.

scheme. Axially symmetric illumination is more convenient for most fusion chamber designs. Further developments in this direction would certainly be appreciated.

In this paper we have also illustrated problems where extensive numerical simulations of plasmas in phase-space play a role in Ion Beam Fusion. We have outlined just a few nonneutral plasma problems arising from HIF alone. (For an outline of issues of HIF beam propagation in gases, see e.g. Ref. 16.) In Section 2 we showed results using our 2.5-dimensional particle in cell code CONDOR. With this method we have made a series of computer simulations of beam compression to achieve peak power similar to those shown in Figs. 2 and 3. Some global characteristics of the runs are summarized in Fig. 4. We have also outlined one of several of the methods that allow the combining of pulse shaping and beam compression to reduce the number of final beam manipulations. This process uses the simple technique of an initial longitudinal velocity tilt in (z, v_z) consisting of several linear pieces as illustrated in Fig. 5. Using this method we can produce pulse shapes similar to those required by a number of targets.

These calculations were performed with grid resolution and particle number determined in part by the geometry of the beam bunch but also largely by the nature of resolving longitudinal pulse compression and shaping. It would certainly be more optimal if we could deviate from the use of a fixed grid to calculate the fields in this type of rapidly compressing Vlasov plasma. This and other related fields would benefit from development of more efficient numerical algorithms to simulate rapidly imploding plasmas in phase-space of 4-6 dimensions. For our mostly electrostatic plasmas with elliptic field solves, it is conceivable (but not tested) that we can save computer time by using a Lagrangian mesh. On the other hand, some specific regimes might allow magnetohydrodynamic (MHD) type of simplification of the physics modelling (and this is being explored for some particle beam problems[17]). However, in some problems such as that illustrated by Section 2, the simulations were initiated specifically to address deviations from a simplified ID version of hydrodynamic models (due to realistic kinetic effects of these beams in $2D$ to $3D$). For this class of problems, we know no immediate numerical algorithms which could give the equivalent order of magnitude reductions in computer time usage we achieved in the target simulations of Section 3 (see Fig. 9).

References
1. J. W-K. Mark, "Near Spherical Illumination of Ion-Beam and Laser Targets" *Phys. Lett.* **114A**, 458 (1986).
2. B. Badger, *et. al.*, "HIBALL-II: An Improved Conceptual Heavy Ion Beam Driven Fusion Reactor Study", Kernforschungszentrum Karlsruhe, West Germany, KfK-3840 (1984); also Fusion Power Associates, Washington, D.C.,

FPA-84-4 (1984), and University of Wisconsin, Madison, Wis., UWFDM-625 (1984).

3. J. Bisognano, E. P. Lee, and J. W-K. Mark, "Numerical Studies of High Current Beam Compression in Heavy Ion Fusion," *IEEE Trans. Nucl. Sci.* NS-32, 2477 (1985).

4. J. W-K. Mark and J. D. Lindl, "Symmetry Issues in a Class of Ion Beam Targets Using Direct Drive Pulses," in *Heavy Ion Inertial Fusion*, AIP Conference Proceedings 152 (American Institute of Physics, New York, N.Y., 1986), p. 441.

5. J. W-K. Mark, D. D-M. Ho, S. T. Brandon *et. al.*, "Studies on Longitudinal Beam Compression in Induction Accelerator Drivers," in *Heavy Ion Inertial Fusion*, AIP Conference Proceedings 152 (American Institute of Physics, New York, NY, 1986), pp. 227.

6. J. W-K. Mark, in *1982 Laser Program Annual Report*, Lawrence Livermore National Laboratory, Livermore, Calif., UCRL-50021-82 (1983), p. 3-16.

7. M. J. Clauser, "Ion-Beam Implosion of Fusion Targets," *Phys. Rev. Lett.* **35**, 848 (1975).

8. J. D. Lindl and R. O. Bangerter, "Low Power Multiple Shell Fusion Targets for Use with Electron and Ion Beams," Lawrence Livermore National Laboratory, Livermore, Calif., UCRL-77042 (1975).

9. M. A. Sweeney and M. J. Clauser, "Low-Z Ablator Targets for Electron Beam Fusion," *App. Phys. Lett.* **27** (9), 483 (1975).

10. R. O. Bangerter and D. J. Meeker, "Ion Beam Inertial Fusion Target Designs," Lawrence Livermore National Laboratory, Livermore, Calif., UCRL-78474 (1976).

11. J. D. Lindl and J. W-K. Mark, "Recent Livermore Estimates on the Energy Gain of Cryogenic Single-Shell Ion Beam Targets," *Laser and Particle Beams*, **3**, 37 (1985).

12. R. O. Bangerter, J. W-K. Mark, and A. R. Thiessen, "Heavy Ion Inertial Fusion: Initial Survey of Target Gain Versus Ion Beam Parameters," *Phys. Lett.*, **88A**, 225 (1982).

13. J. W-K. Mark, "Recent U.S. Target Physics Related Research in Heavy Ion Inertial Fusion," in *Symposium on Accelerator Aspects of Heavy Ion Fusion*, (Gesellschaft fuer Schwerionenforschung, Darmstadt, W. Germany, 1982), p. 454.

14. J. W-K. Mark and Y. L. Pan, "Substantial Reductions of Input Energy and Peak Power Requirements in Targets for Heavy Ion Fusion," in *Heavy Ion Inertial Fusion*, AIP Conf. Proc. 152, (Am. Instit. of Phys., New York, NY, 1986), p. 435; also J. W-K. Mark, in Proc. *1984 Topical Conf. on Phys. of ICF Targets*, Los Alamos Nat'l Lab Report X-1-85-22, LA-CP-85-138 (1985), p. 181.

15. J. W-K. Mark, "Recent Livermore Research on Ion Beam Fusion Targets which: A. Utilize Direct Drive Efficiency while Optimizing Symmetry, B. Utilize Polarized DT-fuel," LLNL Report UCRL-97110 (1987).

16. J. W-K. Mark, "Heavy-Ion Inertial Fusion: Target and Beam Propagation," Lawrence Livermore National Laboratory Report No. UCRL-84821 Abst. and viewgraphs of invited paper, Am. Phys. Soc. Div. Plasma Physics Annual Meeting (1980).

17. J. W-K. Mark and S. S. Yu, "Dynamically Consistent Closure Equations for Hydrodynamic Models of Pinched Beams," *Physics Letters,* **92A**, 179 (1982).

PLASMA PROCESSES RELEVANT TO EXTRATERRESTRIAL PHENOMENA

C. S. Wu

Institute for Physical Science and Technology
University of Maryland, College Park, MD 20742, USA

Recent advances of space plasma physics research in two topic areas are reported. The purpose of this report is twofold; first, to show that research in solar-terrestrial physics is highly interdisciplinary and second, to present a summary of some of the research results attained by us in recent years.

FOREWORD

My research interest shifted abruptly from basic plasma kinetic theory to space plasma physics in the late Sixties. This actually happened after my 1966-67 visit to MIT which was kindly arranged by Professor C. C. Lin who, in those days, motivated me to take an interest in observations and to investigate relevant problems of outstanding scientific significance. In this sense, my visit to MIT was the turning point of my research career. In an effort to express my personal gratitude and to pay a warm tribute to him on this special occasion, I feel it would be appropriate to report and summarize some of the research achievements which I have made in the subject area of theoretical solar-terrestrial physics during the last fifteen years. It is based on this consideration that the present article is written.

1. INTRODUCTION

It is becoming increasingly evident that the physical processes associated with many important and significant phenomena observed in the solar terrestrial environment are peculiar to plasmas. As a result, the consensus established in recent years is: The knowledge of contemporary plasma physics which has been spurred mainly by the research of controlled fusion and extensively developed during the last three decades can serve as a useful tool in the research of extraterrestrial physics and astrophysics. Moreover, the study of observed phenomena in natural plasmas can, in turn, suggest new and important plasma processes which will further enriched our understanding of basic plasma physics.

The objective of the present report is twofold: first, to illustrate that the interplay between plasma physics and space physics is of vital importance in the research of solar-terrestrial physics and, second, to report the findings of my own research program in this interdisciplinary subject area. Of course, it must be stressed that the research achievements mentioned subsequently have been often attributed to joint efforts with many of my colleagues. Scientific collaborations are valuable in all of the works to be described below.

2. RADIO EMISSION PROCESSES FROM EARTH'S AURORAL REGION

Solar system radio astronomy has been advancing very rapidly over the last several decades. The Sun was discovered to be a strong source of radio emissions in the 1940's. Later, in the 1950's, Jupiter's intense decametric radiation was discovered. More recently, the Earth was also found to be a very strong emitter. The phenomenon is usually referred to as the auroral kilometric radiation (AKR) because the characteristic wavelength of the radiation is about 10^3 metres and the source is along the auroral field lines in the polar region of the Earth's magnetosphere. The major observational finding of AKR is presented in an important article by Professor D. A. Gurnett,[1] although there have been numerous subsequent publications by other authors.

AKR has been considered one of the most fascinating phenomenon observed in the terrestrial magnetosphere and the theoretical study of AKR has attracted a great deal of attention over the last ten years for two primary reasons. First, not only is the understanding of the emission mechanism of AKR important to the research of auroral region physics, but it also may lead us to comprehend the other radio emissions processes observed within or beyond the solar system. Second, research of AKR may eventually enable us to better our knowledge of induced radiation processes, a relatively new concept in radiation theory. The general opinion is that the generation mechanism of AKR may have far-reaching implications to the understanding of the radio sources in astrophysical environments, as well as the basic theory of induced radiation processes in general.

The research achievement in this topic area made by my collaborators and me can best be described in a series of articles concerning the generation mechanism of AKR.[2-12] In Ref. 2, we have proposed a very unusual and novel model: That is, mirror-reflected auroral electrons (due to the convergent auroral magnetic field lines and the conservation of the magnetic moment) can possess a loss-cone distribution which can, in turn, lead to an electromagnetic cyclotron instability (also frequently called cyclotron maser instability in the literature in recent years). This instability can amplify RX-, LO-, and Z-modes of radiation and provides an excellent explanation of AKR. Subsequent articles[3-12] were published along the same line, but extended the preliminary analysis presented in Ref. 2 to take into account more elaborate physical models of the AKR source region and a more sophisticated theory of the cyclotron maser instability.

The cyclotron-maser theory of AKR presented by Wu and Lee[2] contains two distinctly interesting features.

1. It is the first time ever that it has been suggested in the literature that mirror-reflected energetic electrons can give rise to a well-formulated idea of a maser "pump".

2. Our theory demonstrated for the first time that the inclusion of both the relativistic effect and the Doppler shift term in the resonance condition leads to a maser instability which amplifies a variety of radiation modes including the RX-mode.

The Wu-Lee theory[2] has been frequently cited in the literature in recent years. The enthusiastic response and growing interest in our work are mainly due to the following reasons. First, the instability discovered by us gives rise to direct amplification of electromagnetic waves (of various modes). Second, the theory is novel and uncomplicated. Third, the instability can occur under fairly general circumstances. Although our discussion has been primarily concerned with the generation mechanism of the auroral kilometric radiation, it is easy for us to envision that similar instabilities may be responsible for many other radio emission processes in astrophysical environments. Recently, it has been suggested in a number of publications that the instability may explain the observed solar microwave bursts, solar continuum radio bursts, Jovian decametric radiation, and radio bursts from flare stars and close binaries. All these discussions have been stimulated by the basic idea discussed in Ref. 2. A comprehensive review of the maser instability is presented in Ref. 11.

3. THEORY OF ION PICKUP PROCESSES IN THE SOLAR WIND

In the solar wind, a variety of neutral atoms and molecules of different origins can occur, for example, neutral atoms and molecules associated with comets and artificially released particles. An issue of outstanding scientific interest is what happens when the neutral species become ionized, say, by solar ultra-violet radiation. The interaction of the newborn ions with the rapidly moving solar wind, which has a typical speed of approximately 400 km/sec, represents one of the important fundamental processes occurring in interplanetary space. In this regard, there are two basic questions:

1. In the absence of collisions, how do the freshly created ions interact with the solar wind?
2. Can the newborn ions be picked up eventually by the solar wind?

It is easily conceivable that the initial distribution function of the newborn ions in the solar wind frame represents a highly nonequilibrium state.[13] When this distribution function has evolved into a nearly spherically symmetric state, we then claim that the newly created ions are assimilated or picked up by the solar wind. Physically, when this state is reached, the newborn ions are practically comoving with the solar wind and "quasi-thermalized".

Because the solar wind plasma is effectively collisionless (i.e., the mean free path is roughly of the order of one astronomical unit (1 AU = the distance between the Sun and the Earth), it is controversial as to whether the solar wind is able to pickup newborn ions and, if pickup is possible, what the typical time scale for such a process would be.

Historically, the effect on the solar wind of helium ions of interstellar origin created in interplanetary space was one of the earliest ion pickup problems. In 1973, it was first pointed out by us[13] that the initial distribution function of the newborn ions is, in general, unstable and can excite hydromagnetic waves and whistlers. The discussion was later elaborated in two subsequent papers.[14,15] Based on these discussions, we proposed a hypothesis that the excitation of waves can consume part of the available free energy and the ensuing turbulence can further result in "thermalization" due to wave-particle scattering. It was emphasized that such collective processes can eventually lead to the picked-up state.[16,17]

For a variety of reasons, the research was discontinued after 1974. However, in 1984, the study was revived due to the joint U. S.-West German "AMPTE" (Active Magnetospheric Particle Tracer Explorers) program. Through a series of computer simulations we have successfully demonstrated that the hypothesis proposed in 1973-74 is indeed correct.[18,19] In 1985, motivated by the forthcoming fly-by of the International Cometary Explorer (ICE) spacecraft near comets Giacobini-Zinner and Halley in 1986, we investigated the effects of newborn heavy ions on the cometary bowshocks. The results of our investigation are reported in Ref. 20.

Among the exciting results acquired with the ICE spacecraft, it has been evidenced that the hydromagnetic waves excited by newborn ions, as predicted in Refs. 13 and 15, are indeed in existence.[21] It was also found that rapid ion pickup processes occur in regions with high levels of turbulence. This finding further stimulated us to conduct a series of studies which aim to understand nonlinear wave-particle interactions. Preliminary reslts are very consistent with ICE observations and are presented recently in Wu et al.[22]

To summarize, our achievements are:

1. It was predicted in 1973-74 by us that newborn can excite hydromagnetic waves.[13,15] The prediction was recently confirmed with the ICE observation near comet Giacobini-Zinner.

2. A hypothesis for an ion pickup process was suggested in 1973-74. Lately, we are able to demonstrate by means of computer simulations that the hypothesis is indeed correct.

4. COMMENTS AND CONCLUSIONS

Because of the restricted length of the article, the presentation is limited to the discussion of research activities in two topic areas only. However, from these efforts and the publications cited, it is shown that knowledge of plasma physics is indeed very important to the research of solar- terrestrial physics. Currently, we are also doing active research in another fascinating topic area, the study of the physics of collisionless shock waves. This is a long-standing research program at the University of Maryland. The results of our investigations are beyond the scope of this report.

In conclusion, I want to once again reiterate that Professor C. C. Lin has influenced my research interests. The numerous stimulating conversations which I not only enjoyed but also benefited from immensely during my visit to MIT in the Fall and Winter of 1966-67 have an intangible and profound effect on my basic attitude toward scientific research. In short, I am very grateful to Professor C. C. Lin.

ACKNOWLEDGEMENTS

The research work mentioned in the preceding presentation has been supported by the National Aeronautics and Space Administration and the Office of Naval Research.

References

1. Gurnett, D. A., *J. Geophys. Res.* **79**, 4227 (1974).
2. Wu, C. S. and Lee L. C., *Astrophys. J.* **230**, 621 (1979).
3. Lee, L. C.; Kan, J. R. and Wu, C. S., *Planet & Space Sci.* **28**, 703 (1980).
4. Wu, C. S., in *Physics of Auroral Arc Formation*, S. Akasofu and J. R. Kan (eds.), *Amer. Geophys. U.* (1981), pp. 418-427.
5. Wu, C. S.; Tsai, S. T.; Xu, M. J. and Shen, J. Wu, *Astrophys. J* **248**, 384 (1981).
6. Wu, C. S.; Wong, H. K.; Gorney, D. J. and Lee, L. C., *J. Geophys. Res.* **87**, 4476 (1982).

7. Wagner, J. S.; Lee, L. C.; Wu, C. S. and Tajima, T., *Geophys. Res. Lett.* **10** 483 (1983).
8. Wagner, J. S.; Lee, L. C.; Wu, C. S. and Tajima, T., *Radio Sci.* **19**, 509 (1984).
9. Wu, C. S. and Qiu, X. M., *J. Geophys. Res.* **88**, 10072 (1983).
10. Omidi, N.; Wu, C. S. and Gurnett, D. A., *J. Geophys. Res.* **89**, 883 (1984).
11. Wu, C. S., *Space Sci. Rev.* **41**, 215 (1985).
12. Omidi, N. and Wu, C. S., *J. Geophys. Res.* **90**, 6641 (1985).
13. Wu, C. S. and Davidson, R. C., *J. Geophys. Res.* **77**, 5399 (1972).
14. Hartle, R. E. and Wu, C. S., *J. Geophys. Res.* **78**, 5802 (1973).
15. Wu, C. S. and Hartle, R. E., *J. Geophys. Res.* **79**, 283 (1974).
16. Wu, C. S.; Hartle, R. E. and Ogilvie, K., *J. Geophys. Res.* **78**, 306 (1973).
17. Hartle, R. E.; Ogilvie, K. and Wu, C. S., *Planet & Space Sci.* **21**, 2181 (1973).
18. Winske, D.; Wu, C. S.; Li, Y.-Y. and Zhou, G. C., *J. Geophys. Res.* **89**, 7327 (1984).
19. Winske, D.; Wu, C. S.; Li, Y.-Y.; Mou, Z. Z. and Guo, S. Y., *J. Geophys. Res.* **90**, 2713 (1985).
20. Omidi, N., Winske, D. and Wu, C. S., *Icarus* **66**, 165 (1986).
21. Tsurutani, B. T. and Smith, E. J., *Geophys. Res. Lett.* **13**, 259 and 263 (1986).
22. Wu, C. S., Winske, D. and Gaffey, J. D. Jr., *Geophys. Res. Lett.* **13**, 865 (1986).

V. GALACTIC ASTROPHYSICS

OBSERVATIONS OF SPIRAL STRUCTURE IN GALAXIES

Ronald J. Allen
Astronomy Department
University of Illinois

Summary

Some of the most convincing evidence for the existence of density waves in galactic disks was provided in the early 1970's by radio astronomers observing the nearby galaxies. At that time, a new generation of radio telescopes had come into operation, exploiting the principles of aperture synthesis, and providing sufficient angular resolution to permit the study of individual spiral arms. C. C. Lin's influence on the development of spectroscopy at the Westerbork Synthesis Radio Telescope in The Netherlands is illustrated with several anecdotes from the author's experience.

In the intervening years, the capabilities of the instruments have steadily improved. As an example of the current state-of-the-art, a summary of several new results on the detailed structure of the spiral arms of M51 and M83 is presented. These data include imaging spectroscopy of atomic and molecular gas, and maps of the thermal and nonthermal radio continuum emission. The main general conclusion of the present work is that the complex nature of the interstellar medium is now becoming evident in the observations of nearby spiral galaxies. The results provide new insights into our understanding of the sequence of events surrounding the star formation process.

1. INTRODUCTION

The origin of spiral structure in galaxies has fascinated astronomers since the Third Earl of Rosse first discovered the spiral nebulae in 1845 using his newly-completed six-foot reflector. By 1850, 14 spiral nebulae were known.[1] Figure 1a shows a remarkably accurate drawing made by Lord Rosse in 1848 from visual observations of one of the first spiral nebulae to be identified. This object is now known variously as M51, NGC5194, or the Whirlpool Nebula.

Since Lord Rosse's discovery, the major advances in the observational study of spiral structure have been closely associated with major advances in astronomical instrumentation. The application of photography to Astronomy towards the end of the nineteenth century provided the ability to

Figure 1a) upper panel. Drawing of M51 made by Lord Rosse from visual observations with his 6-foot reflector in 1848.[1]

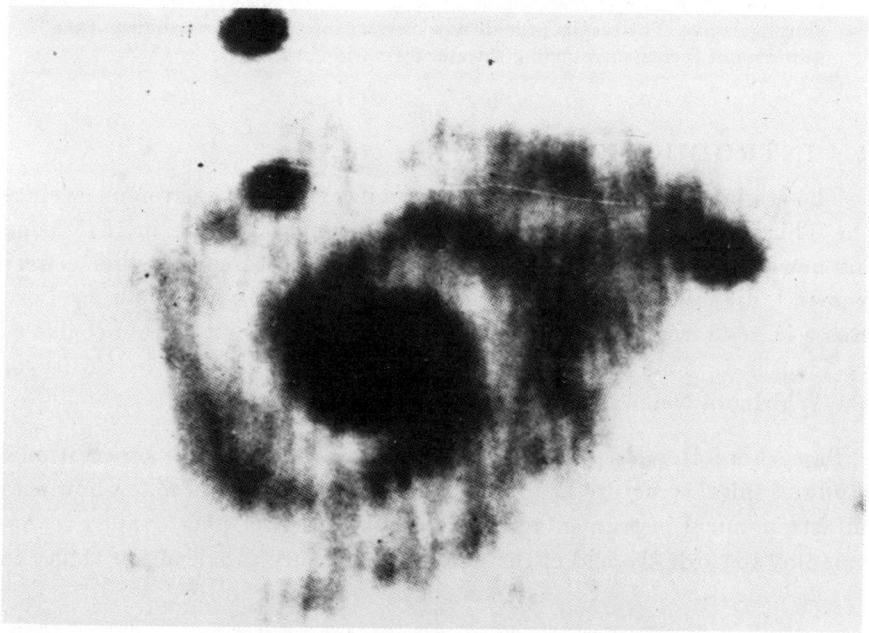

Figure 1b) lower panel. Visual representation of the radio emission from M51 observed with the 4750-foot Westerbork Synthesis Radio Telescope in 1970.[6]

make long time exposures, and Keeler's photographs in 1900 revealed that the spiral forms were common. From about 1912, Slipher used a spectrograph to observe the absorption lines in spiral nebula, revealing both their radial motions and internal rotation as well as confirming their nature as aggregates of stars. Hubble established the distances to the spiral nebulae with the observation of a cepheid variable star in M31 in 1923, and by the end of the 1920's it had become accepted that the spiral nebulae were distant, flattened systems, probably much like our Galaxy, each containing of order 10^{11} stars, and rotating differentially.

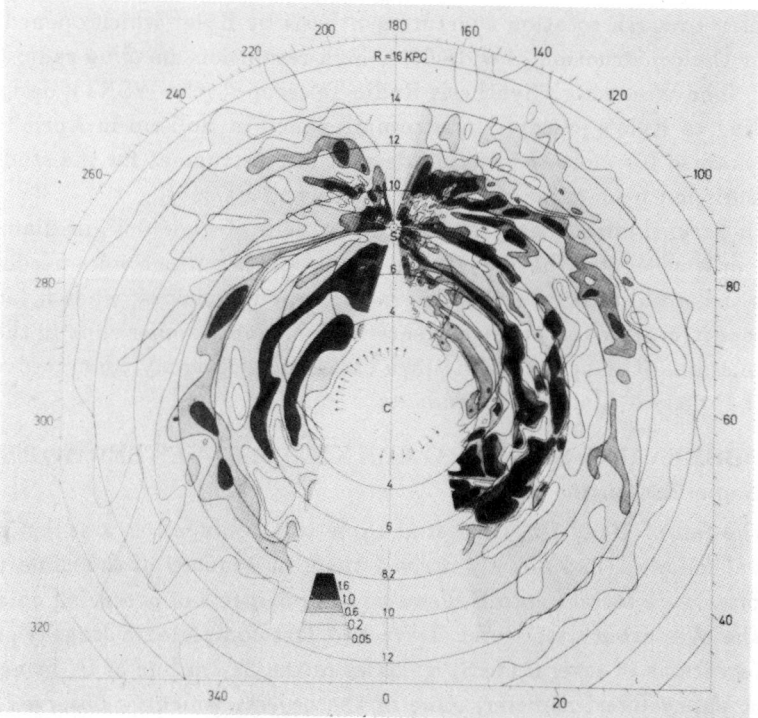

Figure 2. Face-on view of our Galaxy in neutral atomic hydrogen (HI), constructed from radio observations made in the late 1950's in Holland and Australia.[25] This picture is from Burton, who has summarized the reasons why it is probably inaccurate.[26]

Another major advance came with the advent of radio astronomy after the second world war, and in particular with the discovery in 1951 by Ewen and Purcell of the 21-cm line emission from interstellar neutral atomic hydrogen. Detailed and exhaustive studies of the Galaxy in HI using modest radio telescopes in Holland and in Australia provided us with the first view of our own Galaxy as a spiral nebula in 1958 (Figure 2). However, our unique location inside the Galaxy made it impossible to independently establish both the distances and the velocities of the gas clouds. This limitation prevented the use of these observations for a definitive test of the Density Wave Theory when the theory began to take form in the early 1960's. It was clear that observations of galaxies external to our own would provide a better test, but the angular resolution required exceeded by an order of magnitude that which was available with single-dish radio telescopes.

The next major advance in instrumentation had to come in angular resolution. Although interferometry had been applied as early as 1946 in order to obtain high resolution in radio astronomy, it was the extension of interferometry to earth rotation aperture synthesis by Ryle[2] which opened the way for the construction of wide-field, high-resolution, imaging radio telescopes. The Westerbork Synthesis Radio Telescope[3] (the WSRT), designed according to Ryle's precepts, was commissioned in Holland in April 1970, and remained for more than a decade a prime instrument for the study of radio emission from galaxies (see e.g. Allen and Ekers[4]).

With a synthesized beam of 25″ over a field of view 0.5° in diameter at 21-cm wavelength, the WSRT had the properties which were needed to attack the spiral structure problem in the nearby galaxies; what it lacked was a spectrometer to permit measurements of the Doppler shift in the HI line emission. It is here that my own career in astronomy connected with that of Professor Lin and his students.

2. OBSERVING SPIRAL GALAXIES AT WESTERBORK
2.1 Some Tantalizing Results

In the fall of 1969, Dave Rogstad came to Groningen as a senior postdoctoral fellow. I had already arrived there in January of the same year; both of us were attracted to Holland by the prospects of observing galaxies with the Westerbork telescope. Dave had just completed a lengthy series of observations of several nearly galaxies in the 21-cm line of HI using the Owens Valley interferometer. One of the objects which he observed was the giant spiral galaxy M101. Although the 4′ resolution was marginal and the HI distribution (Figure 3a) did not reveal any spiral structure, Dave's reconstruction of the velocity field (Figure 3b) showed large-scale "wiggles" which he suggested could be consistent with density-wave streaming[5]. In this way, Dave introduced me to the theory of spiral structure which C. C. Lin had developed with his student, Frank Shu. It was immediately clear that the Westerbork telescope could provide a definitive test for the existence of density waves in galaxies if we could only get it equipped with a spectrometer. The initial instrumentation at Westerbork was for broadband continuum observations, and the development of a spectrometer was

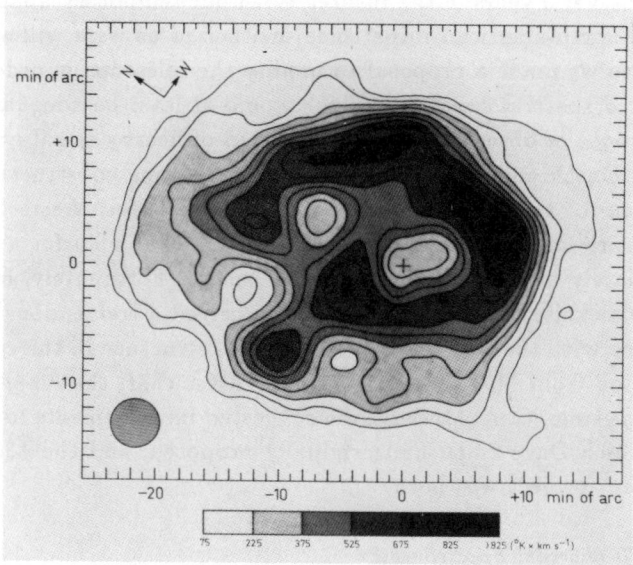

Figure 3. The first radio synthesis observations of the HI in M101, made with the Owens Valley Radio Interferometer.[5] The angular resolution is about $4'$: a) Distribution of column density.

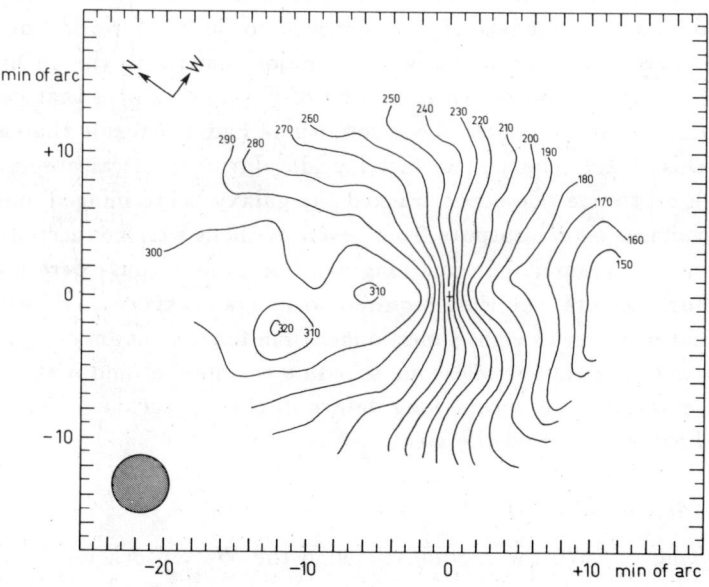

b) Distribution of radial velocity.

planned for several years in the future. With our temporary appointments and typical north-american impatience, neither of us were willing to wait that long. So we made a proposal to modify the telescope in order to provide a limited spectral capability which would at least be adequate for the specific purpose of observing a small selection of nearby spiral galaxies.

Now our Dutch colleagues had spent a lot of time constructing a methodical plan for the development of instrumentation at Westerbork, and it is not surprising that our request to sidetrack that plan for a time was not immediately met with boundless enthusiasm. Fortunately, our efforts to get people excited about the prospects received a tremendous boost by the discovery with the WSRT in 1970 of spiral structure in the continuum radio emission from M51[6] (see Figure 1b). After that, the director of the WSRT at the time, C. A. Muller, even suggested improvements to the modifications which Dave and I had originally proposed, and the "80-channel spectrometer" project was born.

2.2 The 80-channel Spectrometer

The modifications to the Westerbork telescope which C. C. Lin had unknowingly motivated were quite substantial, if not in quantity then certainly in quality[7]. The original arrangement for continuum observations is sketched in Figure 4a. The extent of the modifications was limited by the condition that the 80-interferometer output of the instrument could not be increased to more channels, since there was no budget for additional correlator hardware and no manpower for major changes to the on-line and off-line computer software. Examination of Figure 4b shows that carrying out the modifications under these conditions had the result that a large part of this costly telescope in fact lay idle during spectral observations. Although all twelve telescopes tracked the galaxy being imaged, only half of the available signal outputs from seven of them were connected to the correlator system, and the remaining five telescope outputs were not used at all. Furthermore, the time required to image a galaxy in HI with this instrument was four to eight times longer than for continuum observations, so a major change in the telescope schedule was needed and many continuum observers had to accept long delays in their programs. I sometimes wonder how we ever sold the idea!

2.3 A Pilgrimage

By spring of 1971 the modifications to the Westerbork telescope were well under way, and we began to address the question of which galaxies to choose for our first observing list. This was quite important since it was going to take from two to four weeks of observing time to image any

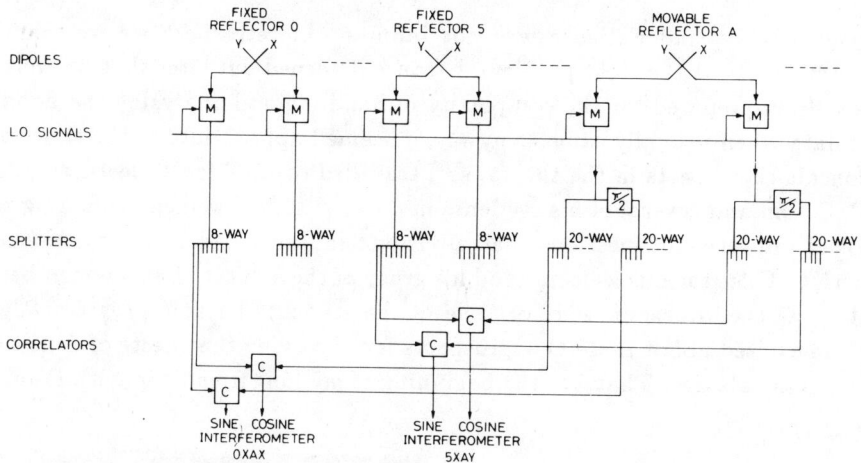

Figure 4. The Westerbork receivers before and after the modifications necessary to enable spectroscopy of HI.[7]:

a) Main components of the Westerbork continuum polarization receiver. Only two of the 80 pairs of correlators are shown, along with two of the ten fixed reflectors and one of the two (now four) movable ones.

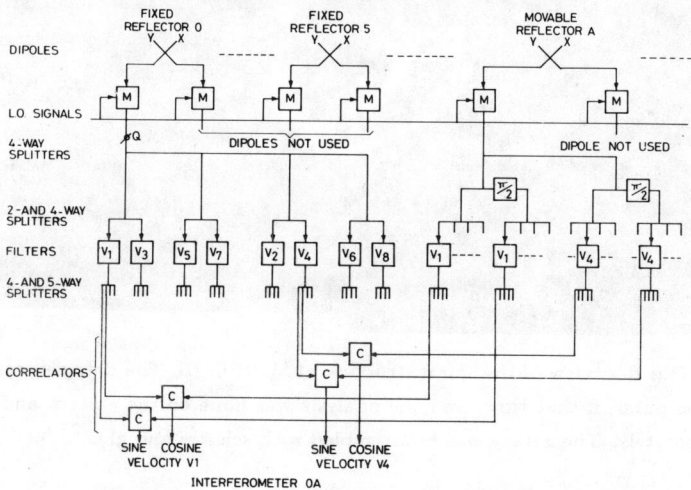

b) The modifications which turn the 10×2-interferometer, 4-polarization continuum system into a 5×2-interferometer, 8-velocity channel spectrometer. The 8- and 20-way splitters have been replaced by combined splitter-filter units and the back-ends of the fixed telescopes are hooked up in pairs.

given galaxy, and patience was not infinite. In retrospect we were not worrying about the right problem, because it turned out later that we were totally unprepared for the complexity of handling and analyzing the flood of data when it finally came. Anyway, it seemed appropriate at the time to consult the experts in the theory, so I travelled to MIT for a meeting with C. C. Lin and several of his students in June of 1971. I remember asking a lot of impertinent questions that afternoon in June, and being impressed with C. C.'s gracious welcome and his grasp of the subject. I also remember that, as the afternoon wore on, one of the students became progressively more excited about these new prospects for observing the effects of density waves in galaxies. That was the beginning of my long friendship with Frank Shu.

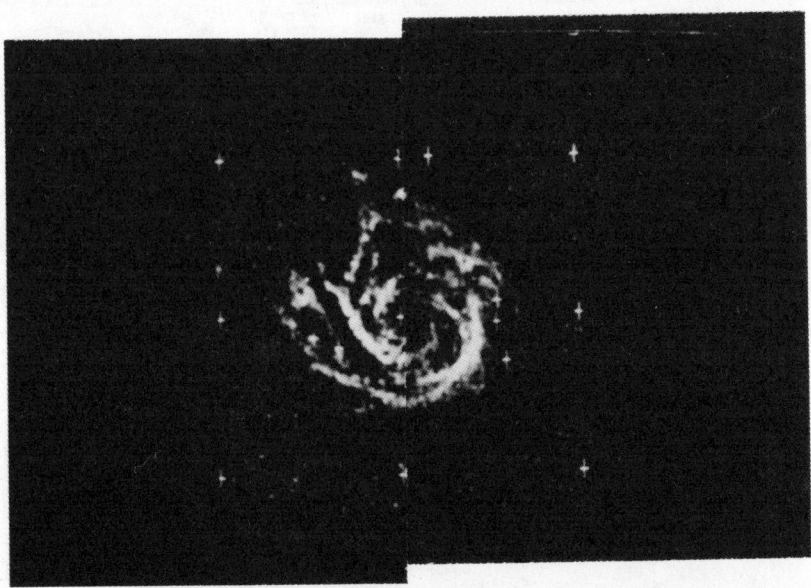

Figure 5. Our first view of the spiral structure of M101 in HI. The data set was too big for the computer of that time, and the analysis was done on the eastern and western halves separately. The galaxy was re-assembled with scissors and glue.[27].

Now I wanted to be able to report in detail on the discussion with C. C. and his students, so I took the trouble of recording the meeting on tape. As I was listening to that old tape recently, I was struck by the foresight which C. C. showed in giving me answers to my questions. First: *Which galaxy should we observe?* My own thoughts were focussed on M101, since that was the original motivation for the whole project. "You should certainly observe M81", said C. C., and history proved him right. We observed M101 too (Figure 5), but at least as far as density waves were concerned this galaxy is probably too face-on for a measurement of the streaming velocities. The second question was: *How much angular resolution is needed?* The answer to this question could make a factor of two difference in the total observing time. Both Frank Shu and I were happy with two beamwidths per spiral arm spacing. C. C. wanted more, maybe even five, and again he was right. The HI images of M81 at 50″ resolution were certainly suggestive of spiral density waves[8], but it was the 25″ images obtained by Rots and Shane[9] and analyzed by Visser[10,11] which provided the unequivocal proof. The final question: *What constitutes a definitive test of the theory?* "The velocities", C. C. answered without hesitation, in particular the radial streaming velocities. And indeed, it was the large-scale wiggles in the M81 velocity field at 25″ resolution, especially around the minor axis, which provided the clearest signature of density-wave-driven streaming in the interstellar gas (Figure 6).

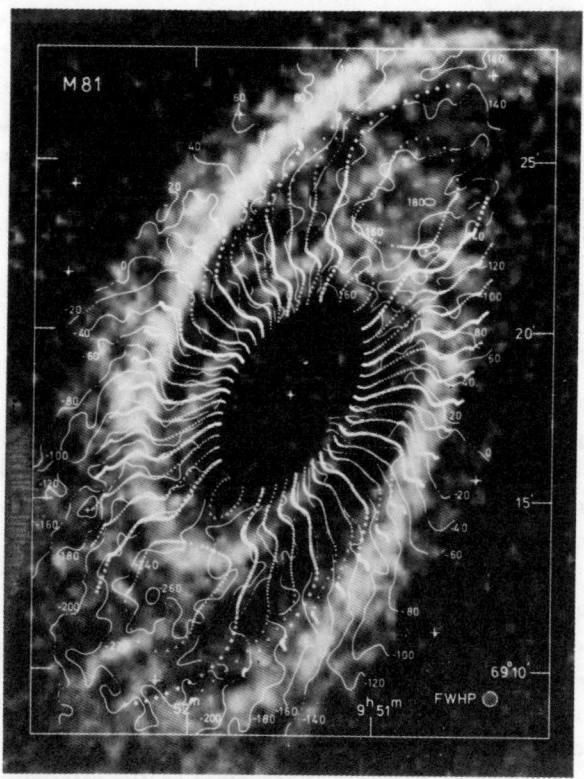

Figure 6. Smoothed radial velocity field of the final density-wave model for M81 (symbols) together with the observed velocity field (full and dashed lines).[11] The angular resolution (FWHP) is 25″. The underlying gray-scale image shows the observed surface density distribution of atomic hydrogen, which is also consistent with the flow velocities.

2.4 Further Developments

During the latter part of the 1970's the HI spectrometers at Westerbork were continually improved, and a stream of papers and Ph.D. theses emerged on a range of subjects such as barred galaxies, interacting galaxies, Z-distributions of HI, galaxy rotation curves and mass distributions, oval distortions, and dark matter. But not much more was done on the study of spiral structure; in fact we had exhausted the angular resolution and sensitivity capabilities of the instrument with the early observations of the nearby large spirals. Following the pattern of discovery started by Lord Rosse more than a century ago, a further advance required further improvements in the instrumentation. By 1981 the size of the Westerbock telescope had been doubled and cooled amplifiers added, and by 1982 the Very Large Array had also been configured for spectral line observations. With these improvements in resolution and sensitivity, a new look at spiral galaxies could be taken. However, spectral imaging data is very voluminous, and the instruments rarely operate perfectly. These features of the observations make the data reduction arduous, and in addition one often has to experiment with various novel methods of analysis. For these reasons it is only recently that results are becoming available. For the remainder of this paper, I will review some of these results on the structure of spiral arms which have been obtained by myself and my collaborators.

3. RECENT RESULTS ON THE STRUCTURE OF SPIRAL ARMS

3.1 Model Shapes for Gaseous Spiral Arms

In order to provide a framework for the interpretation of the new observations to be described below, it is useful to have a picture in mind of the expected cross-sectional shape of a spiral arm as predicted by the early models of gas flow in spiral galaxies. This subject was, of course, first elucidated by another of C. C.'s students, Bill Roberts. Figures 7 and 8 are taken from his seminal paper[12]. Figure 7 shows the density response of a single-cold-fluid model interstellar gas (presumed to be HI in that paper) to an imposed spiral potential perturbation, as a function of spiral phase (which can be roughly taken as either a radial or an azimuthal coordinate). In this picture, the gas density decreases slightly as the gas picks up speed by falling into the spiral potential well from the inside. The flow becomes supersonic and the gas develops a shock close to the minimum of the potential perturbation. The gas density increases sharply there, and star formation is presumably initiated as a consequence of this compression.

The subsequent temporal sequence of events is then stretched out spatially downstream from the shock as the gas flow continues. Figure 8 shows

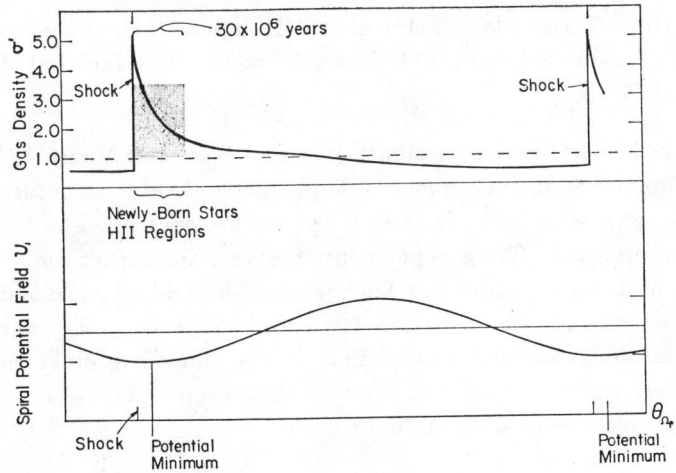

Figure 7. Density distribution of the total interstellar gas along a streamline in the model by Roberts.[12]

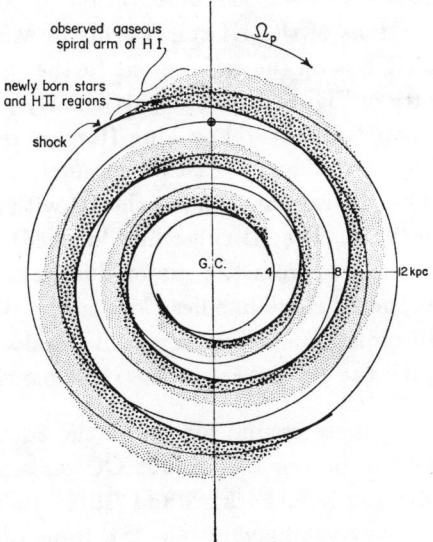

Figure 8. The observable spiral pattern resulting from the density-wave-driven shock in the interstellar gas.[12]

the resulting observable picture in a spiral galaxy. The position of the shock (and thus the maximum in the interstellar gas density) is presumably indicated by long, narrow dust lanes. Star formation is enhanced in a broad band downstream from the dust lanes. The most massive stars will produce giant HII regions after a collapse time of several to ten million years from the time of triggering. These giant HII regions then appear as "beads on a string" along a locus roughly parallel to the dust lane but displaced downstream from it.

Recently Roberts and Hausman[13] (see also Bash and Visser[14]) have refined the model to include a more complicated "cloudy" structure for the interstellar gas, with ballistic as well as pressure-controlled motion of the various components. These refinements are very important for a detailed comparison of the structure and kinematics of individual gas clouds in the arms. However, on the scale of a few hundred parsecs, the main result is to smear things out a bit more than in the single-cold-fluid model. It was therefore quite a surprise to discover that, even on those coarse length scales, real spiral arms look quite different.

3.2 Evidence for the Large-scale Dissociation of Molecular gas in Spiral Arms

The first unexpected result was the observation by Allen, Atherton and Tilanus that large sections of the HI spiral arms in M83 are not situated on the dust lanes at all, but on the HII regions further downstream[15]. The observations are shown in Figure 9 for one particularly prominent example; this piece of spiral arm is about 7 kpc long (for an assumed distance of 9 Mpc to M83) and located about 2' east of the nucleus close to the minor axis of the galaxy. The central panel shows the HI with a resolution of 10" as observed with the VLA. The HI ridge line (dashed) is displaced about 700 pc further out in radius from the major dust lane (thick line). The left panel shows that the HI ridge line lies close to the string of HII regions revealed by their Hβ emission. However, the HI peaks on that ridge line are not coincident with the Hβ peaks, as the right panel shows.

The interpretation of these results hinges on the additional observation that, as indicated by the presence of bright CO emission, M83 is apparently rich in molecular gas (e.g. Rickard and Blitz[16]) although the details of its distribution are not yet known. On the basis of this fact and the unexpected morphology described above, we have suggested[15] that *a substantial fraction of the HI in the spiral arms is a dissociation product of the star formation process*, and not a precursor to it. In this scenario, the gas which falls into the spiral shock (arrow # 1 in Figure 9a) is predominantly molecular. The star formation process is enhanced in the shock, although the gas remains in its mostly-molecular state. Some time further downstream (along arrow #2), HII regions form; the HI then appears near the HII regions as a dissociation product. As the HII regions die out and the gas moves further into the cooler interarm areas, the HI recombines to H$_2$

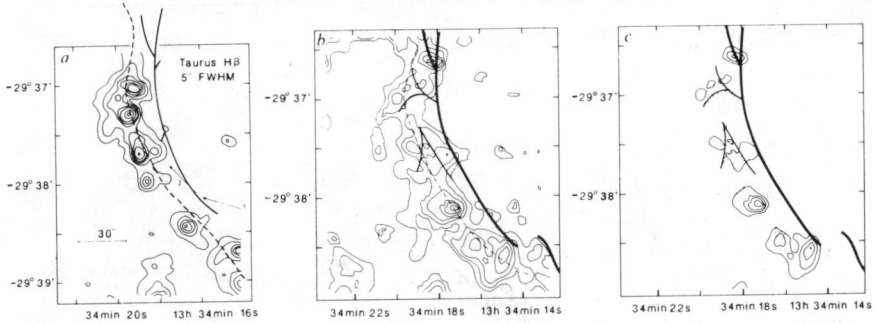

Figure 9. A portion of spiral arm in M83 showing:[15]
a) left panel: Contours of the Hβ emission at 5" resolution. The solid line is the approximate locus of a prominent dust lane. The dashed line marks the ridge of the HI emission at 25" resolution.
b) middle panel: Contours of the HI emission at 10" resolution. The dashed curve is the 25" ridge line, as in the left panel.
c) right panel: The peaks of Hβ (dashed) and the HI (solid) from the other two panels, showing a general anti-coincidence.

on the dust grains and the cycle is ready to be repeated in the following spiral arm.

There is much to be questioned about this interpretation, and the data are still being analysed in detail in order to further test the picture. In particular, the velocity information both for the HI and the Hβ is available, and will tell us whether the gas is really streaming in the way we have assumed. There are many other parts of M83 where the displacement of dust lanes and HI arms does not appear so clearly; we would expect that in those arms the streaming velocities are too small to spread the components far enough apart to be resolved with our 10" beam. If the spiral shock region is as rich in molecules as the above scenario would have it, then observations with sufficient resolution in molecular lines such as CO and other species would provide a wealth of information on the temperature and density structure of the interstellar gas in spiral arms.

A new HI synthesis observation of M51 with the extended WSRT shows features very similar to those seen in the VLA HI image of M83 described above. Figure 10 is a preliminary HI image of M51 which we have recently obtained[17] with a resolution of 12"×18". The major dust lanes are indicated by thick lines. The spiral arm which starts from the southern side of the nucleus and winds out towards the northeast shows a systematic displacement between the HI ridge and the dust lane which persists over nearly 180° in azimuth. The displacement is about 10" or 450 pc in the radial direction, somewhat less but of the same order as in M83. Molecules are also abundant in M51, and very recently we have produced[18] a synthesis image in the CO(1-0) line with the Hat Creek millimeter interferometer

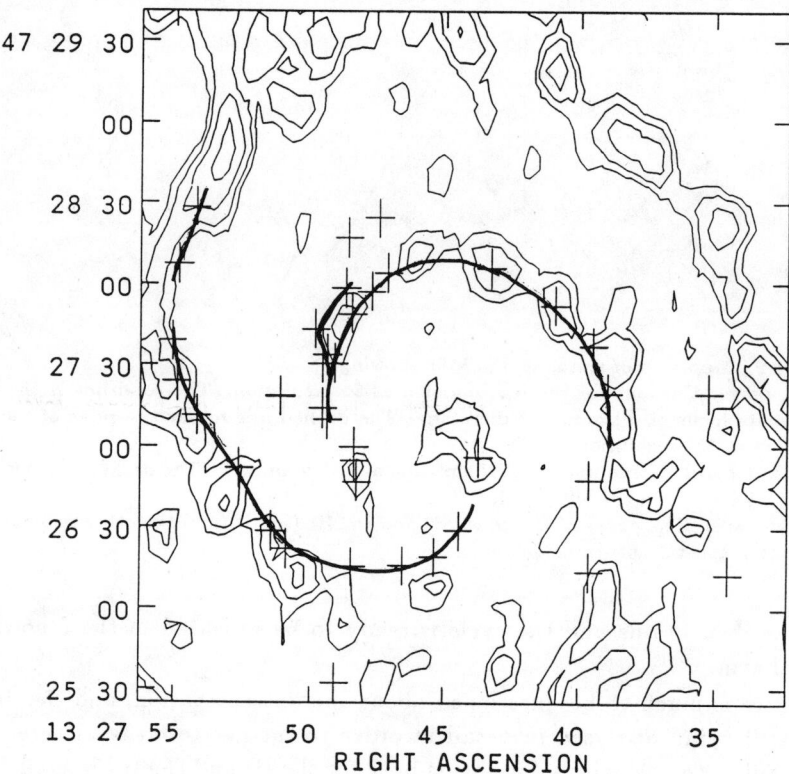

Figure 10. Preliminary HI image of M51 made with the WSRT at $12'' \times 18''$ resolution.[17]

at a resolution of $7'' \times 10''$. A *very* preliminary CO picture is shown in Figure 11; the two fields observed cover (unevenly) a major section of the southern spiral arm. The strong variations in the CO surface brightness suggest a temperature effect, so that the conversion to molecular column densities will not be straightforward.

3.3 Evidence for a Kinematic Separation of the Interstellar Gas in Spiral Arms

Another unexpected result is that the observed cross-sectional shape of the nonthermal emission from spiral arms is just opposite to that predicted by the single-cold-fluid model of the gas flow described previously. In that model, the cosmic ray electrons and magnetic field are convected along with the rest of the gas, and the synchrotron emissivity is monotonic in the total gas density (e.g. Mouschovias, Shu, and Woodward[19]). In regions of low gas density, and in particular just inside the shock, the synchrotron emission should therefore be faint. At the dust lane the nonthermal surface brightness should increase sharply as a consequence of the compression, followed by a slow decrease further downstream from the shock. The general

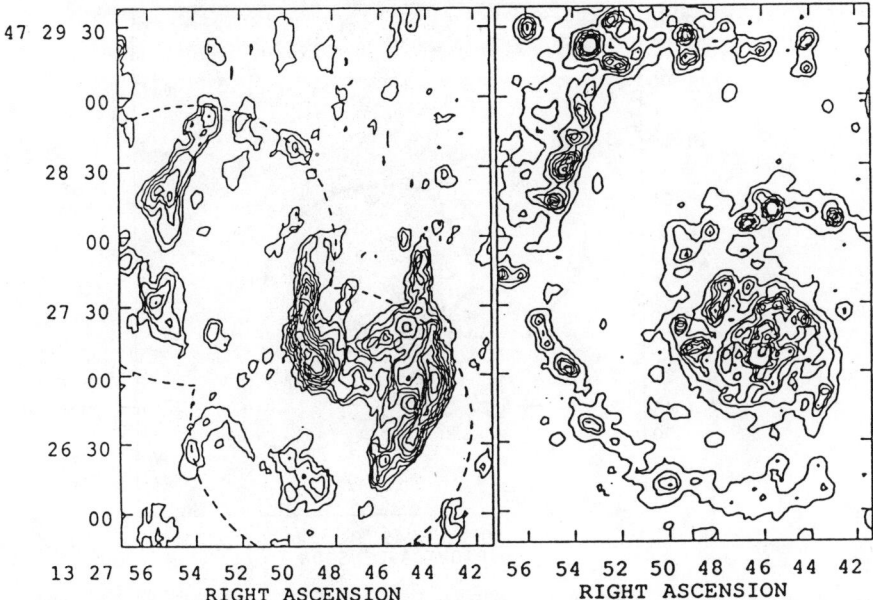

Figure 11. Left panel: Very preliminary map of the CO(1-0) brightness distribution in M51.[18]. Right panel: Hα distribution in M51.[20]

cross-sectional shape of the nonthermal arms should be similar to Figure 7.

Figures 12 and 13 show the distribution of thermal and nonthermal radio emission in M51 at a resolution of 8″, as we have recently established from new VLA observations.[20] The thermal component in Figure 12 shows peaks of emission just outside the prominent dust lanes. The peaks correspond closely with the Hα from filter photographs, as we would have expected since the emission mechanism is basically the same. The morphology of the thermal component agrees with the picture sketched in part (1) above. However, the surprise is the detailed shape of the spiral arms in the nonthermal emission, as shown in Figure 13. Over substantial lengths of both the northern and southern spiral arms the nonthermal emission lies on the *inside* of the dust lanes instead of on the outside. The section of arm located about 1′ to the northwest of the nucleus shows the effect particularly clearly; the dust lane, indicated by a series of crosses and a thick line drawn through them, marks the *outer boundary* of the smooth synchrotron arm. This latter arm has a somewhat asymmetric profile transverse to its long axis; the emission generally falls off more slowly towards the inside of the arm, and more rapidly towards the outside through the dust lane. Further downstream beyond the dust lane we eventually reach the HII regions with their thermal radio continuum emission as shown in Figure 12. This general morphology can be recognized in the southern arms as well, where the

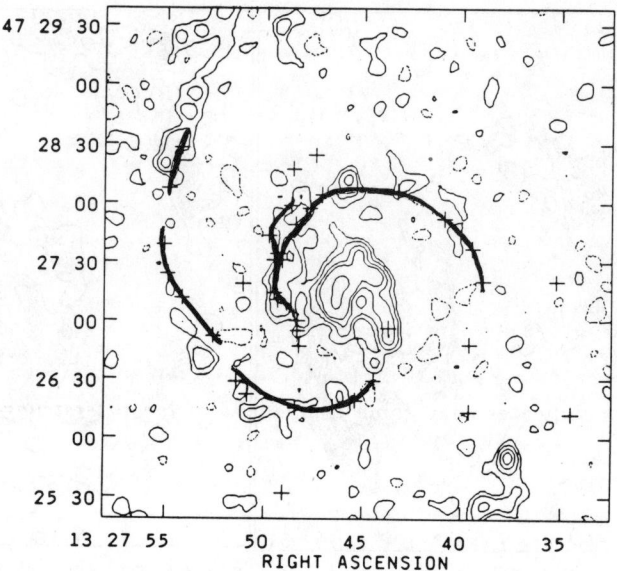

Figure 12. Thermal radio emission from M51 as derived from VLA observations at 8″ resolution.[20]

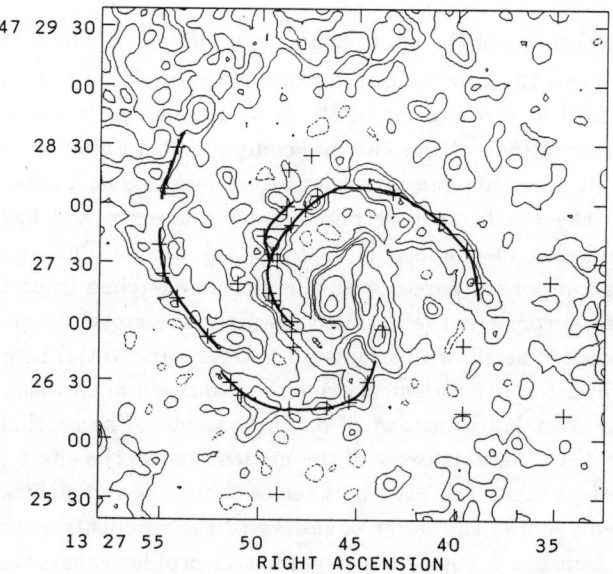

Figure 13. Nonthermal radio emission from M51 as derived from VLA observation at 8″ resolution.[20]

loci of the dust lanes and the nonthermal continuum are separated from the thermal/Hα by a gap with an average width of about 10″ (450 pc for a distance of 9.3 Mpc) from p. a. −180° to −270°, and the nonthermal arm is asymmetric in cross-section.

The conclusion seems inescapable that *the synchrotron-emitting component of the gas is not behaving as a cold fluid.* The gradual increase in nonthermal surface brightness as one approaches the dust lane from the inside indicates that this gas has a higher effective sound speed, and is not shocking. This would be the case if the appropriate speed was the Alfven velocity, i.e. if the magnetic field was having an important influence on the gas dynamics. On the other hand, the presence of the dust lane further downstream indicates that *some component of the interstellar medium is indeed suffering a shock.* This must be the cool molecular cloud component discussed in the previous section, which apparently separates from the synchrotron-emitting gas as the latter begins to slow down on its trajectory into the inside of the spiral arm. The cool clouds then run away to collide with other cool gas already present in the dust lane.

As was the case in the previous section, this description leaves many questions unanswered and many details to be completed. Work is progressing in this exciting subject, and we can expect more surprises in the future.

3.4 Conclusions

From the observations of the Galaxy we have known for many years that the Galactic interstellar medium is complicated and consists of at least several distinct components or "phases". The main general conclusion of the present work is that this complex nature of the interstellar medium is now becoming evident in the observations of nearby spiral galaxies. In particular:

—There is strong evidence that much of the atomic neutral hydrogen found in the inner arms of some spiral galaxies is a product of the dissociation of molecular gas by the activity of recently-formed stars. Furthermore;

—The one-component fluid models currently used to describe the motion of the interstellar gas in spiral arms are inadequate to explain the observations; the composition of the gas must be more complex. Finally;

—The results suggest that magnetic fields are affecting the gas dynamics in spiral arms, and models of gas flow should include them in the future.

4. ACKNOWLEDGEMENTS

Professor Lin, I know you will agree with me that the study of spiral structure in galaxies is exciting and fun. Besides my own work and that of my colleagues outlined above, you have stimulated many other observers into providing detailed results which are relevant to the theory. Some further examples include the detection of the underlying smooth spiral arms in the old stellar disks by Schweizer[21], the correlation of the flow velocity normal to the arm with the luminosity class of the galaxy by Roberts, Roberts, and Shu[22], and observations of multi-armed spirals by V. Rubin (see e.g. Haass, Bertin, and Lin[23]).

It is a great tribute to you personally that you have not only pioneered the development of the theory, but that you have also had the patience to talk with observers like me about its implications. In this way you have motivated the expenditure of a considerable amount of resources by myself and my colleagues and students since 1970. Expensive radio telescopes have been butchered, many months of precious observing time have been allocated, and endless man-years devoted to the reduction and analysis of mountains of data, and all because you thought that spiral structure could be described simply as a bunch of *waves*!

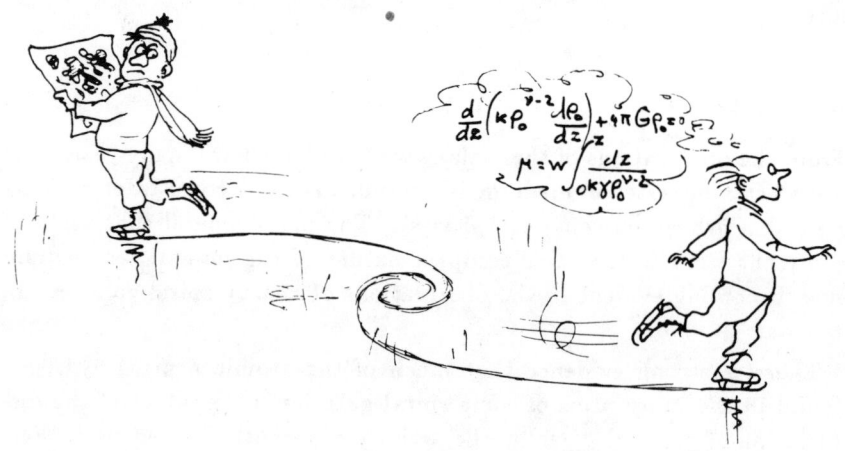

Illustration de J.P. PETIT, participant au Colloque

Figure 14. The spiral skaters, by J. P. Petit.[24]

In 1973, we both attended a delightful conference at the French Institute for Advanced Study at Bures-sur-Yvette. It was intended to be the first major confrontation between theory and observations of spiral structure in galaxies. The dialogue was not easy; we had to learn each other's language.

At the conclusion of that meeting, the situation was succinctly illustrated in a cartoon (Figure 14) drawn by one of the participants[24]. I have frequently identified myself with the exasperated observer on the left side of this figure, with his hands full of messy galaxies and nothing to compare them with but pages of equations. However, the recent work on the theory of spiral modes by you and your colleagues gives me hope that *images* may soon become our common *lingua franca*. And to keep the challenge going, I want to say that I intend to continue observing spiral galaxies, with instruments of ever higher angular resolution, ever better sensitivity, and ever greater wavelength coverage. C. C., you have led me on a merry chase for over 18 years. I have enjoyed it immensely, and I am ready for more. Thank you, Professor Lin!

References
1. Rosse, "Observations on the Nebulae", in *The Scientific Papers of William Parsons, Third Earl of Rosse 1800-1867*, ed. C. Parsons (Lund, Humphries & Co., London), 109-124 (1926).
2. Ryle, M., *Nature* **194**, 517-518 (1962).
3. Baars, J. W. M., van der Brugge, J. F., Casse, J. L., Hamaker, J. P., Sondaar, L., Visser, J. J., and Wellington, K. J., *Proc. IEEE* **61**, 1258 (1973).
4. Allen, R. J., and Ekers, R. D., "Ten Years of Discovery with Oort's Synthesis Radio Telescope", in *Oort and the Universe*, ed. H. van Woerden, W. N. Brouw, and H. C. van de Hulst (Reidel; Dordrecht), 79-110 (1980).
5. Rogstad, D. H., and Shostak, G. S., *Astron. Astrophy.* **13**, 99 (1971).
6. Mathewson, D. S., Van der Kruit, P. C., and Brouw, W. N., *Astron. Astrophys.* **17**, 468-486 (1972).
7. Allen, R. J., Hamaker, J. P., and Wellington, K. J., *Astron. Astrophys.* **31**, 71-78 (1974).
8. Rots, A. H., *Astron. Astrophys.* **45**, 43 (1975).
9. Rots, A. H., and Shane, W. W., *Astron. Astrophy.* **45**, 25 (1975).
10. Visser, H. C. D., *Astron. Astrophys.* **88**, 149-158 (1980).
11. Visser, H. C. D., *Astron. Astrophys.* **88**, 159-174 (1980).
12. Roberts, W. W., *Ap. J.* **158**, 123 (1969).
13. Roberts, W. W., and Hausman, M. A., *Ap. J.* **277**, 744 (1984).
14. Bash, F., and Visser, H. C. D., *Ap. J.* **247**, 488-498 (1981).
15. Allen, R. J., Atherton, P. D., and Tilanus, R. P. J., *Nature* **319**, 296-298 (1986).
16. Rickard, L. J., and Blitz, L., *Ap. J. Lett.* **292**, L57-L60 (1985).
17. Tilanus, R. P. J., and Allen, R. J., *Ap. J. Lett.* (in press) (1988).
18. Lo, K. Y., Tilanus, R. P. J., Allen, R. J., Wright, M. C. H., and Jackson, J., "Gaseous Spiral Structure in M51", in *Molecular Clouds in the Milky Way and External Galaxies*, ed. R. Dickman, R. Snell, and J. Young (Springer-Verlag, in press 1988).
19. Mouschovias, T. Ch., Shu, F. H., and Woodward, P. R., *Astron. Astrophys.* **33**, 73 (1974).
20. Tilanus, R. P. J., Allen, R. J., Van der Hulst, J. M., Crane, P. C., and Kennicutt, R. C., *Ap. J.* (in press; July 15, 1988).

21. Schweizer, F., *Ap. J. Supps.* **31**, 313-332 (1976).
22. Roberts, W. W., Roberts, M. S., and Shu, F. H., *Ap. J.* **196**, 381-405 (1975).
23. Haass, J., Bertin, G., and Lin, C. C., *Proc. Nat. Acad. Sci. USA*, **79**, 3908 (1982).
24. Petit, J. P., in *La Dynamique des Galaxies Spirales*, ed. L. Weliachew (CNRS Colloquium No. 241, Paris), frontispiece (1974).
25. Oort, J. H., Kerr, F. J., and Westerhout, G., *M.N.R.A.S.* **118**, 379 (1958).
26. Burton, W. B., *Proc. Astron. Soc. Pacific* **85**, 679 (1973).
27. Allen, R. J., Goss, W. M., and Van Woerden, H., *Astron. Astrophys.* **29**, 447-451 (1973).

MODAL THEORY OF GALACTIC SPIRAL STRUCTURE

Giuseppe Bertin

Scuola Normale Superiore,
Pisa, Italy

A long-standing problem in astrophysics has been the existence of coherent grand-design spiral structure in galaxy disks, which are known to be characterized by strong differential rotation. In the early 60's the possibility of Quasi-Stationary Spiral Structure (QSSS) was recognized. The collective nature of the pattern maintenance was stressed and QSSS was set as a working hypothesis in a semiempirical approach. A natural dynamical implementation of this program is found in the modal theory of spiral structure. QSSS is identified as a situation where one or a small number of global spiral modes dominate the scene. The modal theory, originally started in the context of tightly wound spiral structure, has now grown to cover most spiral types as found along the morphology classification of spiral galaxies.

I. WAVES AND MODES

In this paper I shall try to outline the key concepts and the main results of the *modal* theory of spiral structure in galaxies, as emerging from research efforts mostly concentrated in the last fifteen years. The focus of the theory is on grand design spiral galaxies, but less regular objects can easily be interpreted within the modal framework.

Basic issues are the origin and the persistence of coherent, usually bisymmetric, spiral structure in a differentially rotating disk. Earlier, in the 60's, a major turning point had been the realization that collective phenomena, dependent on the amount of self-gravity, differential rotation, and random motions, play a crucial role in the dynamics of galaxy disks and can change the picture provided by single-star kinematics altogether. Thus distinct physical scenarios converged into some general agreement that departures from axisymmetry in galaxy disks naturally occur in terms of large scale *density waves*. In particular, quasi-stationary patterns, regenerative spiral features, or even transient and fast evolving disturbances can all be labeled density waves, although they refer to quite different physical situations and they indeed may require or emphasize different physical mechanisms. Each scenario actually corresponds to a distinct perception of the way galaxy disks are structured. The relevant systems are not directly accessible (in contrast to most laboratory physics) and this is one reason for the interest in dynamical theories (since they can provide important constraints on the structure of galaxies) and probably the main reason for the persistence of some controversies (since definitive tests or proofs are extremely hard to conceive).

Especially because spiral structure appears to be correlated with a number of physical properties that are likely to change slowly in time, the most appealing procedure for astrophysical applications is to develop a semi-empirical approach based on the hypothesis of Quasi-Stationary Spiral Structure (QSSS). While material arms would quickly wrap up because of the differential rotation, density waves need not wind up. Thus, for a density wave, QSSS in a differentially rotating disk is obviously a possibility, even though QSSS is not at all a necessary implication. In fact, in the highly inhomogeneous and dispersive environment of galaxy disks, one may expect waves to evolve and to propagate, thus facing a more subtle kind of wind-up problem for spiral structure due to convection.

Particularly in view of the fact that the main focus is on grand design spiral structures that involve the whole disk, sometimes over huge scales (e.g. UGC 2885), there are at least three good reasons for advocating the QSSS hypothesis. The first, primarily empirical, reason is the fact that it is very simple and it can be easily quantified in terms of predictions, and tested. Secondly, for many astrophysical applications, such as shocks and star formation processes, the relevant time scales are short enough, so that we can work *as if* the spiral structure is steady, even when it is evolving on a time scale close to the dynamical time scale. On the basis of these considerations, a large number of important calculations and successful observational tests of the "density wave theory" were performed in the 60's and in the early 70's, so as to claim "consistency with the observations". For this purpose, the mathematical basis was a local *pattern* theory, the so-called dispersion relation for tightly wound spirals, where the pattern frequency is a free parameter to be constrained by the observational data. The third good reason for advocating the QSSS hypothesis is basically of a theoretical nature and it is a crucial point particularly if one feels optimistically about the possibility of formulating a dynamical classification of galaxies. This is the fact that, since large scale coherent phenomena are involved, boundary conditions play an important role in the density wave problem. Thus it is very natural to imagine that when we look at a beautiful grand design spiral, such as M81, NGC 5364, NGC 1300, or even M51, we are indeed looking at internal discrete notes of the system, i.e. *global modes* of the disk, that are *composed of waves* that may satisfy locally a dispersion relationship. Fortunately, we need not worry excessively about excitation processes, because the differential rotation is an abundant and readily available energy source for the modes (see following Sec. III).

Once the modal approach is accepted, mastering the relation between the properties of the basic states and the properties of the relevant spiral modes and identifying realistic basic states by discussing the processes of evolution and self-regulation can lead to a unified view of morphologies and of astrophysical processes in spiral galaxies. Even if the concept of global modes was already present, albeit lacking proper calculation and demonstration, in the 60's and early 70's, the major turn from pattern theory to modal theory occurred only in the mid-70's. Now the modal theory has grown to cover the key morphologies that are found along the Hubble classification scheme.

Certainly the density wave theory is not only *about* cooperative phenomena but is *itself* some sort of a cooperative effect. In this Symposium we honor C. C. Lin who gave constant inspiration and drive to the field from the very beginning by promoting the QSSS hypothesis and later by stimulating its natural dynamical implementation as a modal theory. As a participant in this fascinating and challenging program, I shall try to present a summary of the key points of this research. Mathematical details and reference to individual contributions can be found in the literature.

II. MODAL APPROACH

If a galactic disk is essentially dominated by one (self-excited) spiral mode, the resulting spiral structure is expected to be very (unusually) regular. If a small number of modes dominate the scene, the structure is expected to be less regular and slowly evolving due to the beating of the various mode frequencies. If many modes are present, the appearance of the disk may be flocculent, as in NGC 2841, unless some external agent, such as a tidal encounter, occasionally organizes the system into a more coherent form. The resulting spiral structure would then evolve, but with no essential change of morphological type. A situation where all modes are damped could also be considered as a case for flocculent galaxies. In this latter picture tidal driving is expected to produce spiral structure subject to a somewhat rapid change of morphological type, unless a single damped mode stands out. Thus one important role of the dynamicist in this research field is to identify and to quantify the above-described situations.

As far as the number of modes is concerned, already at the linear level one can realize that a situation where one or a few modes dominate is not at all unreasonable. Many arguments show that the spectrum of growing spiral modes is likely to be discrete, not only mathematically, as expected for an effectively finite system when the possible singularities of the model are removed, but also physically, judging from the separation of the relevant eigenvalues. In addition, it is found that the total number of modes is severely restricted by the possibility that a mode resonates with the frequency of epicyclic motions around the circular orbits of the star guiding centers. Modes with multiple arms or with very low pattern frequency suffer most from this resonant interaction and tend to be damped. Thus a situation where only one or a small number of growing modes dominate can be demonstrated.

The concept of mode dominance is usually related to the size of the growth rate, which should yield acceptable amplification over a cosmological time scale, and to the radial extent, which should be comparable to the size of the optical disk. Non-linear phenomena may also be associated to the concept of mode-dominance, but these have not been quantified yet in the context of spiral structure.

Most of the theoretical work in the dynamics of disk galaxies has been carried out on simple infinitesimally thin one-component systems. Of course, the presence of a bulge-halo spheroidal system is usually taken into account, but only as providing a rigid background force field. On the other hand, at least in a local sense, many "complications" of the real system have been studied in some detail. We can list among these the role of thickness, effects of disk-halo interaction, and the various roles of gas and other "cold" material. These studies show that cooperative phenomena are very sensitive to geometry and to the existence of separate components. Thus the use of infinitesimally thin one-component systems can be very valuable, only provided they are seen as *equivalent* systems and their parameters are carefully chosen so as to best represent the actual three-dimensional multiple-component galaxy. For example, in many cases a relatively flat density distribution for the *active participating mass*, with a hole in the middle, may be the best way to represent the galactic disk for dynamical purposes, even when observed photometries tend to point to exponential disks. In fact, the hole in the middle is a simple way to include the three-dimensional effect of the transition from disk geometry to bulge geometry in the inner regions. In addition, the thin Population I disk is dynamically more active than the thicker Population II disk and it has a flatter mass distribution.

For astrophysical applications the attention should be focused on basic models characterized by modes with *moderate growth rate*. Models subject to violent instabilities would rapidly evolve and essentially correspond to the wrong choice of the basic state. Thus one important stage of the modal studies is to identify moderately unstable basic states within the constraints provided by observations. This procedure actually discards a large number of models as astrophysically uninteresting.

Two important issues can be raised at this point. (i) If the system is initially *inside* the regime of moderate instability, are there physical mechanisms that can prevent it from evolving away from such a regime? (ii) If the system is initially *outside* the regime of moderate instability, are there physical processes that can make it evolve into it? The answer to these questions has been called *self-regulation*. If sufficient cold interstellar gas is available, mechanisms of self-regulation are quite natural. (Fortunately smooth-armed galaxies, i.e. grand design spirals without gas, form just a statistical minority). The gas is known to be subject to shocks which dissipate and saturate at small amplitudes the otherwise exponentially growing modes. Gas-rich galaxies, like Sc's, are likely to control modes with relatively large growth rate. In addition, the torques associated to a spiral mode do not seem to induce rapid evolution, so that, as an answer to (i), models that are moderately unstable are expected to evolve only slowly. Another worry could derive from the fact that the collisionless stellar disk, when any kind of evolution is induced, is likely to increase steadily its amount of random motions. This is, as a general trend, an evolution towards more stable configurations. However the gas component is expected to act like a *thermostat*, since it is subject to internal cooling. Thus if the gas disk is not stirred by instabilities, it tends to cool and the composite disk is brought back to instability. On the other hand, if the initial system is too cold, rapid stirring ensures evolution towards moderate instability levels. Therefore the answer to point (ii) is also thought to be positive. The fact that much of the self-regulation relies on gas is not at all surprising, since we know that spiral structure is mostly associated with and traced by the so-called Population I layer.

Many of these arguments are supported by detailed calculations, others still await more quantitative analysis. This discussion shows that the modal approach goes well beyond the mere calculation of linear modes in a thin self-gravitating disk, since it is concerned with the various astrophysical processes that participate in and result from QSSS. It also shows that the choice of the basic state is a crucial and subtle issue. One can be easily misled by the wrong choice of the basic state.

III. THEORY OF MODES

In general, a disturbance on an axisymmetric self-gravitating disk can be separated into an odd and an even part with respect to the (reflection symmetry) equatorial plane. Odd perturbations bend the disk and are observed in many galaxies as warps (mostly affecting the outer (lighter) gaseous regions, but sometimes detected in the (heavier) optical disk). Even perturbations are associated with density enhancements in the disk, such as spiral arms. These disturbances can be seen as characterized by a density perturbation and by a disturbed gravitational potential. In the absence of external driving, density and potential must be consistent with each other. In the following we shall refer to even disturbances, i.e. to density waves.

In galaxy disks most stellar orbits are confined to narrow annuli. However, the relation between density and potential has an integral character, since, for the long-range gravity problem, the field on a point results from the sum of the contributions of the density at every other point of the disk. On the other hand, if one considers perturbations with fast spatial variations, i.e. with large *total* wave number, an approximate *local* relation can be derived. The total wave number is a vector composed of a radial part (associated with the radial spacing between spiral arms) and an azimuthal part (essentially the number of spiral arms). Thus, two safe limits for this approximate description are the regime of tightly wound spiral arms or the case of multiple-armed spiral structure. (But asymptotics often extends generously its domain of application.) This discussion usually goes under the name of potential theory and it is at this stage that the most delicate mathematical assumptions are made. In order to describe the inhomogeneous problem, the radial wave number is allowed to change as a function of the galactocentric distance.

Under the above assumptions, in a one-component fluid model of an infinitesimally thin disk embedded in a bulge-halo inactive spheroid, a simple local dispersion relation has been derived that connects the so-called (dimensionless) Doppler-shifted frequency ν to the magnitude of the total (dimensionless) wave number K. The relation is a cubic in K which depends on two parameters Q and J that characterize the basic state:

$$\frac{Q^2}{4} = \frac{1}{K} - \frac{(1-\nu^2)}{K^2 + \frac{J^2}{(1-\nu^2)}} .$$

The parameter J, which is proportional to the disk density, measures the amount of self-gravity of the disk, relative to the total gravitational field. In addition, J depends on the local shear rate, i.e. the differential rotation, but does not depend on the local thermal energy. On the other hand, Q is proportional to the local equivalent sound speed. Therefore J essentially measures how massive the disk is, and is a primary instability parameter, whereas Q measures how hot the disk is, and is easier to change as the result of dynamical processes. In general J and Q are functions of the galactocentric distance. Note also that for a given value of the rotation frequency of the spiral pattern, the frequency $\nu = (\omega - m\Omega)/\kappa$ is generally a monotonic increasing function of the radius; it is negative inside the corotation circle and positive outside.

Following the above (approximate) description, for a given galaxy model and for each pattern frequency one can construct the propagation diagram which identifies at each radial location the allowed wave-branches and the associated pitch angles of the related density waves. In addition, all the various properties, like group velocity, density of wave action, angular momentum and energy flux, which define these dispersive waves, can easily be derived and calculated. In particular, the density of wave action is *negative* inside the corotation circle and positive outside it, because of the monotonically decreasing character of the differential rotation.

The (J, Q) plane is divided in two regions by a transition line (see Figure 1). At low values of disk mass (low J), the cubic dispersion relation admits three real roots for K. Above the transition line (high J) the cubic admits only one real root. Thus a global mode can be maintained by different waves and wave-cycles according to the parameter regime that is considered. Two simple limits can be defined. In the (low J) regime of tightly wound spiral structure, modes can be maintained by superposition of two different wave-branches (short and long) of the trailing type (observed spiral structure is trailing, i.e. winding in the "natural" direction with respect to the differential rotation). In contrast, in the (high J) regime of high disk mass only one wave-branch is available so that a global mode must be composed of waves with opposite winding, i.e. leading and trailing waves. Each of these two simple limits, involving *two* kinds of wave, can thus be reduced mathematically to a second order differential equation for the perturbed enthalpy in the radial variable. This is the differential counterpart of the local dispersion relation. In each case, the differential equation corresponds to a *two-turning point problem* where the pattern frequency and the growth rate of the mode are the eigenvalue, to be determined by imposing the appropriate boundary conditions (see Figure 2).

The inner turning point is simple and essentially describes a *feedback mechanism* for incoming signals that are refracted outwards from the central (bulge) region of the disk. A regularity condition at the center selects the evanescent wave inside such turning point. The outer turning-point is double and is located at the corotation circle. Both sides of this latter turning point are propagation regions. The best way to represent resonant, and possibly turbulent, absorption outside corotation (as suggested by stellar dynamic and physical arguments), is to impose a *radiation boundary condition* outside, so that a wave is selected that transfers energy and angular momentum outwards. For growing modes this condition is consistent with the requirement of vanishing amplitude at large radii. Therefore a "quantum condition" is determined by proper matching of solutions. This yields not only the pattern frequencies of the discrete spectrum, but also the growth rates of the modes. A positive growth rate, which corresponds to self-excitation, is the global counterpart of an *overreflection* process that is realized at corotation. Indeed the transfer of wave action across the corotation circle outwards ensures amplification, because energy is extracted away from a region of negative energy density and transferred to a region of positive energy density. The distributed source for amplification is the differential rotation of the disk, but the values of J and Q (at corotation) determine the amount of energy that is tapped, and therefore the growth rate of the mode. In Figure 1 the regime of moderate amplification is indicated as a dotted strip. The radiation boundary condition is consistent with the trailing nature of observed spiral structure.

This discussion and the related physical picture can be confronted with numerical mode calculations that include the potential theory in its exact integral form. Only in certain regimes, as indicated earlier, does the neat picture of the above two cases apply in detail. In practice, many interesting basic states belong to intermediate regimes and, in addition, as suggested by the cubic dispersion relation itself, more complicated wave-cycles could be considered. However it is delightful to find that the overall picture and even quantitative results are well covered by simple extrapolation of the above two basic cases.

IV. MORPHOLOGY OF SPIRAL MODES

The two simple representative cases described in the previous section correspond to two major categories of spiral morphologies. In fact, in the low J regime the relevant wave-cycle is composed of short and long waves of the trailing type; thus the resulting spiral mode has a relatively tight normal spiral appearance. In contrast, in the high disk mass regime the mode is composed of leading and trailing waves; thus, for the astrophysically interesting case of moderate growth, it has an open barred appearance, possibly with a mild trailing disturbance around the outer neighborhood of the corotation circle. (Note that normal spiral modes of the Sc type can also be obtained in the high J regime, but they are subject to very high growth. This can be regulated only when a large amount of gas is present). Thus, it seems that the key-categories of the morphology classification of spiral galaxies are at hand and indeed easily recognized along the narrow strip of moderate instability in the proper parameter space.

A detailed concrete appreciation of all the modal morphologies in moderately unstable disks, and especially of those corresponding to regimes of the transitional type, can be obtained from extensive numerical surveys

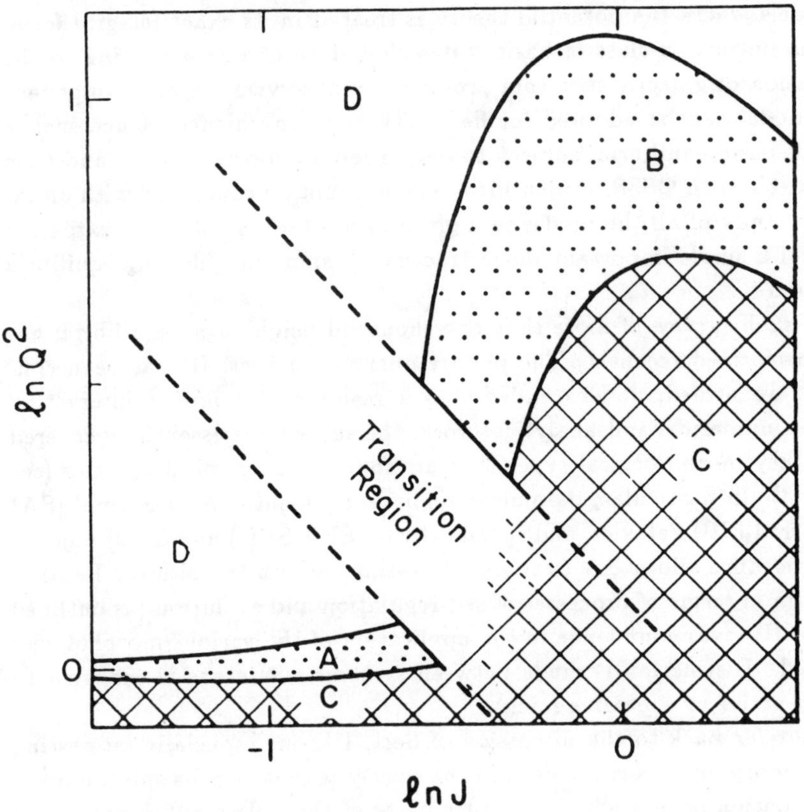

![Figure 1](parameter regimes)

Figure 1. Parameter regimes in parameter space. The dotted area represents the strip of moderate growth. The hatched area C corresponds to violent instability. A transition region separates the two simple regimes of normal (A) and barred (B) spiral appearance.

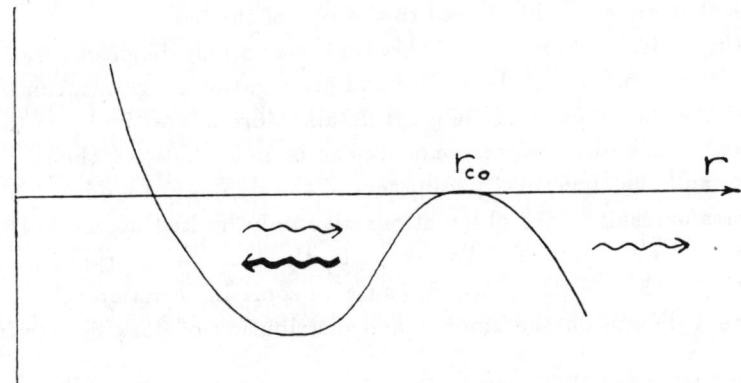

Figure 2. Sketch of the two turning-point problem, indicating feedback, overreflection, and radiation boundary condition.

of models where the potential theory is treated in its exact integral form. To this purpose a class of basic states should be chosen according to the perception of galactic structure provided by observers. A one-component fluid model can be adopted for flexibility and convenience. Once such a class of basic equilibria, subject to only a few important modes and thus compatible with QSSS, is identified, one need not be concerned with an intensive study of all the modes in a given basic state. Rather, an extensive study of a single important mode tracked on many neighboring equilibria is desired.

One such survey, of more than one thousand neighboring equilibria, has been performed recently. The picture outlined in Sect. III has been confirmed, quantified, and extended. As a result of continuous interaction between numerical and analytical work, the survey has essentially covered all the key morphological types that are found among spiral galaxies (see Figure 3). In particular, parameter regimes are identified for normal (SA) and barred (SB) galaxies, and specifically for SB0, SB(r) and SB(s) objects. The procedure allows one to track interesting smooth transitions. Reinterpretation in terms of processes of self-regulation and evolution (as outlined in Sect. II) gives confidence in the application of the various morphologies obtained. Distinction is made between systems where gas is expected to play an active or a passive role.

If we refer back to the discussion of Sect. III, one especially interesting class of morphologies recognized in the survey is that of SB-s spiral modes. This situation occurs when the inner parts of the galactic disk are in the high density regime and the outer regions are in the low density regime and kept cool by the presence of abundant gas. Thus, modes can be found that are characterized by a long bar extending up to a sizeable fraction of the corotation circle; the bar then turns sharply into relatively tight trailing spiral arms that are smoothly joined to the tips of the bar.

On the theoretical side, surveys of this kind, via various diagnostic tools, such as Fourier transform of the modes and propagation diagrams, can be used to test the theory of modes in great detail. More interestingly, on the astrophysical side, a direct correspondence can be made between the spiral morphology and the structural parameters of the dynamical model. This is an important result, since observations give only limited access to the physical properties of galaxies. For example, the observed morphologies, if interpreted in the modal approach, can give some information relevant to the current debate on the amount and distribution of dark matter in galaxies.

V. BEYOND THE PRESENT LINEAR THEORY OF MODES

This exposition has tried to outline the major steps on the long road

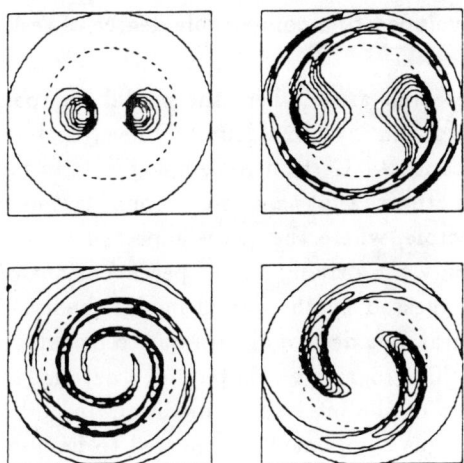

Figure 3. Key prototypes of spiral modes (after R. Thurstans, Ph.D. Thesis 1987). Three modes (a normal SA, a barred SB0, and a barred SB-s) have moderate growth rate. The open spiral mode on the lower right is violently unstable.

that follows the hypothesis that grand design spirals are long lasting. This road naturally leads to the view that *the majority* of spiral galaxies can be interpreted in terms of the modal approach. Such perception is quantified in a framework that relates detailed dynamical mechanisms to various astrophysical processes. A large number of specific issues has been faced and clarified, thus strengthening considerably the modal theory of spiral structure. Some of these issues are the character of the discrete spectrum of modes, refraction and feedback mechanisms, role of resonances, excitation processes, appropriate boundary conditions, and self-regulation. Many points have been addressed in significantly deep technical detail, as can be found in the literature. Still the overall picture is amazingly simple and can almost be told in a few words.

Obviously, even in the modal approach it is recognized that galaxies are complex systems and in specific objects spiral structure may originate in a number of ways. In a statistical minority of cases one should not exclude the role even of some scenarios, such as that of stochastic and shock induced star formation, that do not rely on density waves at all. Other more transient density waves are also expected to operate occasionally; for these cases the modal representation, even if, strictly speaking, still applicable, would be less convenient. However, for the statistical majority of spiral galaxies, an alternative theory has apparently not yet been formulated, or at least not yet developed to a comparable degree of testing and quantitative analysis.

Many are the research areas where the modal approach awaits new efforts. Multiple component systems definitely need better understanding in the global modal context. Certain morphologies may be properly understood only through studies that take into account two or more components explicitly. For example, where the gas is expected to play a fairly passive role, one should study the driving due to perturbations of the stellar component that are suggested by the global modal theory. Most of the ring morphologies in spiral galaxies could turn out to be categorized within this approach. Clearly this program would require a detailed study of the structure and the physics of the interstellar medium. In addition, much of the reaction of the cold gas component is expected to be non-linear.

Within the stellar-dynamic modal theory more work is required to cover the regime of high disk density. Furthermore, now that the linear theory of modes is in a reasonable shape, it would be time to make progress in the case of nonlinear regimes. Here too one could take advantage of the knowledge of linear modes in order to construct collisionless spiral structure at finite amplitudes. For example, current efforts aimed at identifying self-consistent structures with large amplitudes by superposition of single star-orbits might be initialized with the morphologies suggested by a linear modal theory. These self-consistent studies of purely stellar spiral arms would be of interest to the so-called smooth armed spirals, but they could also shed light on possible mechanisms of non-linear growth and saturation that are presently overlooked. Mechanisms of non-linear excitation or non-linear dominance of modes might be of great astrophysical interest.

Another research area where more work would be desired is a quantitative realistic analysis of tidal interaction, starting out with various conditions for the galactic disk subject to tidal driving, according to the scenarios outlined in Sect. II. These and other issues could be proposed and discussed further. They show how far the long road of the modal theory might continue. But, in closing, I would like to comment on another aspect of the theory of spiral structure.

Beyond the face-value of the results obtained, the theory of spiral galaxies is a gold-mine for many research fields. For the applied mathematician, it provides the opportunity to develop asymptotic methods, to work out complex turning point problems, to discuss subtle issues of boundary conditions, but especially to develop a good model for complex three-dimensional inhomogeneous systems. These points often require the use of sophisticated analytical and numerical techniques. The systems involved are characterized by a highly complicated phase-space structure: this offers a tremendous challenge to scientists who aim at producing theoretical experiments in terms of N-body simulations.

Another general field of interest is theoretical mechanics. The study of stellar orbits, resonances, and responses to a symmetry break can lead the way to problems that are regarded as fundamental issues in theoretical physics. One example is the process of transition through deterministic equations to stochasticity, which can be witnessed in many phenomena relevant to galactic dynamics. In fact, some of the relevant concepts originate from scientists active in the dynamics of spiral galaxies.

As should be already clear from the discussion of Sect. III, a number of analogies can be drawn with phenomena that belong to hydrodynamics, meteorology, and geophysics, mostly as a result of the application of the subtle art of wave motion in systems characterized by shear. The study of collective phenomena and the correspondence between rotation and magnetic field is a source of fascinating connections with the world of electromagnetic plasmas, especially of high temperature plasmas. A number of specific examples can be given where basic concepts, that are used in the theory of spiral structure, have a translation in the theory of laboratory plasmas, e.g. in the theory of magnetic toroidal devices. Clearly, a general common ground should be the emphasis on the modal approach and on the role of inhomogeneities. Besides, it is likely that detailed phenomena, like large scale odd (bending) and even (lumping) modes in self-gravitating disks, have a direct counterpart in current disks, which seem to be a common feature in plasmaspheres of astrophysical interest.

For the astrophysicist probably the main food for thought derives from the basic strategy suggested by the semi-empirical method and by the focus on morphologies and observed phenomena. Therefore, beyond the present linear theory of modes, I see primarily a delightful field for learning.

OBSERVATIONAL EVIDENCE ON THE DENSITY WAVE THEORY — AN INTENSIVE STUDY OF M81

Frank Bash
The University of Texas at Austin

Michele Kaufman and the author have made an intensive study of the spiral galaxy M81 using V.L.A. maps of the galaxy and in collaboration with several other observers. Our continuum V.L.A. maps have been used to measure the width and location of the spiral arms as well as the H II regions near the arms. We have used the V.L.A. 21-cm maps of M81, made by Hine and Rots, to observationally locate the two-armed spiral shock wave. We have been able to compare the observed locations of the H II regions, containing newly formed stars, with their expected locations based on the assumption that their formation was triggered in the spiral shock and that they move subsequently according to the predictions of the density wave theory. In addition we have compared our results with optical data gathered by other collaborations. This study probably represents the most stringent and most complete observational test of the density wave theory.

1. INTRODUCTION

The galaxy M81 is perhaps the single best example of density waves in a spiral galaxy. Herein we describe a portion of a major study of M81 by the author, Michele Kaufman (Ohio State) and various collaborators. We do not claim that M81 is typical of all spiral galaxies and thus, by implication, that all spiral galaxies contain density waves. Instead we use M81 as an example of a simple, two-armed galaxy which happens to contain a density wave (at least a shock-like velocity signature) in the 21-cm H I gas along the spiral arms. In a separate project with W. Roberts, we are comparing the observed density wave to predictions of the density wave theory which has been modified to include a cloudy interstellar medium. (eg. Hausman and Roberts, 1984).

Here we use observations of the locations of giant H II regions, the light of the spiral arm stars and the dust lanes in M81 as compared to the *observed* location of the shock wave in order to try to deduce the displacement between them. If the shock wave plays a role in triggering star formation and if we know the initial velocity of the clouds which have been triggered, then by integrating the cloud orbits we can compute the time delay from the triggering to the formation of massive stars which power the H II regions and cause the lanes of blue light along the arms.

It is important to point out that the results of the study reported here do not, to first order, depend on any predicted location or strength of the density wave. We use the *observed* location of the shock wave, the *observed* location of H II regions and we assume that the clouds which gave birth to the H II regions were triggered in the shock and, in this paper, move on circular orbits at the local circular velocity. In subsequent work the observed post-shock velocities will be used as initial velocities of the clouds.

2. THE OBSERVATIONS

Bash and Kaufman (1986) have published B, C and D configuration V.L.A. maps of M81 made at λ 6 cm and 20 cm. These observations allow the detection of the giant H II regions in M81 and the faint non-thermal spiral arms. The typical synthesized beam size is $10'' = 160$ p.c. at a distance to M81 of 3.25 Mpc. In addition our collaborators Hine and Rots (Hine, 1984) have made V.L.A. maps of the λ 21 cm radiation from H I in M81 at similar angular resolution. Our λ 20 cm map of M81 (turned face-on) is shown in Figure 1. The brightest areas are giant H II regions but the faint non-thermal radiation along the arms can also be seen.

We also have the Hα maps of M81 made by Hodge and Kennicutt, the digitized B and I plate photographs by D. Elmegreen and the high-precision Hα radial velocities of the bright H II regions made by Levreault.

We can observationally:

1. Locate the spiral arms in the blue light of the young stars (B-plates) in the Hα radiation from the H II regions, and in the thermal radio radiation from the H II regions (V.L.A. maps),
2. Locate the linear density wave in the old disk stars on near-infrared photographs (I-plates) and future λ 2μm observations,
3. Locate the spiral arms through the compressed gas and fields giving rise to enhanced synchrotron emission (V.L.A., λ 20 cm maps) and
4. Locate the spiral shock by observing the velocity discontinuities in the H I, 21-cm velocities along the arms (V.L.A. H I map).

We plan to use these data to see if the density wave theory fits what we see. Also we can actually measure the location of the young stars with respect to the shock wave and ask how they got there and how long it took. If stars form in dense molecular clouds which are unaffected by external magnetic

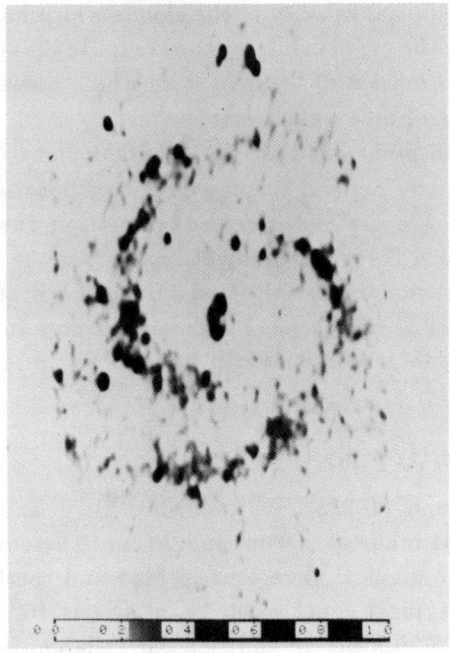

Fig. 1 Face-on λ 20 cm V.L.A. map of M81 from Bash and Kaufman (1986).

or ram-pressure forces for at least several tens of millions of years then we can integrate their (purely ballistic) orbits using the observed post-shock velocity as their initial velocity.

Figure 2 shows a set of four observed maps of M81 (turned face-on) with the *observed* location of the H I spiral shock drawn-in. In these face-on pictures the major axis is horizontal and north is on the left. The east arm is on the right and lower; the west arm is on the left and upper. The first results of the analysis of these data are given in the next section. It seems to be obvious from examining Figure 2 that the shock wave lies along the inside (upstream) edge of the blue spiral arms, the H II regions, the H I concentration and even the broad, faint non-thermal arms.

3. THE RESULTS (SO FAR)

In the first results (Bash and Kaufman, 1986) we found that the spiral arms are too broad to correspond with classical density-wave theory. That is, they are too broad to be the width of the predicted T.A.S.S. shock, but they do agree with the predictions of "cloudy models" eg. Hausman and Roberts (1984). We also found that the Inner Lindblad Resonance

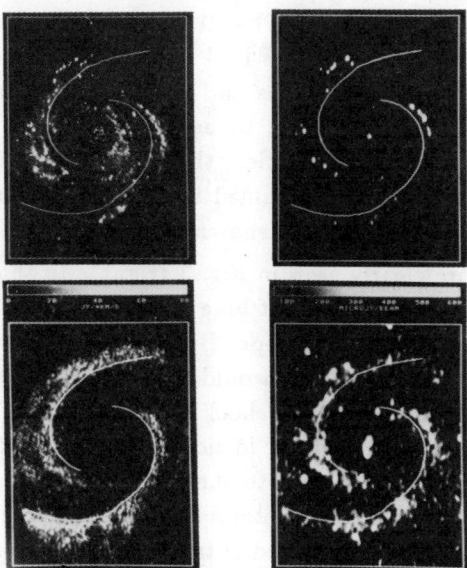

Fig. 2 Four face-on maps of M81 all to the same scale with the observed location of the shock wave drawn in. Upper-left: Blue-light photograph which has had a near-infrared photograph subtracted. Upper-right: Giant H II regions. Lower-left: The H I column density. Lower-right: The λ 20 cm V.L.A. map seen in Figure 1.

may be at larger radii than previously assumed. The spiral arms in M81 are approximate logarithmic spirals at large radii but they differ from each other in pitch angle. The arm nearest M82 also seems distorted. The second paper, Kaufman, Bash, Kennicut and Hodge (*Ap. J.* in press) deals with the H II regions in M81.

Our next task is to examine the location of the observed shock wave with respect to the young stars which may have been triggered to form by the shock wave. Figure 3 shows the location of the giant H II regions in M81 with respect to the shock wave. Assuming that the giant H II region forms in a cloud which moves on a galactocentric circle at the local circular velocity we can compute how long it takes to get that far from the shock. The shock is assumed to move at the spiral pattern speed. We took the rotation curve and pattern speed from Bash and Visser (1981). A future, more elaborate, calculation will use the measured post-shock velocity for the initial velocity and we will actually integrate the orbits.

Figure 3 (left), on which is marked lines for 10^7 and 4×10^7 years, shows that most of the giant H II regions are found between 10 and 40 million years downstream of the shock. (The 10 and 40 million year lines cross at corotation.) The 11 most luminous H II regions are circled and, especially outside of the circle of H II regions and young stars near R = 5 kpc, the most luminous H II regions seem to be seen at about 40 million years downstream of the shock. In addition, the whole set of giant H II regions, at all radii, does seem to be contained in the region bounded by the 10^7 yr. line and a 5×10^7 yr. line (not shown).

This is even more clearly seen in Fig. 3 (right). In that figure we show the locations of the sharp ridges of blue starlight. Again those stars form a nearly continuous ring at R~ 5 kpc. But outside of R = 6 kpc the upper envelope of points falls with R as would be expected if they die at a fixed time (45 million years?) from the shock wave at all R. This suggests that star formation is triggered to occur in clouds by the shock wave or by some process located at a place parallel to the shock wave. By seeing where the massive stars die, then with those stars' lifetimes we can compute how long star formation takes once the cloud is triggered (assuming that the shock is the cause).

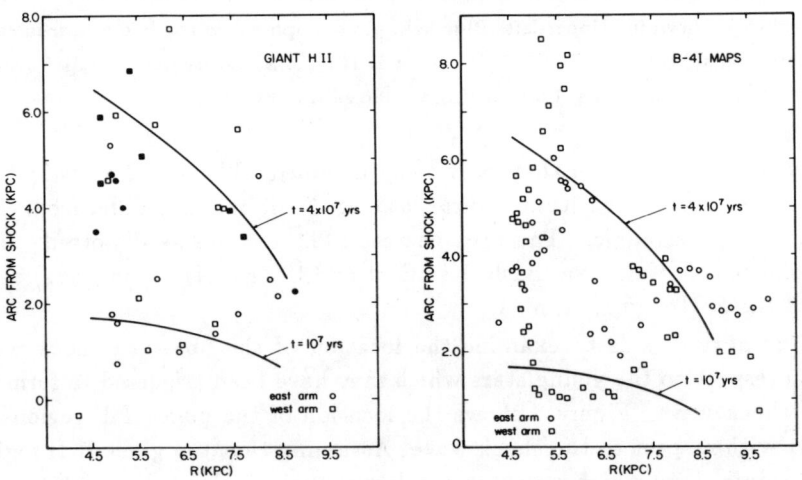

Fig. 3 (Left). The location of the giant H II regions in M81 with respect to the location of the shock wave. The H II regions of each arm are plotted with different symbols and the 11 most luminous H II regions are shown filled in. Each H II region's distance from the nearest shock is shown in terms of the distance along a circle centered on the center of the galaxy and its distance from the center of M81. Positive arc distances are those "downstream" of the shock in terms of galactic rotation. (Right). Same as on the left, except we show the ridges of blue light along the arms. The blue photograph has had an enhanced (by 4 times) near infrared photograph subtracted from it in order to remove the smooth underlying disk of old stars.

4. CONCLUSIONS

We and our collaborators have collected a large amount of observational data on the galaxy M81. Those data seem to allow us to test the detailed predictions of the density wave theory, to examine the role of the density wave in triggering the formation of massive stars and also to look at how long the star formation process takes.

I would like to acknowledge the support of NSF Grant AST-8611784 and its predecessors.

References
1. Bash, F. N., and Kaufman, M. 1986, *Ap. J.*, **310**, 621.
2. Bash, F. N., and Visser, H. C. D. 1981, *Ap. J.*, **247**, 488.
3. Hausman, M. A., and Roberts, W. W. 1984, *Ap. J.*, **282**, 106.
4. Hine, B., 1984, M. A. thesis, University of Texas.

COLOR IMAGING OF SPIRAL GALAXIES

Debra Meloy Elmegreen, Bruce G. Elmegreen, and Philip E. Seiden
IBM Watson Research Center
Yorktown Heights, NY 10598

1 INTRODUCTION

The revolutionary theoretical work on galactic spiral structure by Professor Lin and his colleagues has inspired many astronomers to seek evidence for spiral density waves. Computerized color imagery using B and I passbands may be an important new tool for recognizing these waves. Computers can superimpose the blue and near-infrared intensity images of a galaxy, and then realign the result to a face-on orientation for a better view of the spirals, or subtract the underlying disk light to emphasize only the azimuthal variations. The computer images show the shapes and color distributions of spiral structure better then conventional photographs. Spiral arms can often be traced for more than a complete turn around the nucleus, inner ovals are sometimes revealed, and blue patterns of star formation can be identified. This paper summarizes some of the results obtained from a preliminary study of our total sample of approximately 150 galaxies.

2 OBSERVATIONS

The observational material was obtained between 1978 and 1981. The majority of the data consists of photographic plates in the B and I passbands taken with the Palomar 1.2m Schmidt telescope; other plates are from the Palomar 5m, CTIO 0.9m Curtis-Schmidt and 4m, and Las Campanas 2.5m DuPont telescopes. Plates in five passbands (UBVRI) were taken for $\sim$ 15 galaxies. An additional 50 galaxies were observed in v, g, and r passbands with the direct-SIT vidicon camera on the Palomar 1.5m telescope.

The photographic plates were scanned with the IBM microdensitometer (Angilello *et al.* 1984), and the plate densities were converted to relative intensities with sensitometer wedges. The direct-SIT vidicon images were already linear in intensity. Sky and fog (or dark count) corrections were made in the usual way. The B and I (or v and r) images were then aligned and normalized with respect to each other, and a fake G image was produced from their average. The resulting B, G, and I images were combined to produce a color composite, with blue for B, green for G and red for I.

The normalization of the blue and red intensities was done by equating the total light in the two passbands that comes from the region between 1/3 and 2/3 of the optical radius at 25th magnitude per square arcsecond (visual band). This gives an attractive color image of the galaxy, with the blue enhanced just enough to show most of the large regions of star formation. A true color image, made with a blue-to-red normalization derived from the actual B-V and V-I colors of the galaxy, gives a rather uniform, dull, yellow image without much color contrast in the regions of star formation. The solar-like, yellowish color of the old stellar disk dominates the light in all but the brightest regions of star formation.

Two additional color images were made for each galaxy: a rectified image, in which the galaxy is artificially deprojected by stretching the image along the minor axis, and a high contrast image, in which the average radial profile is removed from the rectified version.

3 RESULTS

Figure 1 shows a deprojected color image of NGC 598 (M33) with the radial profile subtracted. The level of detail shown by this figure is much greater than in conventional photographs. A small oval distortion appears in the center, and the outer arms contain many blue regions of star formation.

NGC 598 is an example of a galaxy with a weak density wave and numerous large patches of star formation inside and between the spiral arms. The density wave nature of the arms has been determined from a comparison of azimuthal intensity profiles in the blue and near-infrared passbands (Elmegreen and Elmegreen 1984; hereafter EE84) and from the observation of streaming motions of gas in the arms (Newton 1980). It has an Arm Class 5 (Elmegreen and Elmegreen 1987; hereafter EE87) in a classification system that distinguishes between various levels of spiral arm continuity and symmetry. Approximately 25% of galaxies with bisymmetric spirals ("grand design galaxies") are fragmentary and patchy like NGC 598.

Figure 2 shows a deprojected color image of NGC 3031 (M81) with the radial profile subtracted. This galaxy, along with other well-studied galaxies such as M51 and M100, are examples of Arm Class 12, which is for spiral patterns that are continuous and symmetric over much of the optical disk. Approximately 25% of all grand design spirals are of this type. The arms are stellar density waves as in Arm Class 5, but they are stronger and more regular.

Figure 2 indicates that the spiral structure in NGC 3031 covers only the outer half of the galaxy; it terminates near the bulge. This may be an example where a wave excited in the outer part of the disk, possibly by the companion galaxy M82, is absorbed by an inner Lindblad resonance or reflected by a Q-barrier produced by the bulge.

Figure 3 shows a multiple-arm galaxy, NGC 6946, deprojected with the average radial profile removed. This galaxy is similar to many other multiple arm galaxies, including NGC 5457 (M101) and IC342, and is designated as Arm Class 9. Approximately 34% of grand design galaxies are in this class. They usually have an inner two-arm symmetry, but the multiple outer arms are full of branches and spurs. The inner pattern is often not continuous to the outer region.

NGC 6946 has an inner oval, as shown in Figure 3. It also has an elongated bar-like region of CO emission in the nucleus (Zaritsky and Lo 1986). IC 342 and NGC 5457 have no obvious bars or ovals in optical photographs, but IC 342 has a CO bar in the nucleus (Lo *et al.* 1984). IC 342 is relatively isolated; its nearest neighbor is a small spiral 3° away at a similar velocity. For an assumed heliocentric distance of 3 Mpc, the two galaxies are separated by at least 160 kpc. This is not atypical of the average separation among group galaxies, although it is not clear whether the spiral structure in IC342 is induced by a companion. NGC 5457 has a companion, NGC 5474, which could conceivably have influenced its spiral structure, but there is no particular reason for believing that it actually did (van der Hulst and Huchtmeier 1979).

Figure 4 shows an example of a flocculent spiral galaxy, NGC 5055, which is an Arm Class 2. As in the other figures, this image is rectified, with the average radial profile subtracted. The blue patches in NGC 5055 seem to be strung out in long thin arcs, as if the star formation pattern has long-range order. This resembles computer simulations of stochastic self-propagating star formation (Seiden and Gerola 1982), and may be an example of spiral-like structure that is not related to a density wave.

Approximately 2/3 of all spiral galaxies are flocculent. This fraction, and the other fractional abundances of spiral arm types discussed here, is based on the following data. Arm classes were determined (EE87) for all

08 spiral galaxies in the Second Reference Catalogue (de Vaucouleurs et al. 1976) that are larger than 1 arcmin and less inclined than 55°. Out of this sample, 2/3 are grand design and 1/3 are flocculent. However, the flocculent galaxies are 50% smaller in physical size than the grand design galaxies, based on their average radial velocities and angular sizes. Thus, the angular size-limited survey covers a smaller volume of the sky for the flocculents than for the grand design spirals. The factor of 1.5 size difference implies a sampling volume 3.3 times larger for the grand design galaxies than for the flocculent galaxies. When the number of flocculents is corrected by this factor, the flocculent galaxies account for approximately 2/3 of all spiral galaxies (barred and non-barred) in a given volume.

NGC 5055 and other flocculent galaxies tend to be more homogeneous in the I band than in the B band. This is very different than for grand design galaxies, which appear similar in the two bands. The difference is that most of the structure in the blue band image of NGC 5055 is from star formation, and so it is very weak in the I band. Indeed, the patches look very blue in Figure 4. Most of the large-scale structure in B images of grand design galaxies is from stellar density waves, which look about the same in the B and I bands because the stellar population is compressed somewhat uniformly with respect to age (and color).

There are exceptions to these generalizations, however. NGC 2841 is a flocculent galaxy (Arm Class 3) with short reddish arms that may be small-scale density waves, or ripples (EE84). NGC 628 is a grand-design, multiple-arm galaxy (Arm Class 9) with at least one long blue arm, which may be pure star formation with no underlying wave in the old stellar population (or it could be strongly triggered star formation with an imperceptibly weak wave).

Color images reveal such peculiarities immediately. It is important to remember that all of the galaxies are processed and normalized in exactly the same way, so color differences in the final images correspond to real stellar population differences in the galaxies.

Each computer-enhanced image of a spiral galaxy seems to show something new. For example, there are occasionally weak oval-like structures in the centers of SA galaxies, including one in NGC 5194 (M51). These ovals are more easily seen after rectification. The distributions of spiral arms in the outer regions of multiple arm galaxies, such as NGC 5457, NGC 628, and NGC 4254, are often asymmetric. This is particularly clear after the average radial profile has been subtracted. Spiral arms in some grand design galaxies can also be traced for as much as 540° around the nucleus. This is evident on radially subtracted images because these show structure in both the inner and outer regions of a galaxy without saturation of the nucleus.

The color images are not only interesting from a scientific point of view but they are also quite beautiful. Perhaps this may explain, more than anything else, the attraction to galactic spirals that we share with Professor Lin.

References

Angilello, J., Chiang, W.-H., Elmegreen, D.M. and Segmuller, A. 1984, Goddard Astronomical Microdensitometry Conference, ed. D.A. Klinglesmith, NASA Conference Publication 2317, 229.

de Vaucouleurs, G., de Vaucouleurs, A. and Corwin, H. 1976, Second Reference Catalogue of Bright Galaxies, Univ. of Texas Press, Austin, Texas.

Elmegreen, D.M. and Elmegreen, B.G. 1984, *Astrophys. J.Suppl.*, **54**, 127 (EE84).

Elmegreen, D.M. and Elmegreen, B.G. 1987, *Astrophys. J.*, **314**, 3.

Lo, K.Y., Berge, G.L., Claussen, M.J., Heiligman, G.M., Leighton, R.B., Masson, C.R., Moffet, A.T., Phillips, T.G., Sargent, A.I., Scott, S.L., Wannier, P.G., and Woody, D.P. 1982, *Astrophys. J. Letters.*, **282**, L59.

Newton, K. 1980, *Monthly Not. Roy. Astron. Soc.*, **190**, 689.

Schweizer, F. 1976, *Astrophys. J. Suppl.*, **31**, 313.

Seiden, P.E. and Gerola, H. 1982, *Fundamentals of Cosmic Physics*, **7**, 241.

van der Hulst, J.M., and Huchtmeier, W.K. 1979, *Astron. Astrophys.*, **78**, 82.

Zaritsky, D. and Lo, K.Y. 1986, *Astrophys. J.*, **303**, 66.

Figure captions: 1.(top) M33 from a composite of B and I band photographic plates, rectified to face-on orientation, and with the average radial light profile subtracted. The vertical lines represent 100 pixels (left) and the radius at 25 magnitudes per square arcsecond surface brightness in the B band. 2. (bottom) M81, as in Figure 1.

Figure 3 (top) NGC 6946, and figure 4. (bottom) NGC 5055, as in Figure 1.

RADIAL AMPLITUDE VARIATIONS
IN SPIRAL ARMS

Preben Grosbøl[*]
European Southern Observatory,
Karl-Schwarzschild Straße 2,
D-8046, Garching, Fed. Rep. of Germany.

The amplitudes and phases of the spiral patterns in the two late type spiral galaxies NGC 1980 and NGC 1376 were obtained from CCD images in the V and i filters using Fourier transform techniques. The radial variation of these quantities was analyzed and compared with the density wave theory. The amplitude ratio between the $m=2$ and $m=4$ components of the spiral pattern in NGC 1080 indicates that its density wave show non-linear effects. In NGC 1376 two separate spiral modes were found.

1. INTRODUCTION

The density wave theory by Lin and Shu (7) has given the dynamic explanation of the grand design spiral patterns seen in many galaxies. Observational evidence for the existence of density waves in spiral galaxies was given *e.g.* by Schwiezer (8) studying the surface brightness of spiral galaxies and by Visser (9) analyzing the velocity field of M 81.

New linear detectors like CCD's have made it possible to study the surface brightness on galaxies with higher precision compared with photographic techniques In the present paper the CCD surface photometry of two late type spiral galazies is analyzed to obtain the amplitudes and phases of their spiral patterns as function of radius. The azimuthal shape of the spiral arms can indicate if non-linear effects as discussed by Contopoulos and Grosbøl (2) are present. Further, the data can show the existence of different spiral mode in the galaxies.

2. DATA

Two Sc-Scd galaxies NGC 1080 and NGC 1376 were chosen for the investigation. The first galaxy is a grand design spiral while the latter has a more flocculent pattern. This makes it possible to study typical azimuthal intensity profiles of spiral arms for these spiral types. Two color bands

---

[*] Based on observations at the ESO La Silla Observatory.

Galaxy	NGC 1080	NGC 1376
Position angle	$-77°$	$-73°$
Inclination angle	$33°$	$21°$
Radius $R(I_c=23.0^m)$	$31''$	$58''$
Pitch angle i_2	$-28°$	$21°$
Interarm $(V-I)_c$	1.12^m	1.12^m

Table 1. Derived parameters for observed galaxies.

were used, namely the i filter (Wade *et al.* (10)) mainly showing the old disk population and the V filter (Bessell (1)) which is more affected by the young stars in the arm region.

2.1 Observations

The observation were made with a RCD CCD at the danish 1.5m telescope at ESO, La Silla. Three exposures of 10 min. in each filter were compared and added in order to remove artifacts and cosmic ray events from the final intensity maps. The V and i intensities were transformed into the VI system of Cousins (3) using standard stars in the E-regions.

2.2 Analysis

Although the intensity in I_c mostly originates from old disk stars, it is also effected by a significant amount of flux from young stars. In order to do a zero-order correction of this effect it was assumed that the color variation across arms originated from a change in the ratio between a young and an old stellar population each with constant color index and that internal absorption was insignificant. Taking $(V-I)_c = 0.00^m$ for the young population and measuring the color index of the old stars in the interarm regions the corrected surface brightness in I_c is :

$$F_I^* = \frac{r-R}{r-1} F_I \qquad (1)$$

where F_I is the observed I_c intensity. The intensity ratios are defined as $r = (f_v^a/f_i^a)/(f_v^d/f_i^d)$ for the stellar populations and $R = (F_v^a/F_i^a)/(F_v^d/F_i^d)$ for the observed surface brightness where v and i give the color while a and d denote arm and disk/interarm, respectively.

The position and inclination angles of the galaxies were determined by fitting ellipses to the outer isophotes of the I_c images (see Grosbøl(5)). Amplitudes and phases of the $m = 2$ and $m = 4$ components were computed for each radii by Fourier transforming the azimuthal intensity profiles (Grosbøl (4)) of the deprojected maps. A radial step of $2''$ was used being slightly smaller than the seeing. The analysis was done out to a radius corresponding to the 23^m isophote in I_c in order only to use high signal-to-noise data. The derived quantities for the two galaxies are given in Table 1.

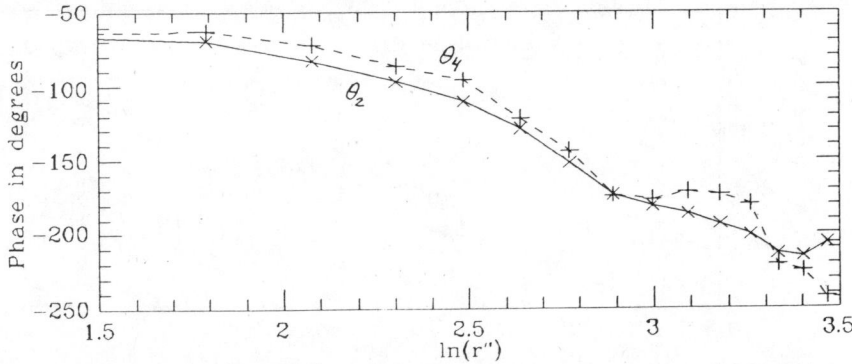

Fig. 1. Phases of the θ_2- and θ_4-components of the spiral pattern in NCG 1080 as function of radius ($\times\times\times$ θ_2-component; $+++$ θ_4-component).

1. RESULTS AND DISCUSSIONS

The phases of the θ_2- and θ_4-components (*i.e.* $m = 2, 4$) for NGC 1080 are given in Fig. 1 as function of radius. A logarithmic scale is use so that logarithmic spirals will appear as straight lines in the plot. In the inner 7" the phase is roughly constant due to the bulge which in the deprojected map appears as a bar like structure. The phase variation in the region outside 7" can be approximated by a logarithmic spiral with a pitch angle of $\approx -28^c$. The deviation from a straight line around $\ln(r) = 2.9$ is likely caused by errors in the projection angles. The small phase difference between the θ_2- and θ_4-component indicates that the $m = 4$ component is associated to the main two armed spiral. Since the two components are in phase, the intensity profile of the spiral is more peaked than a single sinusoidal wave.

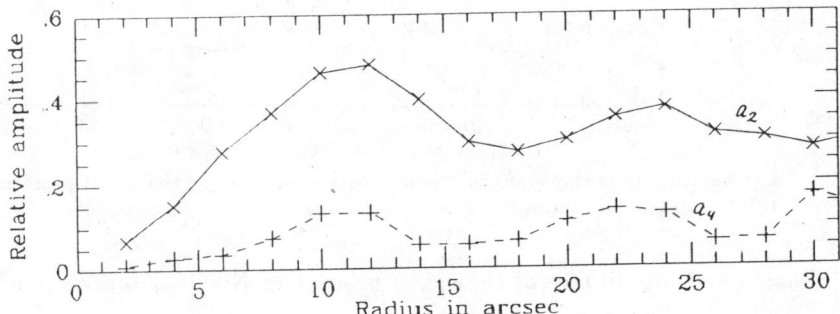

Fig. 2. Relative amplitudes of the θ_2- and θ_4-components of the spiral pattern in NCG 1080 as function of radius ($\times\times\times$ θ_2-component; $+++$ θ_4-component).

The corresponding relative amplitudes a_2 and a_4 for NGC 1080 are shown in Fig. 2. Except for the inner region the amplitude a_2 is in the range of 0.3-0.4 of the axisymmetric intensity in I_c whereas a_4 is significantly smaller.

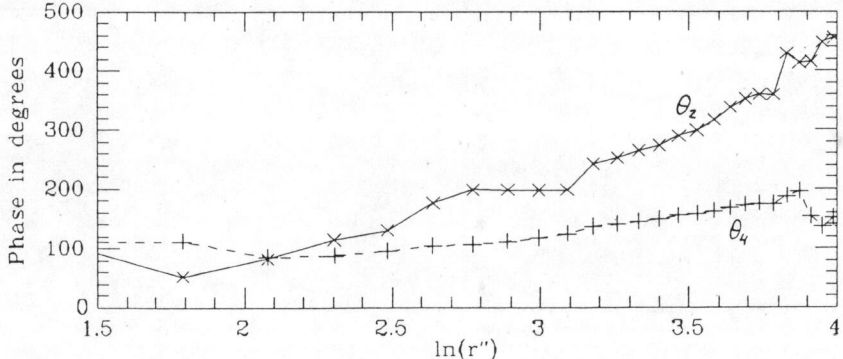

Fig. 3. Phases of the θ_2- and θ_4-components of the spiral pattern in NCG 1376 as function of radius ($\times \times \times$ θ_2-component; $+ + +$ θ_4-component).

The ratio a_4/a_2 is roughly 0.2 in the outer part of the galaxy. Since I_c intensity has been corrected for population effects this indicates that also the surface density of the two armed spiral is peaked. The amplitude ratio between a_2 and a_4 is of the same order as that found by Contopoulos and Grosbøl (2) for a model galaxy with a strong spiral perturbation. This suggests indicate that the spiral pattern in NCG 1080 is so strong that non-linear effects in the density wave start to be significant.

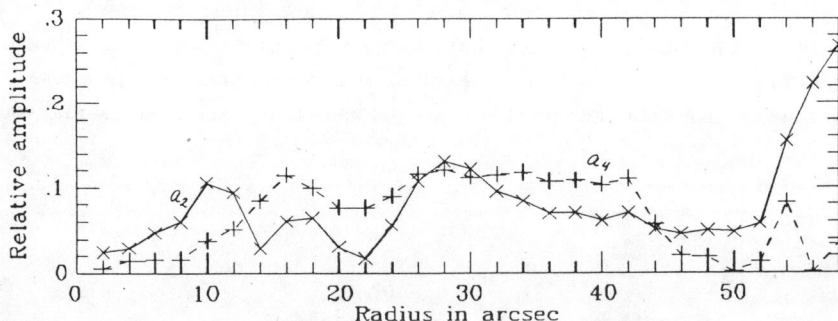

Fig. 4. Relative amplitudes of the θ_2- and θ_4-components of the spiral pattern in NCG 1376 as function of radius ($\times \times \times$ θ_2-component; $+ + +$ θ_4-component).

Phases and amplitudes of the spiral pattern in NGC 1376 are given as function of radius in Fig. 3 and 4, respectively. The phase variations for both the θ_2- and θ_4-component are well represented by logarithmic spirals, however, with significantly different pitch angles. The θ_2 mode is more tightly wound ($i_2 \approx 21°$) while the $m = 4$ component is quite open with $i_4 \approx 48°$. This difference strongly suggests that the two components in this galaxy are unrelated and represent separate spiral modes (see Lau, Lin and Mark (6)). In fact, NGC 1376 also shows a significant $m = 3$ component.

The well correlated phases of both θ_2 and θ_4 further emphasis that the spiral pattern consists of several mode of a density wave and not random spurs of arms.

The spiral amplitudes in NGC 1376 shown in Fig. 4 are much smaller than in NGC 1080 and of the order of 0.1. The two armed spiral is dominant only in the inner 12" while the four armed pattern is as strong or stronger in the outer parts. The strong increase of the θ_2 amplitude beyond 50" is caused either by a warp in the disk or small errors in the projection angles.

1. CONCLUSIONS

The grand design spiral galaxy NGC 1080 has a strong two armed pattern with a relative amplitude of the order of 40% in population corrected I_c color. The θ_2- and θ_4-components are in phase and the ratio a_4/a_2 between their amplitudes is ≈ 0.25. This suggests that the spiral perturbation in NGC 1080 is so strong that non-linear effects in the density wave start to be important.

The flocculent spiral galaxy NGC 1376 has much weaker spiral arms. Both a two and four spiral structure is present in this galaxy. These two patterns have different pitch angles indicating that they represent two separate modes of a density wave.

Referrences
1. Bessell,M.S., *Pub.A.S.P.* **88**, 23 (1976).
2. Contopoulos,C., Grosbøl,P., *Astrom. Astrophys.* **155**, 11 (1986).
3. Cousins,A.W.J., *Mem.R.A.S.* **81**, 25 (1976).
4. Grosbøl,P., in *Astronomical papers dedicated to Bengt Strömgren*, Eds. A. Reiz and T. Andersen, Copenhagen, 397 (1978).
5. Grosbøl,P., *Astron. Astrophys. Suppl.* **60**, 261 (1985).
6. Lau,Y.Y, Lin,C.C., Mark,J.W.-K., *Proc. Natl. Acad. Sci. USA* **73**, 1379 (1976).
7. Lin,C.C., Shu,F.H., *Astrophys. J.* **140**, 646 (1964).
8. Schweizer,F., *Astrophys. J. Suppl.* **31**, 313 (1976).
9. Visser,H.C.D., *Astron. Astrophys.* **88**, 149 (1980).
10. Wade, R.A., Hoessel, J.G., Elias, J.H., Huchra, J.P., *Pub. A. S. P.* **91**, 35 (1979).

DISTRIBUTION FUNCTIONS FOR TRIAXIAL GALAXIES

C. Hunter
Department of Mathematics, Florida State University
Tallahassee, Florida 32306, U.S.A.

The problem of constructing self-consistent stellar dynamic models of triaxial galaxies is outlined. Methods of finding models for which the gravitational potential is of Stäckel form, and the manner in which these are non-unique, are then described.

1. STATEMENT OF THE PROBLEM

The basic aim of the research that is described here is to understand the dynamics of isolated elliptical galaxies. An elliptical galaxy is primarily a collection of stars. The scale of the galaxy and the number of stars are both so large that the stars can be treated as a continuum, describable by a density distribution function $f(\mathbf{r}, \mathbf{v})$. This function gives the density of stellar matter in the six dimensional phase space of position $\mathbf{r}$ and velocity $\mathbf{v}$. Because collisions between individual stars are infrequent on the time scale of the age of the universe, f satisfies the collisionless Boltzmann equation

$$\mathbf{v}.\nabla f - \nabla V . \frac{\partial f}{\partial \mathbf{v}} = 0. \tag{1}$$

Here $V(\mathbf{r})$ is the gravitational potential in which the stars move.

For a self-consistent model of an elliptical galaxy, the potential V must be that produced by the stars. Self-consistency can be ensured by selecting, at the outset, some spatial density field $\rho(\mathbf{r})$ for which the gravitational potential V is known, and then requiring that the solution of equation (1) satisfies also the equation

$$\rho(\mathbf{r}) = \int f(\mathbf{r}, \mathbf{v}) d^3\mathbf{v}. \tag{2}$$

Equation (1) allows the distribution function f to be a function of any time-independent integral $I_j(\mathbf{r}, \mathbf{v}) =$ constant of the equations of motion

$$\frac{d\mathbf{v}}{dt} = -\nabla V. \tag{3}$$

However, it is also necessary to require that f depends on the phase-space coordinates $(\mathbf{r},\mathbf{v})$ through isolating integrals of the motion only[6,9]. The energy per unit mass $E = \frac{1}{2}\mathbf{v}^2 + V$ provides one isolating integral, whatever the potential, and other isolating integrals exist for restricted forms of V[10].

Elliptical galaxies were once thought to be oblate spheroids, whose oblateness is due to rotation. The component of angular momentum L_z about the z-axis of symmetry is then a second isolating integral, and any $f = f(E, L_z^2)$ satisfies equation (1). Equation (2) then gives a double integral equation

$$\rho(\varpi, z) = \int f(E, L_z^2) d^3\mathbf{v}, \quad \varpi = \sqrt{x^2 + y^2}, \qquad (4)$$

that can generally be solved uniquely to give f for some given density field ρ[8]. Here and subsequently, the solution for f is of interest only if it is non-negative everywhere.

There is now considerable evidence, both observational and theoretical, that elliptical galaxies rotate only slowly and that more complicated models with three isolating integrals and triaxial density distributions are needed to describe them[1]. The first such non-rotating model was constructed numerically by Schwarzschild[11]. Schwarzschild did not solve equation (1) directly. Instead, he computed a large number of orbits of equations (3) in an assumed triaxial potential, and found how long each orbit spends in each of the cells into which he had subdivided the mass. Then he solved a finite difference analog of equation (2) to obtain a non-negative combination of his orbits that correctly reproduces the mass in each cell. Although Schwarzschild's potential is unlikely to possess any exact isolating integral other than the energy E, his computations showed that his equations are near-integrable and possess three effective integrals of motion. Therefore he did implicitly compute a solution of equation (1) that depends on three isolating integrals.

2. STÄCKEL MODELS

It has recently been realized that further analytical progress can be made by seeking distribution functions for a class of non-rotating gravitational potentials for which three exact isolating integrals are known explicitly[2,7]. These are the Stäckel potentials[12], which have the symmetric form

$$V = -\frac{(\lambda - a^2)(\lambda - c^2)G(\lambda)}{(\lambda - \mu)(\lambda - \nu)} - \frac{(\mu - a^2)(\mu - c^2)G(\mu)}{(\mu - \nu)(\mu - \lambda)} \\ - \frac{(\nu - a^2)(\nu - c^2)G(\nu)}{(\nu - \lambda)(\nu - \mu)}, \quad (5)$$

in terms of ellipsoidal coordinates λ, μ, ν. The ellipsoidal coordinates of any point in space are the three roots for τ of the cubic equation

$$\frac{x^2}{\tau - a^2} + \frac{y^2}{\tau - b^2} + \frac{z^2}{\tau - c^2} = 1. \quad (6)$$

where $a^2 \geq b^2 \geq c^2$ are positive constants, with the roots being labelled according to their magnitudes

$$\lambda \geq a^2 \geq \mu \geq b^2 \geq \nu \geq c^2. \quad (7)$$

Stäckel potentials can represent realistic mass distributions. One example is the "perfect" ellipsoid[4,7] of mass M with density

$$\rho = \frac{M}{\pi^2 abc(1 + x^2/a^2 + y^2/b^2 + z^2/c^2)^2}, \quad (8)$$

whose equidensity surfaces are similar ellipsoids. Its potential is given by

$$G(\tau) = \frac{\gamma M}{\pi} \int_0^\infty \frac{du}{u + \tau} \left[\frac{u + b^2}{(u + a^2)(u + c^2)}\right]^{\frac{1}{2}}. \quad (9)$$

where γ is the gravitational constant.

The dynamics in a Stäckel potential is simple because separate equations of motion are obtained in each ellipsoidal coordinate. Specifically, each generalized momentum p_τ satisfies the equation

$$p_\tau^2 = \frac{(\tau - a^2)(\tau - c^2)[E + G(\tau)] - (\tau - c^2)I_2 - (\tau - a^2)I_3}{2(\tau - a^2)(\tau - b^2)(\tau - c^2)}, \quad (10)$$

where E is again the energy and I_2 and I_3 are two other independent isolating integrals of the motion. Equation (2) is now a triple integral equation from which the distribution function $f(E, I_2, I_3)$ must be determined for some given density $\rho(\lambda, \mu, \nu)$, such as (8). But, unlike the oblate spheroidal case (4), f is not now fully determined by ρ. The reason why many different distribution functions can be found for triaxial Stäckel models is a consequence of the fact that, unlike the oblate spheroids, they allow more than one kind of orbit.

3. THE ELLIPTIC DISK

The simplest illustration is provided by the extreme case of the "perfect" elliptic disk[3] in which motion is confined to the x-y plane ($\nu = c^2$), $I_3 = 0$, and the limit $c \to 0$ is taken in the volume density (8) to give a surface density

$$\Sigma(\lambda, \mu) = \int \rho dz = \frac{M}{2\pi ab(1 + x^2/a^2 + y^2/b^2)^{3/2}}. \tag{11}$$

Coordinates λ and μ in the x-y plane are given by the roots of the quadratic that is obtained from (6) when the z^2-term is dropped. Curves of constant λ are ellipses, while those of constant μ are hyperbolae. All curves have the points $x = 0$, $y = \pm(a^2 - b^2)^{1/2}$ as foci. The portion of the y-axis between the foci is the degenerate ellipse $\lambda = a^2$, whereas the portion outside the foci is the degenerate hyperbola $\mu = a^2$. (See Figure 1).

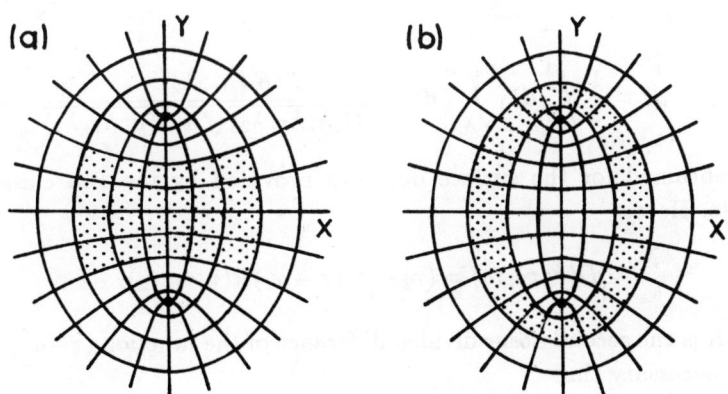

Fig. 1. Coordinate curves in the x-y plane. The foci are marked by large dots. Parts (a) and (b) show examples of regions covered by a box and a tube orbit respectively.

The requirement that $p_\tau^2 \geq 0$ for both $\tau = \lambda$ and μ allows two possible kinds of bound ($E \leq 0$) motion. The first, for which $I_2 \leq 0$, is a libration in both coordinates which is confined by turning points $\lambda_0(E, I_2)$ and $\mu_0(E, I_2)$ to the region

$$a^2 \leq \lambda \leq \lambda_0, \quad b^2 \leq \mu \leq \mu_0. \tag{12}$$

This is called a box orbit. The second, for which $I_2 > 0$, is called a tube orbit because the motion in μ is an unrestricted rotation, while the libration in λ confines the motion to the region

$$\lambda_1(E, I_2) \leq \lambda \leq \lambda_2(E, I_2), \tag{13}$$

between two ellipses. (See Figure 1).

It is mathematically more convenient to use the turning points just defined, rather than E and I_2, as integrals of the motion. Also, we shall seek distribution functions of the form $F_b(\lambda_0, \mu_0)$ and $F_t(\lambda_1, \lambda_2)$ for the box and tube orbits respectively. The functions F give mass densities in the spaces of their respective arguments, and differ from our earlier f by Jacobian factors. Two separate integral equations

$$\Sigma_b(\lambda, \mu) = \frac{1}{2} \int_\mu^{a^2} d\mu_0 \int_\lambda^\infty d\lambda_0 \frac{F_b(\lambda_0, \mu_0)}{\sqrt{D(\lambda; \mu_0, \lambda_0)}\sqrt{-D(\mu; \mu_0, \lambda_0)}}, \tag{14}$$

and

$$\Sigma_t(\lambda, \mu) = \frac{1}{2} \int_{a^2}^\lambda d\lambda_1 \int_\lambda^\infty d\lambda_2 \frac{F_t(\lambda_1, \lambda_2)}{\sqrt{D(\lambda; \lambda_1, \lambda_2)}\sqrt{-D(\mu; \lambda_1, \lambda_2)}}, \tag{15}$$

are obtained for the surface densities provided by the two classes of orbits. Here

$$D(\tau; \tau_1, \tau_2) = (\tau_2 - \tau)(\tau - \tau_1)R(\tau, \tau_1, \tau_2), \tag{16}$$

and R is the second order divided difference of the function $(\tau - a^2)G(\tau)$. It is necessary that

$$\Sigma_b(\lambda, \mu) + \Sigma_t(\lambda, \mu) = \Sigma(\lambda, \mu). \tag{17}$$

Because both kinds of orbits are able to provide density over the whole plane while only their sum (17) is specified, a considerable variety of distribution functions is possible. The most straightforward way of constructing one is governed by the fact that (14) is a Volterra integral equation for determining F_b from a known Σ_b, whereas equation (15) for an unknown F_t is not of Volterra type. Another consideration is that $\Sigma_b(\lambda, a^2) = 0$ if F_b is to be finite, and hence the density on the y–axis outside the foci $y = \pm(a^2 - b^2)^{\frac{1}{2}}$ must be provided by the tube orbits that cross it, rather than by the edges of marginal box orbits. We can therefore proceed as follows: first select some non-negative tube distribution function $F_t(\lambda_1, \lambda_2)$ for which $\Sigma_t(\lambda, a^2) = \Sigma(\lambda, a^2)$, and then solve equation (14) for F_b with the now known $\Sigma_b(\lambda, \mu) = \Sigma(\lambda, \mu) - \Sigma_t(\lambda, u)$. The determination of the unique solution of (14) is formally straightforward, except near the foci, because (14) is a generalized Abel equation. Numerical methods are needed to compute specific solutions and to verify that they have non-negative F_b, as required. Near the foci where $\lambda = \mu = a^2$, both D's in the denominator of (14) contain two factors that can vanish, and (14) is no longer of generalized Abel type. Analysis of this more complex region shows that $F_b(\lambda_0, \mu_0)$ is finite but discontinuous, because multi-valued, at the extreme corner point $\lambda_0 = \mu_0 = a^2$ of the envelopes of box orbits.

4. ELLIPSOIDS

The coordinate surfaces of the full three dimensional problem are ellipsoids ($\lambda = $ constant) and hyperboloids of both one ($\mu = $ constant) and two sheets ($\nu = $ constant). Besides box orbits for which there is now an extra condition $c^2 \leq \nu \leq \nu_0$ added to those of (12), three kinds of tube orbits are possible. As well as tubes that encircle the short z–axis as with the elliptic disk (librations in λ and ν, rotation in μ), there are two kinds of tubes that encircle the long x–axis. Both involve rotations in ν, but their librations in the other coordinates differ. For the outer long axis tubes, equations (10) cause the motion to be confined to a part of the region $\lambda_1 \leq \lambda \leq \lambda_2, \mu_1 \leq \mu \leq a^2$, between two ellipsoids, whereas inner long axis tubes are confined to a region of the form $a^2 \leq \lambda \leq \lambda_2$, $\mu_1 \leq \mu \leq \mu_2$ between two one-sheeted hyperboloids.

With four kinds of orbits, a great variety of solutions becomes possible. There are now two restrictions provided by the four integral equations that are the analogs of equations (14) and (15) of the disk if infinite distribution functions are to be avoided. They are that the density on the surface $\nu = b^2$, a degenerate hyperboloid of two sheets, must be provided by long axis tubes while the density on the degenerate one-sheeted hyperboloid $\mu = a^2$ must be provided by short axis and outer long axis tubes. Once some combination of tube orbits that satisfies both these restrictions has been chosen, there remains a three variable Volterra equation to be solved for the box orbit distribution function. This equation is again of generalized Abel type except on the two curves $\mu = \nu = b^2$ and $\lambda = \mu = a^2$, which form the edges of the two degenerate hyperboloids mentioned earlier. At these edges, the three dimensional box distribution function has the same kind of discontinuities that the two dimensional distribution F_b of §3 has at its focus.

5. CONCLUDING REMARKS

The dynamic properties of the many models that can be constructed by the methods described here are not yet fully understood. They need to be compared more fully with the models that have been computed by Teuben[14] for elliptic disks (11) and by Statler[13] for ellipsoids (8), but using numerical methods that are more closely related to the method of Schwarzschild[11] than to the methods described in §§3 and 4 here. They need also to be related to observations, to elucidate which of the possible models are favored by real galaxies.

6. ACKNOWLEDGMENTS

The results described in §§3 and 4 were obtained in collaboration with Tim de Zeeuw and Martin Schwarzschild[3,5]. My work has been supported in part by NSF grant DMS-8420624.

My interest in astrophysical problems began in the early 1960's when I had the good fortune to be a new Ph.D. at M.I.T. It was an exciting time because C.C. Lin was then beginning to master a new field and to lay the foundations for his work on spiral wave theory. I learned a great deal from C.C., both then and subsequently, and am delighted to be able to contribute this paper in his honor.

References

1. Binney, J.J., *Comments on Astrophys.* **8**,27 (1978).
2. de Zeeuw, P.T. *Mon. Not. R. Astron. Soc.* **216**, 273 (1985).
3. de Zeeuw, P.T., Hunter, C., and Schwarzschild, M., *Astrophys. J.*, **317**, 607 (1987).
4. de Zeeuw, P.T., and Lynden-Bell, D., *Mon. Not. R. Astron. Soc.*

215, 713 (1985).
5. Hunter, C., and de Zeeuw, P.T., work in progress.
6. Jeans, J.H., *Mon. Not. R. Astron. Soc.* **76**, 71 (1915).
7. Kuzmin, G.G., in *Dynamics of Galaxies and Clusters*, ed. T.B. Omarov, (Alma Ata, Akademiya Nauk Kazakhshoj SSR), 71 (1973).
8. Lynden-Bell, D., *Mon. Not. R. Astron. Soc.* **123**, 447 (1962).
9. Lynden-Bell, D., *Mon. Not. R. Astron. Soc.* **124**, 1 (1962).
10. Lynden-Bell, D., *Mon. Not R. Astron. Soc.* **124**, 95 (1962).
11. Schwarzschild, M., *Astrophys. J.* **232**, 236 (1979).
12. Stäckel, P., *Math. Ann.*, **35**, 91 (1890).
13. Statler, T.S., *Astrophys. J.*, in press.
14. Teuben, P., *Mon. Not. R. Astron. Soc.*, in press.

Shapes of Star-Gas Waves in Spiral Galaxies

Stephen H. Lubow

Space Telescope Science Institute and Johns Hopkins University

Density-wave profile shapes are influenced by several effects. By solving viscous fluid equations, the nonlinear effects of the gas and its gravitational interaction with the stars can be analyzed. The stars are treated through a linear theory developed by Lin and coworkers. Short wavelength gravitational forces are important in determining the gas density profile shape. With the inclusion of disk finite thickness effects, the gas gravitational field remains important, but is significantly reduced at short wavelengths. Softening of the gas equation of state results in an enhanced response and a smoothing of the gas density profile. A Newtonian stress relation is marginally acceptable for HI gas clouds, but not acceptable for giant molecular clouds.

1. INTRODUCTION

The gas gravitational field can have a significant influence on the dynamics of spiral waves. This paper extends the results of a recent study of star-gas density waves.[1] In that work, clouds form a viscous fluid in which the atomic and molecular gas are are treated as a single component which interacts gravitationally with the stars that form a second component. Steady-state, nonlinear viscous fluid equations were derived and solved for a zero-thickness, isothermal gas disk.

An alternative to the approach described here of directly solving fluid equations for the gas is an N-body simulation for clouds.[2] The steady-state fluid model suppresses time-dependent processes, but hopefully provides some significant detail about mean flow properties. The other approach may provide some time-dependent information, but may have some of the usual noise problems associated with N-body calculations of spirals (see discussion of noise in reference 3).

2. SHOCK SMOOTHING VIA GRAVITY

Perhaps the most surprising result of our recent study was that gas gravity actually smooths gas shock profiles (see Fig. 1). This result is actually not very surprising when one considers the fact that for a fixed peak to average gas density ratio at the solar circle *pressure increases the tendency of gas to shock, while gravity opposes pressure to decrease the tendency to shock.* Intuition may suggest that shocks are avoided when pressure is included, because pressure can prevent streamlines from crossing and thus prevent shocks. In local Galactic flow, away from resonances, streamlines do not cross, even in the absence of pressure. Instead, shocks arise as the flow makes a sonic transition from supersonic to subsonic speeds, relative to the wavefront.[4] Therefore, contrary to intuition, raising the sound speed of the gas somewhat actually makes it easier for gas to make a sonic transition from its upstream supersonic velocity, since there is less change in velocity required for a sonic transition and hence for a shock.

Gas gravity pulls gas into the shock front where there is a high concentration of gas. The gravity counters the effect of pressure forces which decelerate the flow through the strong pressure gradient at the front. As a result, the flow profile becomes smoother and thus the gas pressure gradient less. The effect of gravity can be understood mathematically in terms of the inviscid (nonviscous) flow equation[4]

$$\frac{1}{r \sin(i)} \frac{\partial u}{\partial \varphi} = u(2\Omega v + f(\varphi))/(v^2 - c^2) \qquad (1)$$

where u and v are the gas velocities respectively perpendicular and parallel to the spiral wave in the wave pattern frame, and $f(\varphi)$ is the star-gas gravity perpendicular to the wave front at radius r.

Wherever shocks occur, the denominator in the above expression must vanish, which occurs when the gas makes a sonic transition relative to the density wave front. However, the possible singularity in $\frac{\partial u}{\partial \varphi}$ can be cancelled by a simultaneously vanishing numerator. Shocks can occur at a sonic transition *where rotational, $2\Omega v$, and gravitational, f, forces are unbalanced.*

Fig. 2 shows that gas gravity causes the rotational and gravitational forces to be better balanced at the first sonic point, and hence reduces the strength of the shock (Rotational and gravitational forces always nearly cancel at the second sonic point, since this point corresponds to flow from subsonic to supersonic speeds.) This effect is due to lumpy nature of the gas gravity near the density peak. In other words, short wavelength gravitational forces play a critical role in determining gas profile shapes, since those forces must compete with pressure forces, which are intrinsically short wavelength forces. However, short wavelength gravity is most susceptible to finite thickness corrections.

3. FINITE-THICKNESS EFFECTS

Finite thickness corrections most strongly affect the gravitational forces.[5] The strength of finite-thickness effects depends on the product of the wavenumber and layer thickness. The gas contains more density at high harmonics (short wavelengths) than the stars. Although the gas disk is substantially thinner than the stellar, finite thickness corrections for the high order gas harmonics can be as large as the corrections for the stars at its fundamental mode. A full solution for the nonlinear flow with finite thickness corrections involves solving for vertical motions, since the gas disk thickness is likely to vary significantly in phase. Instead, the finite thickness corrections are approximated by suppressing the phase dependence of the vertical disk structure. A further approximation is that the gas and stars as have a constant density within their respective layers.[5] We define thickness
$H_j = 0.5\sigma_j(r)/\rho_j(r, z = 0)$,
for species j. For the gas, the thickness H_g is chosen as 110 pc, a value about midway between that for atomic (150 pc) and molecular gas (65 pc), which contribute about equally to the local surface density.[6] For the stars, a thickness H_* of 700 pc is adopted.[7]

For each harmonic n, the gravitational forces in the star and gas dynamical equations are modified by the correction factor
$\int_{-\infty}^{\infty} \rho_{0i}\phi_{nj}(z)/(\sigma_i\phi_{nj}(H=0))$
where i and j can represent stars or gas. This correction factor works well in the linear flow case.[8]

The gas gravity is significantly softened as a result of this correction. However, our current best model of the solar circle with about 14% gas ($70 M_\odot/pc^2$ total, $10 M_\odot/pc^2$ gas)[9] in Fig. 1 still shows important effects of the gas gravity. If thickness H_g for the gas layer had been over 250 pc, the gas density profile would have been noticeably steeper.

4. Softened Equation of State

Several suggestions exist that the equation of state for a fluid of clouds may be softer than isothermal.[10] An extremely softened model was constructed. This model represents energy dissipation through cloud collisions (proportional to the square of the cloud density times the cube of the random velocity) which balances energy injection by supernova explosions (proportional to the local cloud density).(Actually other effects are also present, such as heating through dissipation of differential rotational energy, see equation (16) in reference 12.) This extreme model assumes that the supernova explosions occur so much later than the cloud collisions which form the supernova precursors, that the local energy injection rate depends only on the local cloud density and not the local rate of cloud collisions. This model yields a polytropic pressure relation of $p \propto \sigma^{1/3}$. (At the opposite extreme is a model in which supernova form immediately after cloud collisions. This model yields an approximately isothermal equation of state.)

Numerical results indicate that, like gravity, the softening of the equation of state results in a smoother gas density profile when comparing models with the same peak to average density ratio and the the mean cloud random velocity. The basic reason for the smoothing is that the sound speed of the gas drops upon compression. A lower sound speed means the gas is harder to shock (see section 2). It can be shown that the sonic condition for a shock can be expressed as

$$\sigma_{crit} = (\nu^2/x_{p0})^{1/(\gamma+1)} \qquad (2)$$

where σ_{crit} is the critical gas surface density (normalized to the circular average gas density) needed to produce a shock, $\nu = m(\Omega_p - \Omega)/\kappa$, $x_{p0} = k^2 c_0^2/\kappa^2$, with sound speed c_0 where the normalized density is unity, and $\gamma = \frac{\sigma}{p}\frac{dp}{d\sigma}$. At the solar circle, for $x_{p0} \approx 0.2$, one finds $\sigma_{crit} \approx 1.6$ for the isothermal case and $\sigma_{crit} \approx 2.1$ for $\gamma = 1/3$.

5. VISCOSITY

Viscosity in disks has some new features.[11,12] Gas in a galaxy is typically nonaxisymmetric and highly perturbed. In such a situation, the usual Newtonian stress relation can be applied only when the collision frequency is much greater than any velocity derivatives in the gas flow. From Fig. 3, we see that this condition is marginally satisfied for HI gas clouds.

For giant molecular clouds (GMCs), the collision frequency is more than 5 times smaller than for HI clouds. The Newtonian stress relation cannot be rigorously applied, since the above condition is not well satisfied. Assuming GMCs are long-lived, rotational effects are likely to be important. The *effective mean free path* is likely to be limited by Galactic rotation, since the epicyclic radius of around 300 pc is smaller than the mean free path for cloud collisions.[13] A proper treatment of this problem would require a solution of the Boltzmann or Krook equation, as has been applied in the context of planetary rings.[14,15] In fact, in this regime, the sign of the effective coefficient of shear viscosity can sometimes even be negative.[15]

Simulations with gas cloud mean free paths that are long compared with the epicyclic radius may not yield simple results. For example, the front thickness will not generally be a few collisional mean free paths. Such an effect is seen in some N-body results.[16]

Acknowledgements

I thank Pawel Artymowicz and Rick White for useful discussions.

References

1. Lubow, S.H., Balbus, S.A. and Cowie, L.L, *Ap.J.*, **309**, 496, (1986).
2. Roberts, W.W., this symposium.
3. White, R.L., preprint (1987).
4. Shu, F.H., Milione, V., and Roberts, W.W., *Ap.J.*, **183**, 819 (1973).
5. Toomre, A., *Ap.J.*, **139**, 1217 (1964).
6. Gordon, M.A. and Burton, W.B., *Ap.J.*, **208**, 346 (1976).
7. Freeman, K. *in IAU Symposium No. 106*, 118 (1983).
8. Vandervoort, P.O., *Ap.J.*, **161**,, 87 (1970).
9. Sanders, D.B., Solomon, P.M., and Scoville, N.Z., *Ap.J.*, **276**, 182 (1984).
10. Cowie, L.L., *Ap.J.*, **236**, 868 (1980).
11. Goldreich, P., and Tremaine, S., *Icarus*, **34**, 227 (1978).
12. Shu, F.H., and Stewart, G.R., *Icarus*, **62**, 360 (1985).
13. Fukunaga, M., *Publ. Astron. Soc. Japan*, **35**, 173 (1983).
14. Shu, F.H., Dones, L., Lissauer, J.J, Yuan, C., and Cuzzi, J.N., *Ap.J.*, **299**, 542 (1985).
15. Borderies, N., Goldreich, P. and Tremaine, S., *Icarus*, **68**, 522 (1986).
16. Roberts, W.W, and Hausman, M., *Ap.J.*, **277**, 760 (1980).

SPIRAL DENSITY—WAVE GAS PROFILE

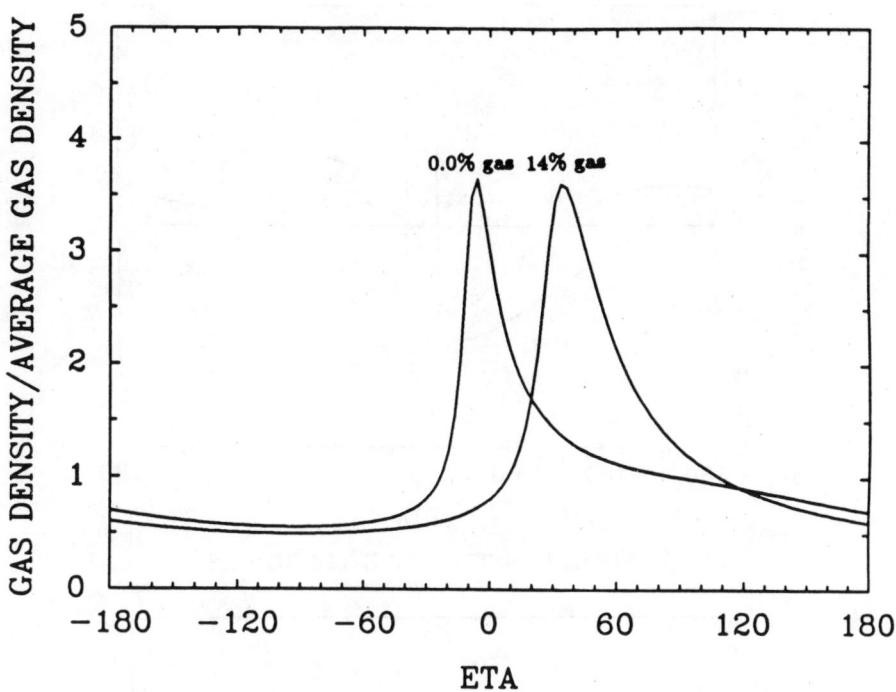

Figure 1. Normalized gas density as a function of phase (ETA = 2φ) along the solar circle for models with 0.01% and 14.3% gas. In these models, gas disk thickness H_g is 110 pc and stellar disk thickness H_* is 700 pc. The stellar wave amplitude was adjusted so that the ratio of peak to average gas density was 3.6 for the two cases.

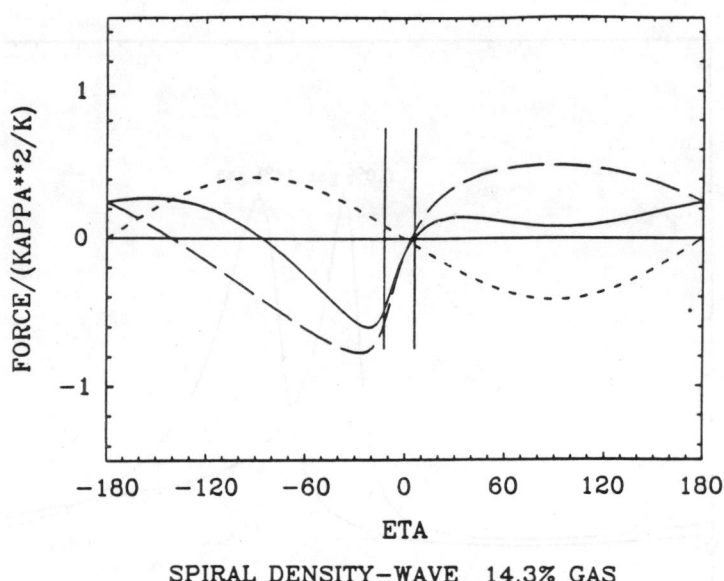

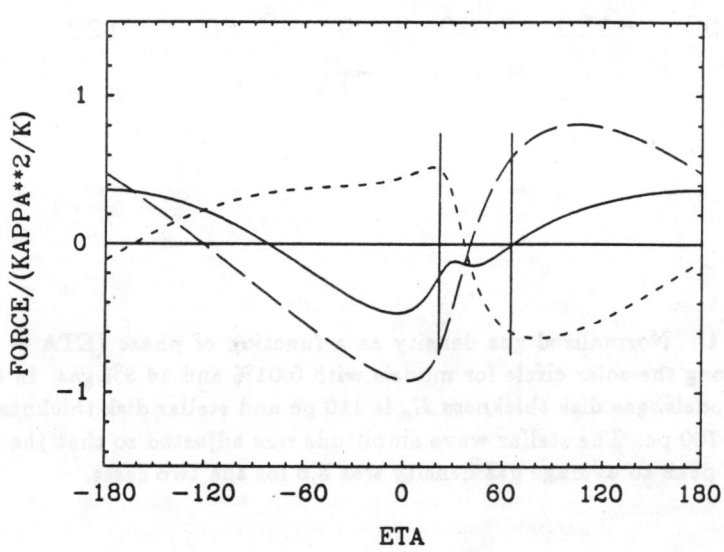

Figure 2. In short dashed lines is the total gravitational force per unit mass perpendicular to spiral arms; in long dashed lines is the perturbed rotational force per unit mass. In solid lines is the sum of the perturbed rotational and gravitational forces. The vertical lines mark the location of the two sonic points of the gas flow relative to the spiral wave pattern.

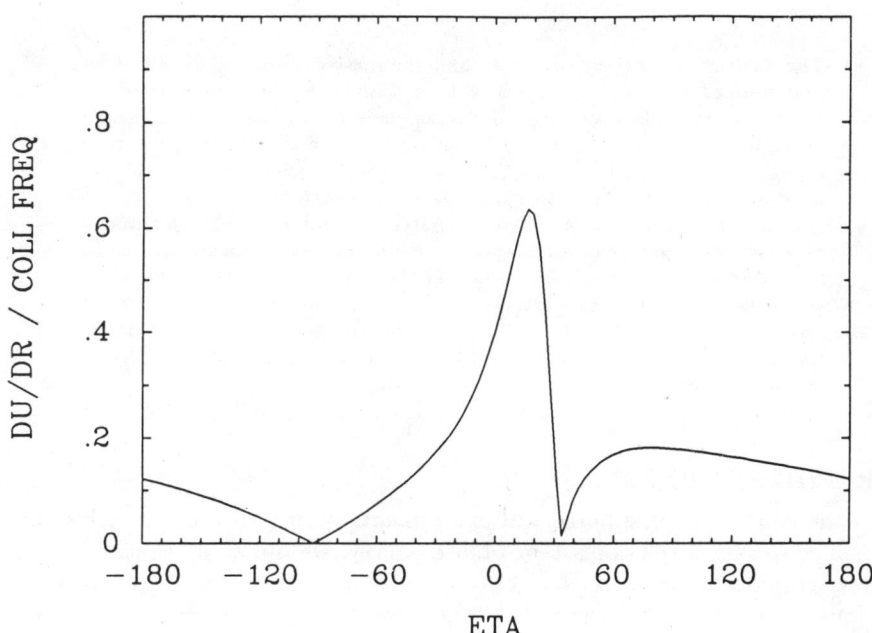

Figure 3. The spatial derivative of the gas velocity perpendicular to spiral arms divided by the local cloud collisional frequency for HI gas clouds is plotted as a function of phase. Mean free path at average surface density is 170 pc; 1-d random velocity is 8 km/sec.

GOINGS-ON AT THE CENTER OF A GALAXY

R. H. Miller and B. F. Smith
University of Chicago and NASA Ames Research Center

The nucleus of a galaxy orbits around the mass centroid. This effect was first noticed in a numerical experiment with a disk embedded within a self-consistent oblate galaxy, but checks show that it has been present in all earlier experiments as well. The effect is physical. Orbital periods are essentially $(G\rho_{\text{central}})^{-1/2}$, and excursions reach a distance from the centroid at which $\rho(r) \sim (1/2)\rho_{\text{central}}$. We use the extremum of the smoothed gravitational potential to define a "potential center," which we take to represent the nucleus of a real galaxy. Numerical experiments are usually started with the artificial condition that the potential center coincides with the mass center. Orbital motions then appear overstable, starting from a seed provided by $\sqrt{N}$ noise. The amplitude doubles in $6 - 10$ orbital periods. It is not clear what physics limits the amplitude. Motion pictures will be shown.

1. THE PROBLEM

In the course of some numerical experiments with a disk of particles embedded within a self-consistent oblate galaxy, we did some checks with a new graphics display at NASA-Ames and zoomed in on the central regions. Much to our surprise, the center was whipping about in a manner that looked erratic on that display. As experimenters, we were immediately concerned lest something had gone awry—in numerical experiments that something is most likely to be a numerical instability. Subsequent checks have convinced us that we were seeing a real physical effect, but that kind of experience is frightening to an experimenter. The motions that trouble us are quite small: the centermost particle moves only as far from the center as the radius of a sphere that would contain about 4×10^{-4} of the galaxy's mass.

We routinely make motion pictures of all our experiments. The picture shown at this Symposium is one of them. It shows nothing unusual going on near the center—but on closer check, the center can be seen to move around with small amplitude. Its motion is not irregular. It is not catastrophic.

A few details of the experimental setup are necessary to set the scene. The "galaxy" consists of 400,000 particles in the form of an $n = 3$ polytrope, squashed to form an oblate spheroid. The entire configuration rotates to encourage it to remain oblate and spheroidal. The rotation axis and the spheroid axis coincide with the $z-$axis. The disk set into this galaxy consists of 1024 particles initially flat, its normal making some angle to the rotation axis. Disk particles are initially in centrifugal balance with the galaxy's radial gravitational force. The balance is not perfect, so some radial oscillations of the disk soon become apparent. There are so few disk particles that the disk acts like a massless tracer. It does little to modify the self-consistent structure of the main galaxy. Disk particles and galaxy particles are identical except that disk particles are yellow and galaxy particles are blue in the motion picture. All 1024 disk particles are in the motion picture, but only 1024 galaxy particles.

2. IS IT PHYSICAL OR NUMERICAL?

One check consisted of looking at motions of the normal to a little patch at the center of the disk. The final scenes of the movie show the motions of the 80 particles closest to the center at the start. That little patch remains remarkably planar through all the whipping. We can define a unit vector normal to the patch, and we can plot the trajectory the tip of that unit vector traces out on the unit sphere throughout the experiment. As part of the process of finding the unit vector, we can explore how fat that little patch is. It fattens very little through the experiment. This part of the movie was made *after* we had run this check, but it confirms that we were dealing with a nearly planar patch.

The trajectory of the normal is shown in Figure 1. It loops and wiggles across the figure, looking like something drawn by a small child. The loops represent a free nutation, like that of the spinning earth. The drift of the loop centers is precession. Note that the loop centers drift at nearly constant latitude, another signature of precession. Apart from increases and decreases in the amplitude of the nutation, it looks like drawings in your mechanics books of a nutating and precessing top. The amplitude changes because of interference with other oscillations that we have not fully identified. Orbital motion of the disk center does not show up in this plot. Figure 1 is not a picture generated by a numerical instability. That should be much more irregular, and it should not look plausible. The disk farther away from the center acts very similarly. There is a fascinating complexity and richness with travelling wavelike patterns, various oscillations, evidence of interference among the oscillations, coupling of radial and vertical oscillations, and so on. We don't have time to go into all this at this Symposium.

3. WHAT CAUSES IT?

Once we had convinced ourselves that we were dealing with a genuine physical process, there arose the problem of exploring that process. This was quite lengthy (it took a couple of years), more a groping than a systematic attack on the problem. We had to grope because we didn't know what we were looking for.

We first checked to see whether we were dealing with a self-consistent effect within the disk by immobilizing the galaxy. A self-consistent disk was run in that potential. Self-consistency effects due to the disk were measurable, but the disk center simply continued to orbit with whatever

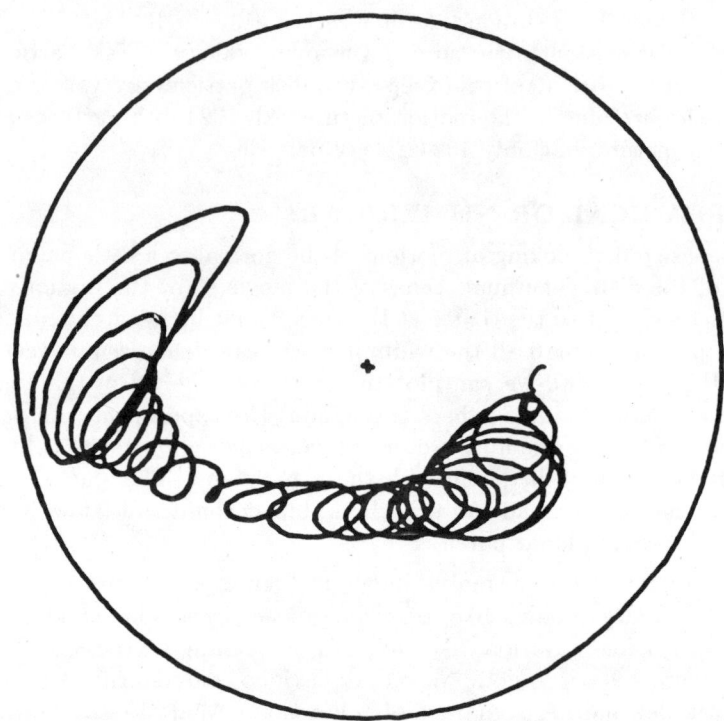

Figure 1. Trajectory of normal to center of disk on the unit sphere as seen looking down the galaxy's rotation axis. The trajectory starts at the right side with small loops and winds around downward and to the left, ending on the left side with large loops.

amplitude we started it with. There were no signs of a growing oscillation. There are some pretty experiments in this set, but they will probably forever remain stacked among lots of other movies we've never fixed up for general circulation.

We next studied a self-consistent galaxy without a disk to see whether the disk was messing up the galaxy somehow. The location of the "potential center," defined by the smoothed galaxy potential, provided a tool for this check. It need not remain fixed, even though the equations of motion require that the mass centroid remain fixed. The potential center started to oscillate even without the disk. So we were dealing with some kind of self-consistent effect within the galaxy.

Next, we made a guess that we might be dealing with an instability at $l = 1$. This kind of oscillation is known in stars (although it is not unstable in stars) and it is reasonable to expect a galaxy to be able to support it as well. We started a self-consistent galaxy with its center displaced relative to its envelope (with the global mass centroid undisplaced), and turned it loose. The $l = 1$ disturbance showed a strongly damped oscillation ($Q \sim 5$ or 6) at a frequency not far from that of the dominant radial pulsational $l = 0$ mode. No instability. The $l = 1$ pattern was imprinted in the velocities rather than in the configuration for a second check. Results confirmed those from the first check. The dominant radial pulsational mode has noticeably higher Q, around 10, although it typically damps fairly strongly as well.

So far we've found several things that did not cause our oscillations. But then, what causes them? We finally nailed this down with a couple of lucky guesses that were confirmed by additional experiments.

4. NAILING IT DOWN

The potential near the center of a galaxy is harmonic—it is the potential of a (possibly anisotropic) harmonic oscillator. A particle, initially at rest at the center, feels tugged first in one direction and then in another, by the wanderings of the potential center. It acts like a harmonically bound particle in Brownian motion. That is a well-known problem. We checked whether this might be the character of the motions we were observing by integrating the response of a harmonically bound particle to forces it would feel due to the wandering center, and found those motions to match the wanderings of the centermost particle of our disk amazingly well.

Some results from this experimental check are shown in Figure 2.

The actual motion of the centermost particle is shown in the topmost trace, and the potential center wandering in the middle trace. Finally, a separate integration of a harmonically bound test particle under the tuggings of the actual potential center wanderings is shown as the bottom trace. Notice the similarity of the top and bottom tracks–even to shapes of individual wiggles. The growth in amplitude with time looks steadier than does an harmonically bound particle driven by Gaussian noise with a similar power spectrum, suggesting that something more is going on. That something can be seen in the center track, where one can imagine that the potential center wanderings near the end are beginning to follow the particle. This similarity has been confirmed by cross-correlating the potential center displacements with the central particle position. They go together (with the appropriate phase for a growing oscillation) with a signal-to-noise of about 3 at the end of the experiment. There is no detectable cross-correlation at the beginning. The system is actually unstable, with feedback coming from the potential center's following the driven particles. But that is a physical, not a numerical, instability.

The effect has been seen with 100,000, 400,000, and with 1,024,000 particles. The growth rate is independent of the number of particles (when

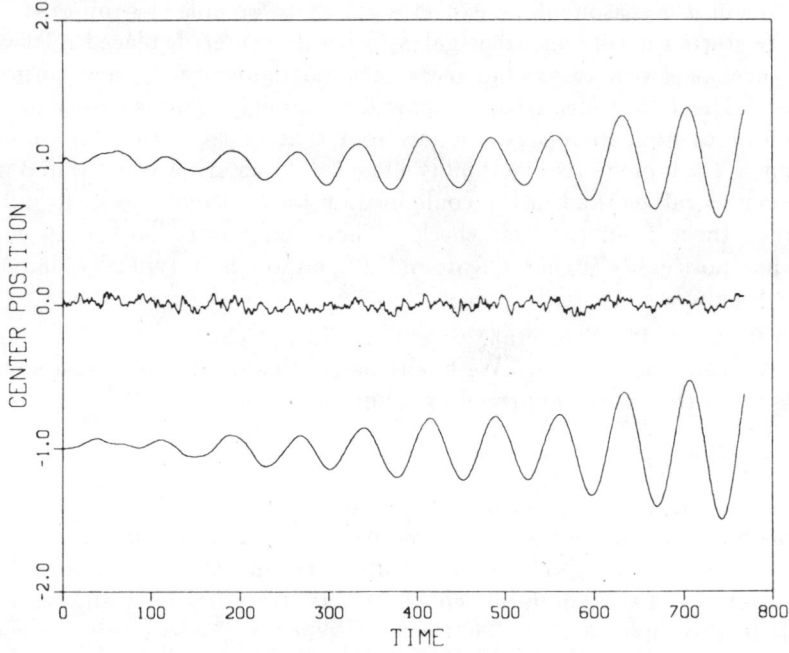

Figure 2. Plots of position *vs* time. Time in the integration is plotted horizontally, the x-component of position vertically. The motion of the centermost particle is shown in the top trace, the potential center wandering in the middle trace, and the motion of a harmonically bound test particle subject to forces produced by that potential center wandering in the bottom track. All three tracks have the same vertical and horizontal scales.

measured in units of the galaxy's crossing time). However, the starting amplitude is about proportional to $1/\sqrt{N}$. Separate checks, with a sequence of initial loads produced by different runs of random numbers, produced a scatter of potential center positions about the same as the early stages of an actual integration with the same number of particles. That confirms that $\sqrt{N}$ noise can cause the potential center to differ from the mass center by the amount observed at the beginning of an experiment. The growth takes over from that seed.

The frequencies of the harmonic potential were first set by counting cycles of particle oscillation. Agreement in plots like Figure 2 was pretty good. But then we trimmed the frequencies a bit (a 5% change). The microscopic details of the plot are quite sensitive to the frequency, although the general character stays much the same. Figure 2 shows our second guess. Agreement is just as good in the other two directions.

We seem to have a density wave orbiting the mass centroid. Professor Lin made density waves popular, and here we have one in rather different circumstances from those in the disks of spiral galaxies. Particles are trapped in the shifting potential minimum near the center, deepening the potential minimum. But the deeper potential minimum can capture a few more particles, and so on. The process feeds on itself. It looks overstable in one component of the motion, but it looks like a growing orbit in three dimensions.

Whatever it is that moves down near the center need not be a density wave, however. A "thing" (e.g., a massive black hole) near the center should show similar motions. The galaxy's potential can be distorted in the neighborhood of the "thing" if the "thing" wanders away from the mass centroid. The distortion of the galaxy's potential by an off-center point mass is overlooked in the usual arguments that a massive black hole cannot wander far from the mass centroid.

Center motions have been noted in $n-$body integrations since the earliest days. von Hoerner, Aarseth, and Wielen, in the 1960's, referred radial dependences to a "density center" as origin. Its wanderings appeared erratic and were considered an awkward feature one had to work around. The present work brings new aspects that show center motions to be physically important. (1) Motions are ordered and follow reasonable trajectories, rather than being random. (2) Amplitudes grow when the system is started with small displacement. This suggests some kind of instability. (3) We see the wanderings in systems with many more particles.

5. WHERE DO WE GO FROM HERE?

Admittedly a start from a nice quiet condition, driven by $\sqrt{N}$ noise, seems kind of artificial. But the real artificiality came when we started our experiment with the potential center atop the mass centroid. Real galaxies are very unlikely to form that way. Instead the most dense region in the galaxy, the "nucleus," and the mass centroid will normally be at different locations. Most of us learned about galaxies by thinking of them as represented in some sort of polar (cylindrical or spherical) coordinate system with a singular line or point at the center. But a physical galaxy doesn't know anything about singular points in coordinate systems. I'm beginning to appreciate why mathematicians prefer coordinate-free spaces. Coordinate systems condition the way you think about a problem.

The two centers may relax toward each other in the more realistic case where they do not coincide at the start. But the overstability says they can't get too close. Some (nonlinear) physical process has to intercede either to limit a growing motion or to damp one that has gotten too large. We have no clue what that physics is.

Yet another concern is that we don't understand the physics well enough to calculate a growth rate. We can't tell how much mass participates in the density wave, or what spatial form it takes. We don't know whether it is a local disturbance or a global mode. We don't have a theory.

And finally we must determine whether this phenomenon is important in real galaxies. Off-center nuclei have been observed in a few galaxies: M33, NGC 3379, and NGC 3389, for example. More galaxies might show off-center nuclei if the effect were systematically sought. Noise in the observations seems to exclude searches in globular clusters, although it is tempting to speculate that this effect is somehow involved in the four out of five globular clusters that do *not* show a cusp at their centers.

The wanderings are also very suggestive of things that happen at the center of our Galaxy. Observations are often interpreted as if the center region were in a stationary steady state equilibrium, but the work reported here shows that assumption to be untenable. Gas, trapped in a potential that sloshes around like this, must do some very interesting things. If a radio-source jet could be nailed to the center, the nutation we've noted would make it appear to precess with a period near $1/\sqrt{G\rho}$, about a quarter of a million years at the center of our Galaxy.

There are lots of fascinating possibilities. We hope some of them capture your imagination.

We thank the organizers of this Symposium, and CC in particular, for the opportunity to show you some of these results and to share these beautiful motion pictures with you. You can imagine our hopes and excitement over future prospects. Many thanks are also due to Dr. Althea Wilkinson of the University of Manchester, who started us down the path that led to these results. This work has been partially supported by Cooperative Agreement NCC 2-265 between NASA-Ames Research Center and the University of Chicago.

N-BODY SIMULATIONS OF THE CLOUDY INTERSTELLAR MEDIUM IN DENSITY WAVE-DOMINATED GALAXIES: ORBIT TRAPPING, SLOSHING, SELF GRAVITY, AND SPIRAL STRUCTURE

William W. Roberts, Jr., David S. Adler, and Glen R. Stewart
University of Virginia

Fluid dynamical problems in galaxies have been of considerable interest to C.C. Lin and a focus of his own research for more than two decades, dating from the early 1960s. The structure and dynamics of galaxies, the grand design spiral structures often observed, and the behavior of density wave modes constitute but a few of C.C. Lin's own research interests and those which he has helped cultivate and stimulate in his students and colleagues over the years. It was C.C. Lin's influence on this graduate student (WWR) in applied mathematics here some twenty years ago which strongly motivated initial research on galaxies and on fluid dynamical phenomena evident in galactic systems.

An N-body, cloud-particle computational code is developed for the purpose of isolating and studying various physical mechanisms and dynamical processes underlying the "cloudy" gaseous interstellar medium in density wave dominated galaxies.[1,2,3,4] The gaseous interstellar medium is simulated by a system of particles, representing clouds, which orbit in a spiral-perturbed galactic gravitational field. Self gravitational effects of the clouds are included via Fourier Transform techniques.[5,6,7] The cloud-particles undergo dissipative collisions with other clouds and experience velocity-boosting interactions with expanding supernova remnants. Associations of protostars form in clouds following such collisions and supernova interactions but take finite times before becoming active themselves and undergoing their own supernovae events.

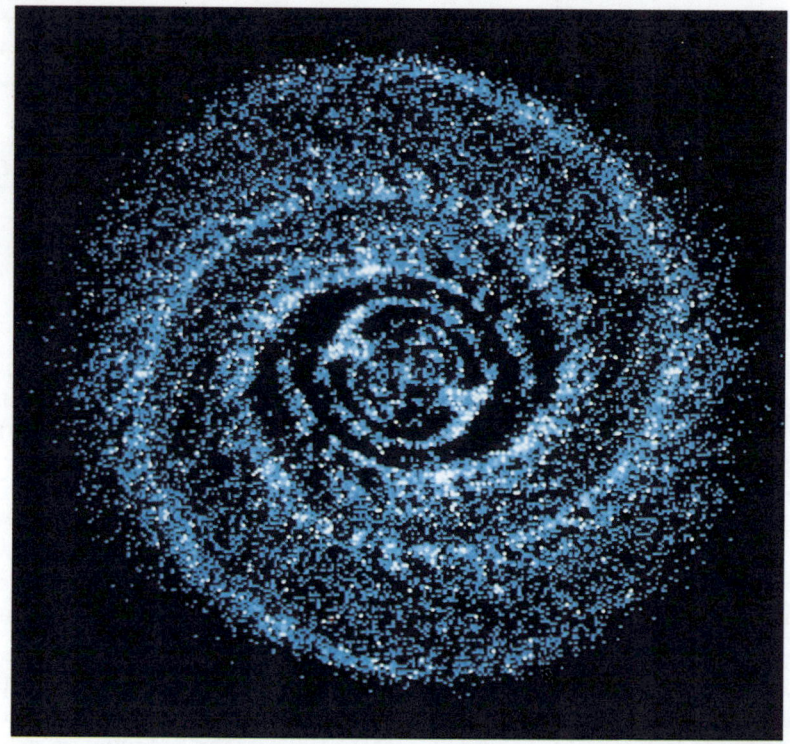

Figure 1.

Figure 1 shows the results of one representative self gravitating simulation in which the gas to stellar disk mass fraction is 10%. Plotted here, in a photographic intensity map at one sample time epoch [900 Myr] during the computations, is the computed global distribution of gas clouds, represented by patches, and young to middle-aged stellar associations active during the past 60 Myr, represented by white dots. The raggedness and patchiness of the global gaseous spiral structure is indeed evident, with massive aggregations of clouds and giant cloud complexes appearing throughout the arm regions. One primary motivation for formulating and developing this type of computational approach is the ability to simulate and study such ragged, realistically-disorderly model "snapshots" of galaxies.

We adopt the philosophy that the results of large-scale, numerical computations can often be best understood by systematically disassembling the model into its basic components so that the main effects of each ingredient on the end result can be separately analyzed and understood. We follow this philosophy by examining and re-examining the present model at substantially reduced levels of complexity through isolation and removal of competing effects. In Figure 2 at the most reduced level of complexity studied, the stripped-down "skeletal" form of the model contains only the gaseous component represented by a cloud system of ballistic particles, driven by the spiral-perturbed gravitational field. Shown in the left frame is the spatial distribution of gas cloud centers. What is striking is the still strong spiral response of the collective system of gas clouds. Thus, dissipative cloud collisions and gas self gravity, both of which are absent, do not constitute the primary factor that underlies the strong global pile-up of clouds in the spiral arms. To appreciate what underlies this pile-up, we display the gas velocity field in the right frame. Dots mark the current positions of the clouds; line segments point along velocity directions. Note the strong convergence of flow in the regions of the spiral arms, followed by divergence, evidenced through the systematic difference between orbital directions of clouds entering the arms and those of clouds leaving the arms. Indeed the strong pile-up of gas clouds in global spiral arms is interpreted to be a manifestation of the collective orbit crowding of cloud-particle orbital trajectories, driven by the spiral perturbed gravitational field. In the presence of a 5% to 10% spiral perturbing force field, the low dispersion cloud system is capable of participating in strong orbit crowding, leading to a strong pile-up of clouds in the arms, even without the aid of dissipative cloud-cloud collisions or self-gravitational effects.

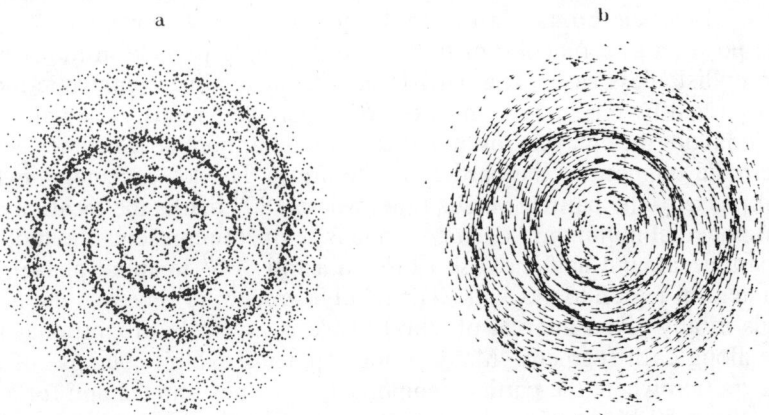

Figure 2.

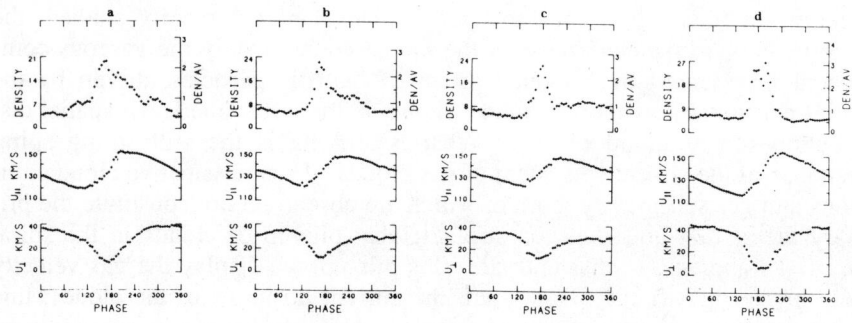

Figure 3.

Plotted in Figure 3 versus spiral phase about a representative annulus are the distributions of number density and the velocity components perpendicular and parallel to spiral equipotential curves for this collisionless gas cloud system at four representative epochs [t = 850, 930, 950, and 1170 Myr]. First and foremost note that the density enhancements attain peak-to-mean values [n_{max}/<n>] up to 3:1 and arm-to-interarm contrasts up to 6:1. The velocity components reflect the magnitude of the gas cloud system's systematic motion. These features "slosh" back and forth in spiral phase and do not reach a "steady state."

These results of the N-body simulations can be understood in terms of individual particle orbits. In Figure 4, the upper panel shows the computed radial position as a function of time for one sample particle in the stripped-down collisionless case. The particle undergoes nonlinear epicyclic motion, characterized by repeatable sequences of radial oscillations. The spiral forcing modulates the epicyclic oscillations in a regular nonlinear manner, leading to periodic trapping of the particle in the arms. The lower panel shows the spiral phase of the particle versus time, where 180^0 marks the location of the spiral potential minimum. Each continuous segment represents the motion of the particle around half the disk (through a full 360^0 in spiral phase), from one interarm region to the next. The relative amount of time that the particle spends in any given interval of spiral phase is inversely proportional to the slope along each segment. Many of the segments exhibit "bumps" of retrograde motion where the particle temporarily becomes trapped and for a short time (about 50 Myr) moves backwards across the arm, before continuing in its forward motion. Such trapping events cause the particle to spend a large fraction of its total orbital time within a spiral arm. It is now clear that the

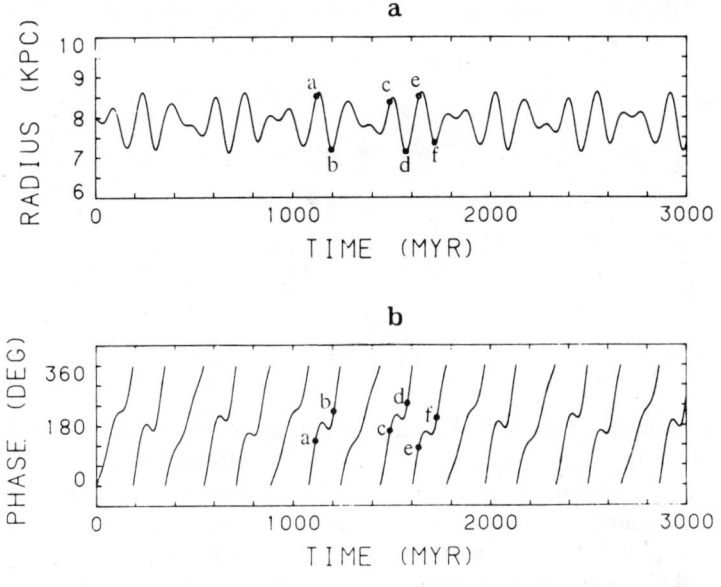

Figure 4.

strong global spiral response of the collective cloud system in the N-body simulations can be largely attributed to this tendency of individual particles to become trapped, and spend a large fraction of their time in the region of the spiral potential minima.

To understand the important role which is played by dissipative cloud-cloud collisions, we now include collisions between clouds. Figure 5 shows the results for a collisional case. Plotted versus spiral phase about the same annulus are the distributions of number density and the perpendicular and parallel velocity components at four representative epochs. Apart from local stochastic variations, a "steady state" has been reached and is being maintained. Stronger density enhancements are exhibited here with peak-to-mean values $[n_{max}/<n>]$ typically of 3:1 and arm-to-interarm contrasts up to 8:1. These steady cloud density enhancements are in striking contrast to those enhancements in the collisionless case which undergo "sloshing." With dissipative cloud collisions present, the $u_\perp$ component exhibits a sharp deceleration from supersonic to subsonic near 180^0 spiral phase and the parallel velocity component exhibits a sharp pointed trough in the region just upstream. These features signify the presence of a galactic shock.

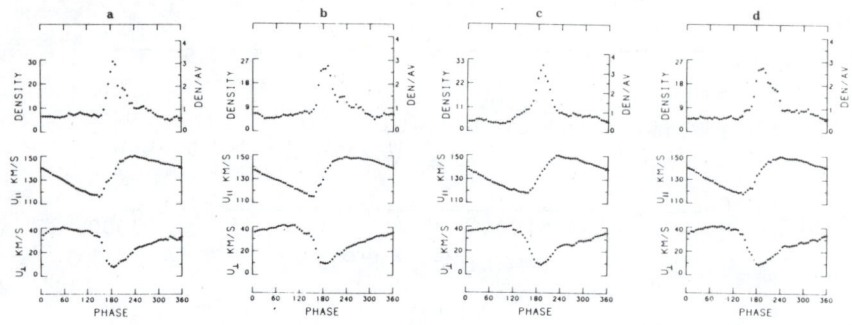

Figure 5.

Figure 6 shows a photographic intensity map of the gas cloud distribution in this collisional case. Dissipative cloud-cloud collisions suppress the sloshing, play an important steadying role for the cloud system's global spiral structure, and damp the relative velocity dispersion of clouds in massive associations, thereby aiding the effective assembling of giant cloud complexes. The assembly of these complexes is evident on kpc scales, producing

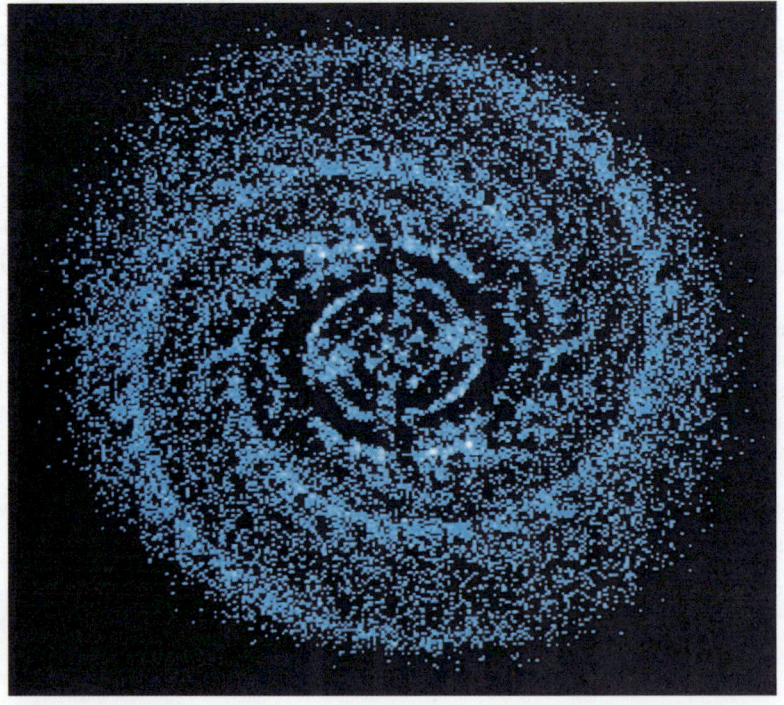

Figure 6.

moderate overall raggedness in the global spiral structure, particularly in the inner half of the disk where the giant cloud aggregations appear to be most prominent.

Figure 7 shows the results when the self-gravitational effects of the gas are put back into the model (10% gas mass fraction). In the preceding case without gaseous self-gravity, the gas response along the global spiral wave arms was only moderately ragged with modest clumping into large complexes, entirely due to the presence of energy dissipating cloud-cloud collisions. Here with gaseous self-gravity, the gas cloud system exhibits a much more ragged global spiral structure. Massive aggregations and large complexes of clouds are much more evident. Of prime importance are the gravitational effects driving the low velocity dispersion gas, including its own self gravity. Gravitationally driven crowding and temporary trapping of cloud orbits in spiral arms, along with dissipative cloud-cloud collisions, underlie the aggregation of the clouds into the giant cloud complexes as well as the organization of the complexes along the ragged global spiral structures. The loosely-associated aggregations and giant complexes of clouds are found to continually disassemble and reassemble rather naturally over time under the influence of the various physical mechanisms and dynamical processes underlying the cloudy ISMs in these model spiral galaxies.

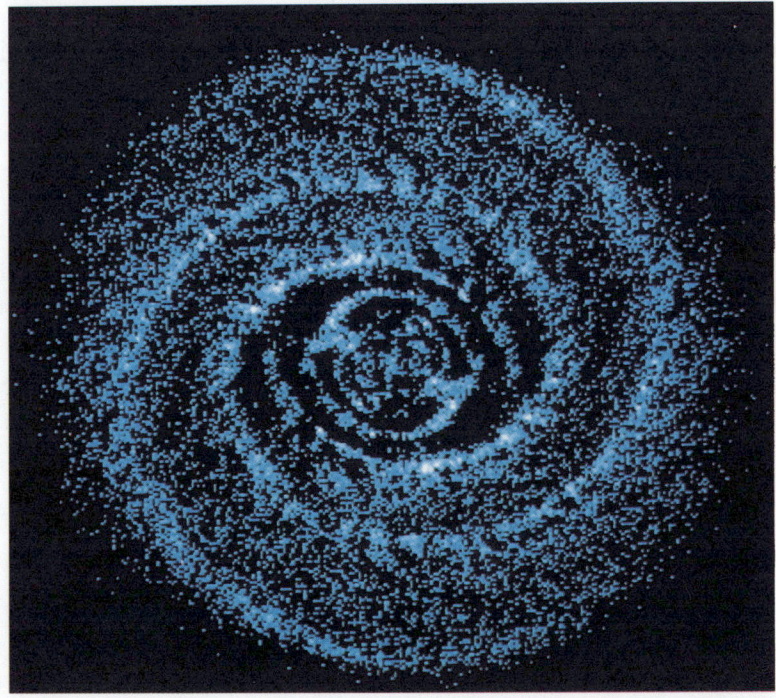

Figure 7.

An intriguing challenge is already emerging within the current theoretical and computational studies. This bears on the important multifold role played by the gas in its self gravitation. As already evident in Figure 7 when self-gravitational effects of the gas are present, the global distribution of the gas becomes quite ragged, with massive aggregations and large complexes of clouds appearing throughout the global spiral structures. For higher gas content, the situation becomes even more interesting. For cases in which the gas mass fraction exceeds a moderate value (25%), what is most striking is the strength with which gas cloud aggregations and complexes interact on local scales and strongly perturb the global two-armed gaseous spiral structure. These perturbations take the form of multitudes of spurs, feathers, and secondary features. The most urgent questions requiring immediate attention in view of these new results are the fundamental questions of stability and persistence. To what extent do these continually-evolving, transient manifestations naturally perturb and rip apart the global spiral structure? To what extent do they enhance the global structure? In the present studies, local spurs, feathers, and secondary features continually break apart and reform as the loosely-associated aggregations and giant complexes of clouds continually disassemble and reassemble over time. Can the global spiral structure persist, in particular for higher gas mass fractions, where there are significant perturbations on local scales? If so, under what conditions?

Acknowledgements

This work was supported in part by the National Science Foundation under grants AST-82-04256 and MCS-83-04459 and the National Aeronautics and Space Administration under grant NAGW-929.

References

1. Roberts, W. W. and Hausman, M. A. 1984, Ap. J., 277, 744.
2. Hausman, M. A. and Roberts, W. W. 1984, Ap. J., 282, 106.
3. Roberts, W.W., and Stewart, G.R. 1987, Ap. J., 314, 10.
4. Adler, D.S., 1987, Ph.D. Dissertation, University of Virginia, in preparation.
5. Miller, R.H. 1976, J. Comput. Phys., 21, 400.
6. Miller, R.H. and Smith, B.F. 1979a, Ap. J., 227, 407.
7. Miller, R.H. and Smith, B.F. 1979b, Ap. J., 227, 785.

THE DISPERSION RELATION AND
THE MASSES OF GALACTIC DISKS

W. W. Shane
Sterrenkundig Instituut
Katholieke Universiteit
Toernooiveld
6525 ED Nijmegen, THE NETHERLANDS

Summary

The dispersion relation, derived from the density-wave theory of spiral structure, relates the surface density of the disk to other structural and kinematical properties of a galaxy. Under favorable conditions this permits an estimate of the disk mass. An application of this method to the Sombrero galaxy is described and possible further applications to other galaxies are discussed.

The problem of the distribution of mass in and around galaxies continues to attract much attention. The mass within any radius can be estimated by measurements of the rotation velocity at that radius, when this information is available, although the results are somewhat model dependent. We find that there is far more mass present than would be suspected from the distribution of light, and locating this dark matter has become an urgent problem. Most of this matter appears to be associated with the galactic haloes. The distribution of this matter can be studied most conveniently in disk galaxies, where the rotation velocities are most readily observed and the geometry is clear. The problem then arises as to what part of the mass resides in the disk and what part in the halo. Sancisi and van Albada (1987) have discussed this problem and drawn attention to the "disk-halo dilemma", which seems to require a conspiracy between the mass distributions in the disk and in the halo such that the rotation curves of most disk galaxies are flat. They consider the extreme cases of the dominating and the insignificant disk. However, a resolution of the problem of the division of mass between these two structures has not yet been found.

Van der Burg and Shane (1986) have resurrected an old idea (see Shu *et al.*, 1971) which may help us to determine the mass of the disk and thus, through the rotation curve, the mass distribution in the halo. If it may be assumed, as is indicated by Wevers (1984), that the mass-to-light ratio in the disk of any galaxy is roughly independent of radius, then the run of surface density with radius and the total disk mass can be determined from photometry and from a single determination of (M/L) at any radius. The only required information is a local value of σ_t, the disk surface density. This can be provided by the dispersion relation from the density-wave theory for tightly wound spirals (Lin et al., 1969 and other references) which may, in simplified form, be written

$$\sigma_t = \kappa^2/\pi G k \chi ,$$

where κ is the local epicyclic frequency, k is the radial wave number and χ is a dimensionless parameter of order one which is a complicated function of the stellar velocity dispersion, the disk thickness and the pattern speed (expressed as the dimensionless frequency with which the material encounters the spiral arms). We restrict ourselves to galaxies in which κ and k are known, thus galaxies with well-defined grand design structures, not to open because of the tightly-wound requirement, and a well-determined rotation curve and inclination. The distance is also required, among other things for the determination of κ. The uncertainties now reside in the evaluation of χ.

The pattern speed, or corotation radius, is never easy to determine, but the requirement that the grand design spiral structure be confined between the Lindblad resonances may lead to a useful limitation in the acceptable values. The isothermal disk approximation leads to a second relation between the velocity dispersion, the disk thickness and the surface density. Given a pattern speed, we are then left with a one-parameter family of solutions relating the surface density to either the disk thickness or the velocity dispersion. We must, of course, not forget to take into account the anisotropy in the velocity dispersion. For any solution the Toomre stability factor Q can be calculated. If the solution is unique (and reliable) this provides a rare opportunity for an empirical determination of Q. If not, then the calculated value of Q may persuade the investigator to prefer certain solutions to others.

This method was applied to the Sombrero galaxy (NGC 4594) by van der Burg and Shane (1986), who restricted their analysis to a zone at about 15 kpc from the center. They had no reliable information on the velocity dispersion or the disk thickness, but they were able to set some limits to the pattern speed. Adopting $Q = 2$, from estimates in other galaxies, they derived a local surface density of 40 to 50 $M_\odot pc^{-2}$ and, from an exponential model of the disk, a total disk mass of 1.1 10^{11} $M_\odot$. These values can be halved by adopting $Q = 3$, which is disturbing, but at least it indicates that a good solution for σ_t will also result in a good determination of Q. In the same zone the HI surface density is $< 2M_\odot pc^{-2}$ (Bajaja et al., 1984) so that it appears that van der Burg and Shane were justified in neglecting the dynamical effect of the gas and using it exclusively as a tracer of the rotation velocity and the spiral arms.

After the apparent success of the method as applied to the Sombrero, it was tempting to look for other possible applications. Reliable estimates of disk thickness and velocity dispersion are scarce, and restricting ourselves to early-type galaxies with well defined spiral structure and reliable rotation curves, we could find no examples as favorable as the Sombrero. An alternative possibility is to select a galaxy in which the exercise can be carried out at two or more radii. If the mass distribution in the disk follows the light distribution then two or more solutions can be used jointly to estimate the unknown quantities. It was not surprising to find that M81, which was the subject of probably the most successful density wave analysis of an external galaxy (Visser, 1980), was a good candidate. A very preliminary analysis at several radii between 5 and 10 kpc yields results with which we are not yet satisfied. Among other things, the calculated values of Q are too small for our prejudices. Part of the trouble is no doubt that we have not yet taken proper account of the gas component. Whereas neglecting the gas may have been permissable in the case of the Sombrero, this is certainly not the case for M81. Even if the method should fail, the exercise will still be of interest in revealing the limitations in the applicability of the elementary density wave theory to real galaxies and may point the way to those improvements which are of the most practical importance.

The work described here was done largely by G. van der Burg and W. Wauben to both of whom I am grateful for stimulating discussions and for providing sometimes very preliminary results.

References
1. Bajaja, E., van der Burg, G., Faber, S. M., Gallagher, J. S., Knapp, G. R., Shane, W. W.: 1984, *Astron. Astrophys.* **141**, 309.
2. Lin, C. C., Yuan, C., Shu, F. H.: 1969, *Astrophys. J.* **155**, 721.
3. Sancisi, R., van Albada, T. S.: 1987, in *Dark Matter in the Universe*, eds. J. Kormendy, G. R. Knapp, IAU Symp. No. **117**, 67.
4. Shu, F. H., Stachnik, R. V., Yost, J. C.: 1971, *Astrophys. J.* **166**, 465.
5. van der Burg, G., Shane, W. W.: 1986, *Astron. Astrophys.* **168**, 49.
6. Visser, H. C. D.: 1980, *Astron. Astrophys.* **88**, 159.
7. Wevers, B. M. H. R.: 1984, A Study of Spiral Galaxies, using HI synthesis observations and photographic surface photometry, Dissertation, University of Groningen.

VI. GENERAL ASTROPHYSICS

SOLITON STARS

T. D. Lee[1]
*Columbia University,
New York, NY 10027*

1. INTRODUCTION

It gives me great pleasure to be here to participate in the celebration in honor of Professor C. C. Lin. In this room, with the exception of Professors Chou Peiyuan, Hsu Hsieh-Hsiu, and a few others, I may well qualify as one of C. C. Lin's earliest admirers.

When Chia-Chiao was a student at Qinghua University before the war, his scholastic achievement was already famous. During the war years, Qinghua joined Peking and Nankai Universities to form the Southwest Associated University of which I was a student. In that sense we are fellow alumni. In Chinese there is a phrase: 先 後 同 學 . Literally translated, it means "Classmates before and after"; Chia-Chiao was before me, and I after him.

For those of you who were not fortunate enough to be students in China in the 30's and 40's, it would be difficult to appreciate the superstar status which the whole university community accorded to a specially accomplished student, sort of like the hero-worship one gives to Babe Ruth, Joe Dimaggio, or other sports greats. As soon as I entered the Southwest Associated University in 1945, I became aware that Lin Chia-Chiao was a name that inspired respect and admiration, even though he had left for the States several years before. Therefore, without Chia-Chiao's knowing it, the retarded potential produced by his worldline had already had a discernible uplifting effect on mine. To our occidental friends, I may point out that C. C. Lin's Chinese name 林家翹 (Lin Chia-Chiao) means "superstar from the Lin family". Chia is "family" and Chiao "superstar". Therefore all this was predetermined by C. C.'s initial conditions.

Back to my story: A year later, in 1946, I also left China to come to the U.S. Before we actually met, Chia-Chiao again made an impact on my life. My first paper was on particle physics, on the universal Fermi interaction

---

[1] This research was supported in part by the U.S. Department of Energy.

and the intermediate boson. My thesis adviser was Enrico Fermi. In 1948 and 1949, Fermi was interested in the origin of the cosmic radiation and nuclear synthesis. At that time, the whole development of renormalization and quantum electrodynamics was about coming to its end, and strange particles were not yet discovered. Fermi himself was not so confident that particle physics had a good future. He thought astrophysics would be a field with a better future and directed me towards that. Through his recommendation, I took up my first postdoc with Chandrasekhar, expecting to work on astrophysics. It turned out, however, that Fermi's optimism about the future of astrophysics was not shared by Chandra, whose interest had shifted to hydrodynamics. So, instead of astrophysics, I wrote several papers on turbulence. Then I sent these papers to C. C. Lin, looking for his encouragement. Chia-Chiao replied quickly. He made some complimentary comments about my papers, but he also made it quite clear that hydrodynamics had very little future. He suggested that I should go back to particle physics. I took his advice and found it to be very, very good.

In appreciation for his excellent advice, I dedicate today's talk to Lin Chia-Chiao. I will begin my discussion by reviewing the well-known Chandrasekhar limit for white dwarfs and the corresponding Oppenheimer-Volkoff limit for neutron stars.

2. THE CHANDRASEKHAR LIMIT

Consider a white dwarf, or a neutron star, of radius R, mass M and fermion number N. The gravitational force is balanced by the Fermi pressure. From the equipartition of energy we expect, for the equilibrium state, the magnitude of the gravitational energy to be comparable to that of the kinetic energy. For ultra-relativistic fermions, we have

$$\frac{GM^2}{R} \sim \frac{N^{4/3}}{R} \qquad (1)$$

where G is Newton's constant. Let m be the effective mass, defined by

$$N = \frac{M}{m}. \qquad (2)$$

For a neutron star, the fermions are neutrons and m is the neutron mass m_N; for a white dwarf, they are the electrons and $m = 2m_N$, since there are two nucleons per each electron. Combining (1) and (2), one sees that a critical mass M_c exists:

$$M_c \sim \frac{1}{G^{3/2} m^2}.$$

Relating (in units $\hbar = c = 1$)
$$G = l_P^2 \tag{3}$$
where l_P is the Planck length, given by
$$l_P \simeq 10^{-33} \text{cm} ,$$
we find (because $m_N^{-1} \sim 10^{-14}$ cm)
$$M_c \sim \frac{m}{l_P^3 m^3} \sim 10^{57} m_N \sim M_\odot . \tag{4}$$

For the white dwarf, this is the well-known Chandrasekhar limit, which is about
$$1.4 \; M_\odot .$$

For M bigger than $1.4 M_\odot$ but less than M_c of the neutron star, white dwarfs cease to exist; instead, one has a neutron star. For the neutron star, because of general relativity[1] and nuclear forces, M_c is somewhat smaller than 4 times the white dwarf limit (as would be indicated by (4)); it is commonly accepted as $\lesssim 5 M_\odot$, depending on the physical assumptions.[2,3]

For M bigger than M_c of the neutron star, the solution becomes singular ($R = 0$). The star collapses into a black hole. This relatively low critical mass $M_c \lesssim 5 M_\odot$ has been used as a criterion for the observation of black holes.

This critical mass M_c for stellar collapse is relatively insensitive to the assumption of the equation of state of matter. For any cold matter (temperature $T = 0$) and assuming the usual thermodynamical limit, the pressure p must be a function of the density $\sim M/R^3$. Take for example
$$p \propto \left(\frac{M}{R^3}\right)^\gamma .$$

By balancing the gravitational force with the force exerted by the pressure, we have, in place of (1),
$$\frac{GM^2}{R^2} \sim pR^2 \propto \left(\frac{M}{R^3}\right)^\gamma R^2 ;$$
i.e.,
$$GM^{2-\gamma} \propto R^{4-3\gamma} . \tag{5}$$

To estimate gravitational collapse we may set R to be the Schwarzschild radius $2GM$. Substituting that into (5), we find
$$G^{-3+3\gamma} \propto M^{2-2\gamma}$$

and therefore the critical mass M_c is always proportional to $G^{-3/2}$, which leads again to (4), independent of γ. (For a relativistic Fermi gas, $\gamma = 4/3$.)

Nevertheless, we would like to ask: can a cold stable star exist with $M > 5M_\odot$, without becoming a black hole? In the following we shall introduce the notion of soliton stars[4-7] whose critical mass M_c can be much larger than that indicated by (4).

3. NONTOPOLOGICAL SOLITON[8,9]

To illustrate the basic mechanism, consider the following example of a nontopological soliton, first without gravity. The theory contains an additive quantum number N (like the baryon number) carried by either a spin $\frac{1}{2}$ field ψ, or a spin 0 complex field ϕ, with its elementary field quantum having $N = \pm 1$. (If one wishes, one may think of ψ as the quark field). In addition, there is a scalar field σ. Take, as a first example, the self-interaction of σ to be the typical degenerate vacuum form:

$$U(\sigma) = \frac{1}{2}m^2\sigma^2\left(1 - \frac{\sigma}{\sigma_0}\right)^2 . \tag{6}$$

We may assign $\sigma = 0$ to be the normal vacuum state, and $\sigma = \sigma_0$ the (abnormal) degenerate vacuum state. (Theories of this type have been studied in the literature, e.g., in connection with the spontaneous T violation,[10,11] the abnormal nuclear model,[8] the bag model[12,13] and the Higgs mechanism.[14]) The soliton contains an interior in which $\sigma \simeq \sigma_0$, a shell of width $\sim m^{-1}$, over which σ changes from σ_0 to 0, and an exterior that is essentially the vacuum. The N-carrying field ψ, or ϕ, is confined to the interior; this produces a kinetic energy E_k (assuming for simplicity that the mass of ψ, or ϕ, is zero when $\sigma = \sigma_0$, but nonzero when $\sigma = 0$)

$$E_k \sim \begin{cases} \frac{N^{4/3}}{R} & \text{for fermions} \\ \frac{N}{R} & \text{for bosons} . \end{cases} \tag{7}$$

In the simplest case of a scalar field ϕ, N conservation requires ϕ to be complex and vary as $e^{-i\omega t}$. Because the σ field changes from σ_0 in the interior to 0 outside, there is also a surface energy

$$E_s = sR^2$$

where s is the surface tension, related to σ_0 and σ-mass m by

$$s \sim m\sigma_0^2 .$$

The radius R can be calculated by minimizing the total energy $E = E_k + E_s$. Setting $\partial E/\partial R = 0$, we have equipartition

$$E_k = 2E_s .$$

Hence, the soliton mass M (which is the minimum of E) can be written as

$$M = 3E_s = 3sR^2 , \qquad (8)$$

the total conserved particle number N is related to M by

$$M \propto \begin{cases} N^{8/9} & \text{for fermions} \\ N^{2/3} & \text{for bosons} . \end{cases} \qquad (9)$$

Very little is known about the mass of such Higgs-like scalars, except that their masses are probably > 30 GeV. For m and $\sigma_0 \gg 1$ GeV, and for a normal nucleus, the above soliton mass M would be much larger than N (the quark number) times $\frac{1}{3}$ of the nucleon mass; therefore, the soliton configuration is unstable when N is small. But because the exponent of N in (9) is < 1, when N is sufficiently large the soliton mass is always less than that of the free particle solution, and that insures its stability against decaying into N free particles (or $\frac{1}{3}N$ free, or nearly free nucleons, in the case of N quarks). But, is there a limit to this stability of very large N?

4. SOLITON STARS

To find the upper limit, we must include the gravitational field. Gravity becomes important when the soliton radius R becomes of the same order as $2GM$. Thus, the critical mass M_c may be estimated by simply setting

$$R \sim 2GM_c ,$$

which leads to, because of (8),

$$M_c \sim \frac{1}{G^2 s} = \frac{1}{l_P^4 s} .$$

A typical Higgs-like field σ may have $\sigma_0 \sim m$ (with m^{-1} much less than l_P); we estimate

$$M_c \sim (l_P m)^{-4} m . \qquad (10)$$

For example, if m is 300 GeV, we have $M_c \sim 10^{12} M_\odot$ and $R \sim 1$ light month. Hence we find the answer to our question: depending on the physical theory, the critical mass for a hadron star to become a black hole can be much greater than $5 M_\odot$. Such cold stable stellar configurations are called *soliton stars*. The estimate (10) holds for both fermion and boson soliton stars.

We note that because of the existence of the surface energy, which leads to (9), the system though large does not have a thermodynamical limit (i.e.,

M is not proportional to N). This is why M_c can have a power dependence on the Planck length, different from -3, as shown by Eq. (4).

Let $n-1$ denote the number of nodes (or, the number of minima at finite radii) of the σ-field. The lowest energy state for a soliton star is always $1s$. An interesting feature is that for a given n, the relation between M and N exhibits a curious behavior, as shown schematically in Figure 1. This zigzag feature is independent of the statistics of the particles that carry the quantum number N, but are characteristics of the type of nonlinear equations with which we have to deal. For each given n, there is a maximum $N = N(ns)$ (with a corresponding mass $M(ns)$), beyond which there is no solution. For $N < N(ns)$, depending on N we may have one solution, or two, or three,..., infinite solutions. Furthermore,[15]

$$M(ns) \simeq \frac{1}{2n-1} M(1s)$$

and (11)

$$N(ns) \simeq \left(\frac{1}{2n-1}\right)^2 N(1s) \ .$$

Thus only the lowest $1s$ solution is stable. For $M > M(1s)$ there is no (spherically symmetric) solution. The critical mass M_c is therefore $M(1s)$.

Another interesting feature is that when $M = M(1s)$, the radius R is $\sim (M_c/s)^{\frac{1}{2}} \neq 0$. Unlike the case of the Chandrasekhar limit, the solution is not singular. Consequently, it is possible to find non-singular solutions for the black holes when $M > M(1s)$ (and $N > N(1s)$).

5. SOLITON BLACK HOLES

It is most convenient to adopt the isotropic coordinates

$$ds^2 = -e^{2u}dt^2 + e^{2v}(dr^2 + r^2 d\theta^2 + r^2 \sin^2\theta d\phi^2) \tag{12}$$

where, for the spherically symmetric solution, u and v depend only on the radius r. Let $2\pi\rho$ be the circumference of a two-sphere. From (12), we see that

$$\rho = re^v \ . \tag{13}$$

The dependence of ρ on r is plotted schematically in Figure 2; the shaded region refers to the star (with nonzero matter density and Higgs field $\sigma \simeq \sigma_0$, the false vacuum). At $r = R + O(m^{-1})$, the σ-field changes from σ_0 to 0 over a distance $\sim m^{-1}$. Outside the surface, $r > R + O(m^{-1})$, the matter density becomes zero, and the metric is determined by the Schwarzschild solution:

$$e^u = \frac{r-a}{r+a}, \qquad e^v = \left(\frac{r+a}{r}\right)^2 \tag{14}$$

where $a = \frac{1}{2}GM$ is the "Schwarzschild" radius in the isotropic coordinates.

For a soliton star, R is $> a$, and ρ is a monotonic function of r, shown in the top drawing in Figure 2.

For a soliton black hole, R is $< a$. Although $d\rho/dr$ is > 0 inside the star, outside the star ρ is no longer a monotonic function. From (13) and (14), we see that outside the star,

$$\rho = a \left(\sqrt{\frac{r}{a}} + \sqrt{\frac{a}{r}}\right)^2 \; ; \tag{15}$$

hence, $d\rho/dr$ is negative for $R < r < a$ (inside the horizon) and positive for $r > a$ (outside the horizon). The horizon is located at $r = a$ (i.e., $\rho = 2GM$). Notice that the inside and outside regions of the horizon are now related *"space-like"* to each other (since the metric is time-independent). This is illustrated by the lower drawing in Figure 2.

The lowest $1s$ configuration of a soliton star (for N sufficiently large, but less than the critical value (10)) is stable.

The soliton black hole is not. The star decays very slowly by sending matter towards the horizon; at the initial stage, the decay (measured by the total particle number N of the star) obeys the usual exponential dependence in time t:

$$N \propto e^{-\Gamma t} \tag{16}$$

with a decay rate

$$\Gamma \simeq (Rx_{\text{in}})^{-1} 8\pi G\sigma_0^2 m\omega (a+R)^2 \exp\left[-2m\left(\frac{a^2 - R^2}{R} + 2a\ln\frac{a}{R}\right)\right] \tag{17}$$

where ω is the rate of t-dependence in the internal symmetry space (given by (1)) and x_{in} is a parameter $O(1)$, determined by the solution. Because this is a black hole solution, R is less than a. For $\sigma_0 \sim m \sim 300$ GeV, since R and a are ~ 1 light month, we find[16]

$$\Gamma \sim (ma)^{\frac{1}{2}} m \exp[-10^{34}]! \tag{18}$$

This very slow time-dependence changes only the solution within the horizon $(r < a)$. Outside the horizon $(r > a)$, the Schwarzschild solution holds at all t.

6. LATITUDE OF THE MODEL

If we take, instead of (6), an MIT-bag-like potential which gives the false vacuum a higher potential than the real vacuum, in the absence of the gravity the soliton mass would be given by, instead of (7)-(8),

$$M = sR^2 + pR^3 + \begin{cases} N^{4/3}/R & \text{for fermions} \\ N/R & \text{for bosons} . \end{cases} \quad (19)$$

Because $\partial M/\partial R = 0$, we have

$$2sR^2 + 3pR^3 = \begin{cases} N^{4/3}/R \\ N/R \end{cases}$$

and therefore

$$M = 3sR^2 + 4pR^3 .$$

Next, we turn on the gravitational field. The critical mass M_c can again be estimated by setting $R \sim GM_c$. This gives

$$1 \sim 3sG^2 M_c + 4pG^3 M_c^2$$

or

$$M_c \sim \frac{1}{3sG^2} \frac{2}{1 + \sqrt{1 + \xi^2}} \quad (20)$$

where

$$\xi = \frac{4}{3s}\sqrt{\frac{p}{G}} .$$

Hence, when $p = 0$ and $s \neq 0$ we have $\xi = 0$ and $M_c \sim 1/sG^2$ as given before by (10). If $s = 0$ but $p \neq 0$ then $M_c \sim 1/p^{1/2}G^{3/2}$, which has the same power of G as the Chandrasekhar limit (4). Consequently, there is an enormous latitude of M_c for a soliton star, which can vary from a galactic mass to a solar mass.

What happens when p and s are both zero? As we shall see, in that case there is an equilibrium solution for any mass (at least classically).

7. MINI-SOLITON STAR

Consider the simple theory consisting of only a "free" spin 0 *complex* field ϕ of mass $m \neq 0$, plus gravity. There is again a conserved additive quantum number N. As shown in (1), for $N \neq 0$, ϕ must be time-dependent:

$$\phi = \sigma(r)e^{i\omega t} .$$

Although this and related problems have been studied in the literature[17,18] there are two new results found in Ref. 3.

(i) Absence of a critical mass for gravitational collapse for the equilibrium solution.

Physically, this can be seen as follows:

Set N relativistic particles in the same orbit in a zero-node s-state of wavelength $\sim R$. Equipartition gives the balance between the kinetic and gravitational energies.

$$\frac{N}{R} \sim \frac{GM^2}{R}. \tag{21}$$

Write $N \simeq M/m$ where m is the mass of the free particle; one sees that there is an upper bound $M(1s)$ for the stellar mass

$$M(1s) \sim \frac{1}{Gm}.$$

(ii) For N relativistic particles in the ns orbit (no. of nodes $= n - 1$), the wavelength is $\sim R/n$, and therefore (1) is replaced by

$$\frac{nN}{R} \sim \frac{GM^2}{R},$$

which gives an upper bound

$$M(ns) \sim \frac{n}{Gm}. \tag{22}$$

The corresponding upper bound on the particle number N is

$$N(ns) \sim \frac{n}{Gm^2}. \tag{23}$$

In Ref. 2, it is shown that this linear dependence of $M(ns)$ on n is quite accurate; it holds to a few parts in 10^4. By increasing the node number n, there is no overall upper bound in the stellar mass M for the equilibrium solution (at least classically). More recently, J. J. van der Bij and M. Gleiser[18] have derived generalization of this result to boson stars with a non-minimal energy-momentum tensor.

2. At any fixed node number n, the M versus N curve has again a zigzag behavior, as in Figure 1. (Here, in contrast to (11), $M(ns)$ and $N(ns)$ both increase with n.) Thus, for a given $N < N(ns)$, there can be more than one solution. The lowest mass branch is very close to the Newtonian approximation. For $N > N(ns)$ [but $< N((n+1)s)$], the equilibrium solution for the ns state ceases to exist and is replaced by the $(n+1)s$ state. In the example of $m \sim 300$ GeV, $M(1s)$ is $\sim 10^9$ kg, the radius is $\sim 6 \times 10^{-17}$ cm and the corresponding density is extremely high, $\sim 10^{43}$ times that of a neutron star! Because of the smallness of its size, we call such a configuration a mini-soliton star.

8. CONCLUSION

The mini-soliton stars, if they exist, may account for some of the dark matter in our universe. The soliton stars can be alternative candidates for quasars. They may also be responsible for some of the gravitational lenses.

Nonlinear field theories have been found to be of importance in all elementary particle interactions: QCD, the electroweak theory, GUT, etc. Many of their physical properties are still in the developing stage. For stellar configurations, although the non-linearity of gravitation is fully recognized through general relativity, that of the matter field is far from adequately explored. The simple examples given here are only meant to illustrate the richness of this exciting new domain.

I hope very much that the physical insight and the mathematical ideas provided by these solutions may also appeal to Chia-Chiao Lin.

References

1. J. R. Oppenheimer and R. Serber, *Phys. Rev.* **54**, 540 (1938); J. R. Oppenheimer and G. M. Volkoff, *Phys. Rev.* **55**, 374 (1939).
2. J. B. Hartle, *Phys. Reports* **46C**, 201 (1978).
3. R. M. Wald, *General Relativity*, University of Chicago Press (1984); S. Weinberg, *Gravitation and Cosmology*, Wiley-Interscience (1972); C. W. Misner, Kip S. Thorne and J. A. Wheeler, *Gravitation*, W. H. Freeman (1973); S. Chandrasekhar, *The Mathematical Theory of Black Holes*, Oxford University Press (1983). See also the references quoted therein.
4. T. D. Lee, *Phys. Rev.* **D35**, 3637 (1987).
5. R. Friedberg, T. D. Lee and Y. Pang, *Phys. Rev.* **D35**, 3640 (1987).
6. R. Friedberg, T. D. Lee and Y. Pang, *Phys. Rev.* **D35**, 3658 (1987).
7. T. D. Lee and Y. Pang, *Phys. Rev.* **D35**, 3678 (1987).
8. T. D. Lee and G. C. Wick, *Phys. Rev.* **D9**, 2291 (1974).
9. R. Friedberg, T. D. Lee and A. Sirlin, *Phys. Rev.* **D13**, 2739 (1976); *Nucl. Phys.* **B115**, 1, 32 (1976).
10. T. D. Lee, *Phys. Rev.* **D8**, 1226 (1973); *Phys. Reports* **9C**, 143 (1974).
11. S. Weinberg, *Phys. Rev. Lett.* **37**, 657 (1976).
12. A. Chodos, R. J. Jaffe, K. Johnson, C. B. Thorn and V. F. Weisskopf *Phys. Rev.* **D9**, 3471 (1974); W. A. Bardeen, M. S. Chanowitz, S. D. Drell, M. Weinstein and T. M. Yan, *Phys. Rev.* **D11**, 1094 (1975).
13. R. Friedberg and T. D. Lee, *Phys. Rev.* **D15**, 1694; **D16**, 1096 (1977).
14. P. W. Higgs, *Phys. Lett.* **12**, 132 (1964); *Phys. Rev. Lett.* **13**, 321 (1964); *Phys. Rev.* **145**, 1156 (1966); F. Englert and R. Brout, *Phys. Rev. Lett.* **13**, 321 (1964); G. S. Guralnik, C. R. Hagen and T. W. B. Kibble, *Phys. Rev. Lett.* **13**, 585 (1964); T. W. B. Kibble, *Phys. Rev.* **155**, 1554 (1967).
15. Y. Pang, to be published.
16. R. Friedberg, T. D. Lee and Y. Pang, to be published.
17. R. Ruffini and S. Bonazzola, *Phys. Rev.* **187**, 1767 (1969) and W. Thirring, *Phys. Lett.* **127B**, 27(1983) have examined the lowest branch of the 1s solution. Similar analysis has been extended by M. Colpi, S. L. Shapiro and

I. Wasserman, *Phys. Rev. Lett.* **57**, 2485 (1986) to include a repulsive quartic interaction.

18. J. J. van der Bij and M. Gleiser (Fermilab preprint) recently examined boson stars, which are like mini-soliton stars, but with a non-minimal stress tensor.

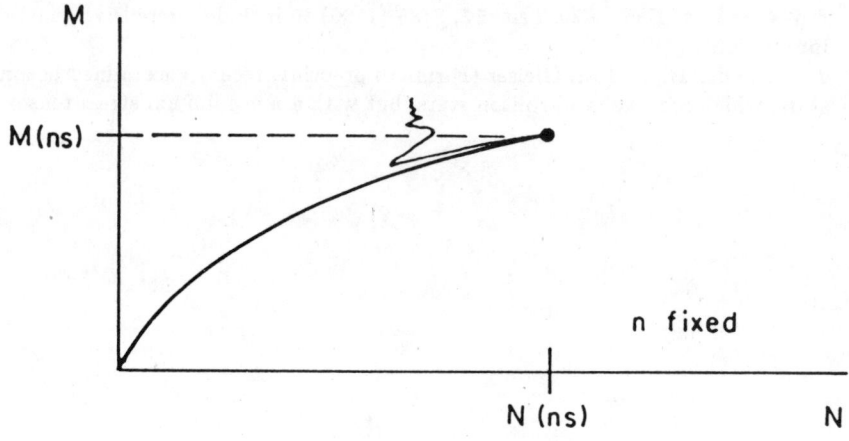

Figure 1 A schematic drawing of mass M vs. particle number N for a soliton (or mini-soliton) star; n denotes the number of nodes of the scalar field.

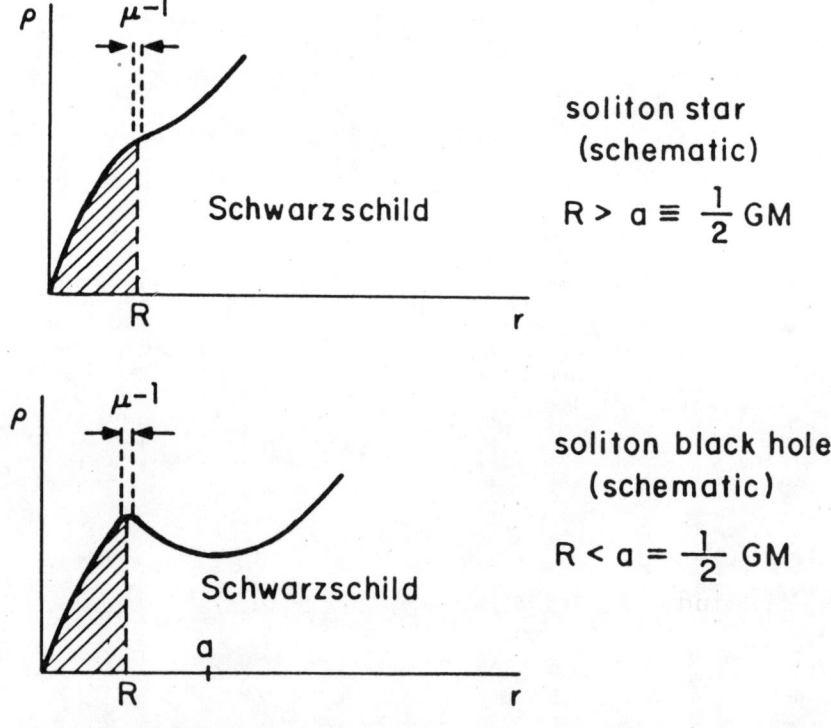

Figure 2 Schematic drawings of the circumference $2\pi\rho$ of a two-sphere vs. the radius r in the isotropic coordinates for a soliton star and a soliton black hole. The shaded region denotes matter in the star and $a = \frac{1}{2} GM$ is the Schwarzschild radius.

THE INTERACTION OF ACOUSTIC RADIATION
WITH TURBULENCE

Peter Goldreich

California Institute of Technology

We are gathered here to honor C.C. Lin who, after a long and distinguished career, has reached the age of retirement. Many of the previous speakers have been former colleagues or students of C.C. and have offered personal reminiscences in addition to delivering scientific talks. Other old friends have presented C.C. with gifts including works of art. My connections with C.C. have not been as close as those of most of you. He has influenced me mostly through his papers on hydrodynamic stability and spiral structure and indirectly through his student, Frank Shu, and his junior colleague, Alar Toomre. Frank originally invited me to speak about spiral density waves in planetary rings but I prefer instead to present a first showing of a new picture, a theory of the interaction of acoustic radiation with turbulence and its application to the excitation of the solar p-modes. This theory is the result of a collaboration with my former graduate student, Pawan Kumar; it is complete and in press in the Astrophysical Journal. However, its application to the excitation of the solar p-modes is still in progress. Given the current state of our work and the strong interest in hydrodynamics of many in the audience, I will concentrate my remarks on the basic theory and only at the end of the lecture briefly indicate how it solves the problem of understanding why the Sun oscillates.

The study of the interaction of acoustic radiation with turbulence is complicated by the absence of a dynamical theory of incompressible turbulence. Turbulence is a difficult problem because of its strongly nonlinear dynamics. The interaction of acoustic radiation with subsonic turbulence is characterized by a small parameter, the turbulent Mach number. The weakness of the interactions allows progress to be made on this problem once a simple physical model has been adopted for incompressible turbulence. The results obtained by this approach are only suggestive and I wonder whether they will be convincing even to applied mathematicians. They do, however, satisfy the standard of rigor in my field of theoretical astrophysics.

Our task is to estimate the equilibrium energies and lifetimes of the acoustic modes in a box filled with fluid some of which is turbulent. For simplicity, we suppose that the turbulence is homogeneous and isotropic within a patch which does not touch the walls of the box. Furthermore, we assume that there are no important mechanisms for the dissipation of acoustic energy other than its interaction with the turbulence. Finally, we adopt a picture of the turbulence as composed of a hierarchy of critically damped eddies. We denote by H and v the linear size and velocity of the energy bearing eddies. These eddies have lifetimes of order H/v and thus couple most strongly to acoustic modes with frequencies $\omega \lesssim v/H$. The fractional density fluctuations associated with the turbulence are of order $\Delta\rho/\rho \sim M^2$ where $M = v/c$ is the Mach number; c is the sound speed in the fluid. You might think that by now we have specified everything which is needed in order to estimate the acoustic interactions with turbulence. However, this is not the case. Our description of the turbulence is seriously incomplete because we have not indicated how it is maintained. Below we distinguish three types of turbulence. Each has its own unique acoustic interactions.

The first type of turbulence is free turbulence, that is, turbulence which is not subject to any external forces. This type of turbulence often arises in the interaction of a jet with ambient fluid as the result of the Kelvin-Helmholtz instability. The other two types are examples of turbulence which is locally forced, namely, turbulence which is maintained by stirring with spoons and turbulent convection.[1]

Now we come to the meat of the lecture, the properties of the acoustic emissivity and absorptivity of the three types of turbulence.

We start with the acoustic emissivity, ϵ, the emission rate per unit volume of acoustic energy. We write

$$\epsilon \sim \frac{\rho v^3}{H} \eta.$$

The first factor is the rate per unit volume at which kinetic energy is dissipated in the turbulent cascade and the dimensionless second factor, η, which we call the acoustic efficiency, is a measure of the fraction of this energy which is radiated in acoustic waves. The appropriate values of η are M^5 for free turbulence and M^3 for both types of forced turbulence. Quadrupoles are the lowest order acoustic multipoles in free turbulence whereas dipoles are created in forced turbulence. This accounts for the lower acoustic efficiency of forced turbulence. The quadrupole nature of the acoustic emission by free turbulence was first established in the papers by Lighthill (1952, 1954) which dominated the subject of the acoustic interactions with turbulence for many years.

The acoustic absorptivity, α, is the coefficient which relates the rate of absorption of acoustic energy per unit volume to the acoustic energy density. It is a more difficult to determine the absorptivity than the emissivity. In fact, there is very little discussion of acoustic absorptivity in the literature. In part this is understandable since most turbulent flows of technological importance are transparent to the acoustic radiation they emit.

---

[1] Strictly speaking, we should digress to discuss an analog of turbulent convection which is homogeneous and isotropic. However, this would take us too far afield in this short lecture so I will just refer to turbulent convection here.

However, I believe that most of the workers in the field were unaware that acoustic emission would be accompanied by acoustic absorption. This type of reasoning does not seem to be common in fluid mechanics. Moreover, the only references we could find to the absorption of acoustic radiation by turbulence, those by Crow (1967) and by Goldreich and Keeley (1977), did not recognize the subtleties described below. We write

$$\alpha \sim \frac{v}{H}\Gamma.$$

The first factor is just the decay rate of the energy bearing eddies. The dimensionless second factor is equal to M^2 for free turbulence, to 1 for turbulence maintained by stirring and to M^2 for turbulent convection. The relative values of Γ for free turbulence and turbulence maintained by stirring are what one might naively expect. After all, quadrupole absorption occurs in the former and dipole absorption in the latter. How then do we account for the absorptivity of turbulent convection which is no larger than that of free turbulence? After all, the emissivity of turbulent convection is enhanced by the existence of acoustic dipoles relative to that of free turbulence. Why then is its absorptivity not similarly enhanced? It took us a long time to crack this puzzle which turns out to have a simple answer. To understand the answer it is first necessary to recognize that classical absorption is really the difference of two terms, true (or quantum mechanical) absorption and stimulated emission. The true absorption by turbulent convection is dominated by acoustic dipoles and is equal to that of turbulence maintained by stirring. However, the classical absorption is much smaller for turbulent convection than it is for turbulence maintained by stirring because the cancellation between the true absorption and the stimulated emission is more complete in the former than in the latter case. I could go on to describe in detail the difference between the two types of forced convection which gives rise to the different absorptivity. However, it is time to move on to another topic so I refer you the my paper with Kumar for further details.

Once we have the values of ϵ and α we can determine the equilibrium energies, E, of the acoustic modes whose frequencies are of order v/H. These are most memorably expressed in terms the mass of the energy bearing eddies, $\mathcal{M} \sim \rho H^3$. We find $E \sim \mathcal{M} v^2$ for both free turbulence and turbulence maintained by stirring and $E \sim \mathcal{M} c^2$ for turbulent convection. Thus the acoustic modes come into equipartition with the energy bearing eddies for free turbulence and for turbulence maintained by stirring. Of course, their lifetimes are longer in the former case than in the latter by a factor M^2. For turbulent convection, the equilibrium energies are higher than the equipartition value. There is really nothing to worry about here. After all, even steady state turbulence is not in equilibrium. Energy must be continually supplied to the largest eddies or else the turbulence will decay.

Finally, we confront our theory with the observations of the Sun's acoustic modes. The most energetic modes have periods of order five minutes and their individual energies are about 10^{28} erg. We are fortunate in that the turbulent convection which interacts most strongly with these modes is probably the solar granulation which is subject to observational scrutiny. The granules are found where the pressure, $P \sim 10^5$ dyne and have typical horizontal dimension $L \sim 10^8$ cm, vertical depth $H \sim 10^7$ cm, and $v \sim 10^5$ cm s^{-1}. Accordingly, we set

$$\mathcal{M} \sim \rho L^2 H c^2 \sim PL^2 H \sim 10^{28} \text{ erg},$$

which agrees nicely with the energy of the most energetic solar acoustic modes.

For completeness, I should mention that Kumar and I have shown that the scattering opacity is much smaller than the absorptive opacity and may be safely neglected in the Sun. We have also evaluated the nonlinear couplings among the modes. Here the situation is more ambiguous. Nonlinear interactions may transfer energy from trapped modes to waves which propagate away from the Sun at a rate which for some modes, especially those with periods shorter than five minutes, is comparable to the rate at which the modes gain and lose energy from the turbulent convection.

REFERENCES

Crow, S.C. 1967, *Phys. Fluids*, **10**, 1587.

Goldreich, P. and Kumar, P. 1987, *Ap. J.*, in the press.

Lighthill, M. J. 1952, *Proc. Roy. Soc.*, A **211**, 564.

Lighthill, M. J. 1952, *Proc. Roy. Soc.*, A **222**, 1.

SUPERNOVA, NEUTRINOS AND COSMOLOGY*

Hong-Yee Chiu
Goddard Space Flight Center
Greenbelt, MD 20771

In this paper we review the physical processes leading to stellar collapse, supernova, and neutron stars and some recent work on the analysis of neutrino detections (the Kamiokande II [KII] and the Irvine-Michigan-Brookhaven [IMB] experiments) from the recent Supernova (SN) 1987A in the Large Magellanic Cloud (LMC). Using a mass correlation method developed for this application, a mass signature was found at 3.6 eV. This mass implies a background neutrino mass-energy density of the order of 3.10^{-30} g cm^{-3}. This energy density is greater than that of ponderable matter ($\sim 5.10^{-31}$ g cm^{-3}) but is still less than that needed for closure (around 10^{-29} g cm^{-3}). Within a decade or two, an accurate determination of the rest mass of the neutrino may probably be achieved in terrestrial experiments such as ^3H decay. If the laboratory value of the neutrino mass is of the order 3.6 eV and not very much smaller, then the laboratory value can be used to redetermine the fundamental distance scale of the universe, i.e., the absolute magnitude-period relationship of Cepheid variables in LMC.

1. INTRODUCTION: SUPERNOVAE, NEUTRINOS AND NEUTRON STARS

The neutron star was postulated over half a century ago by L. Landau, almost as soon as the neutron was discovered. The detailed structure of neutron stars based on the behaviour of an ideal Fermi gas was investigated in 1939 by J. R. Oppenheimer and G. Volkoff, and the collapse of a neutron star exceeding its equivalent Chandrasekhar mass limit was first studied by J. R. Oppenheimer and C. Snyder. While the Chandrasekhar mass limit for the white dwarf star is a generous $1.4 M_\odot$ for composition other than hydrogen, the mass limit for the neutron star composed of an ideal Fermi neutron gas is only around $0.8 M_\odot$. In the next two decades extensive work was carried out to apply theories of properties of nuclear matter to

---

*This summary article, based on the author's recent work in collaboration with Drs. Kwing L. Chan and Yoji Kondo, is dedicated to the celebration of the 70th birthday of Professor C. C. Lin of MIT.

the structure of neutron stars. However, this resulted in more controversies than solutions to the limiting mass. It does seem that no reasonable nuclear matter models can produce a neutron star heavier than say, $3M_\odot$, which may therefore be an absolute upper limit to the limiting mass of the neutron star. In the meantime, the neutron star remained a theoretical object until the discovery of pulsars in 1967 (the first result was published in 1968). The subsequent discovery of the Crab Nebula pulsar, with its rapid period and slow down rate, established that pulsars were the long sought neutron stars. Up to February, 1987, the most recently formed neutron star known was the pulsar at the center of the Crab Nebula.

On February 24, 1987, a new star was discovered near the edge of the Large Magellanic Cloud (LMC) and it was soon recognized to be a supernova.[1] Soon detections of neutrino bursts were reported, by the Kamiokande II group[2] (KII) (in collaboration with the University of Pennsylvania team) and the Irvine-Michigan-Brookhaven group[3] (IMB). The exact times of detection of the KII and the IMB detections are synchronous to within timing accuracy, and preceded the first reported optical discovery by few hours. Although at present it is not directly observed that the remnant star is a neutron star (because of heavy obscuration by the expanding envelope), the detection of neutrino bursts prior to the optical eruption is an anticipated indication of the core collapse of a massive start.[4–11]

Since the late 1940's it was recognized that neutrino emission processes have important consequences during the late stages of stellar evolution. Work during the 1960's showed that when the core temperature exceeds $5.10^8 K$, neutrino emission dominates the energy dissipation over surface radiation and other processes. As stellar evolution advances, the importance of neutrino emission grows while stellar density and temperature increase. Stars with different composition and mass may evolve along different paths, but a plausible scenario is as follows: When the degenerate core of the star exceeds the Chandrasekhar mass limit, the density will increase indefinitely and with increasing density the Fermi energy of the electrons also increases, inducing the inverse beta reactions:

$$e^- + (Z, A) \to (Z - 1, A) + \nu . \qquad (1)$$

In addition, at this stage the temperature of the core may be very high so that photodisintegration of nuclei may take place,[12] transforming nuclei (predominantly ^{56}Fe) into more elementary forms such as p and n, which then undergo reaction (1). Both processes may cause the star to collapse, for the following reasons: Firstly, the removal of electrons via neutrino emission and photodisintegration causes a rapid dissipation of stellar energy. Secondly, as the electron degenerate pressure is already dominant, the removal of electrons will decrease the source of pressure thus producing instability. Indeed, detailed calculations showed that there is no stable theoretical stellar configuration when the central density exceeds 10^9 g cm^{-3}. A gravitational collapse thus takes place.

The time scale of collapse is essentially the free fall time, which, at this stage, is $<< 1$ sec. During collapse, the Fermi energy of the electrons increases with the density, causing reaction (1) to proceed even more vigorously, producing a runaway implosion. Finally, when the Fermi energy (per nucleon) of the electrons exceeds the nuclear binding energy per nucleon (around 8 MeV), all nuclei dissolve into protons p and neutrons n. Since the mass-energy difference of the neutron and the proton is much smaller than the Fermi energy of the electrons, almost all protons will be converted to neutrons:

$$e^- + p \to n + \nu . \qquad (2)$$

The reaction cannot be completely unidirectional, since the neutron is unstable. To maintain stability a small fraction of matter must remain in the form of protons and electrons to satisfy the equilibrium condition:

$$E_F(e^-) = (m_n - m_p)c^2 + E_F(n) \qquad (3)$$

where E_F is the Fermi energy and m_p and m_n are the masses of protons and neutrons respectively. Typically around 1% of matter is in the form of p and the rest is predominantly n at a typical neutron star density of 10^{14} g cm^{-3}.

Although during gravitational collapse the dissipation of energy via neutrino emission is strong, the neutron matter can still heat up to rather high temperatures. The internal temperature of the star just after collapse may be as high as 10^{11} K. The temperature will drop quickly because of neutrino emission, as discussed below.

If we define $T_e = m_e c^2/k = 5.93.10^9$ K, then for $T >> T_e$ the electron gas in relativistic and behaves almost like radiation. Electron pairs can exist in equilibrium with radiation. Even in the presence of degeneracy a substantial population of e^+ can remain. The chief neutrino processes are:

$$e^- + p \to n + \nu \qquad (4)$$

$$e^+ + n \to p + \overline{\nu} \qquad (5)$$

$$e^- + e^+ \to \nu + \overline{\nu} \qquad (6)$$

Although these equations appear symmetrical, the partition of energies is not equal between ν and $\overline{\nu}$. Because of neutron degeneracy, the neutron created in (4) must have energies close to or above the Fermi energy for the neutron gas (typically 50 MeV after the formation of the neutron star), resulting in the suppression of the energy in ν. This is not the case for reaction (5), since the proton is not degenerate. The energies of ν and $\bar{\nu}$ in reaction (6) are about equal. Overall, the production of $\bar{\nu}$ is energetically favored over ν.

The following scenario is envisaged: First a collapse takes place during which reactions (4), (5), and (6) dissipate stellar energy. Next the newly formed star rapidly cools down. Two neutrino emission time scales then apply: a fast one during stellar collapse when the medium is still not overwhelmingly degenerate and a slower time scale during the subsequent cooling. The details of neutrino emission during stellar collapse are under study at GSFC. The following discussion of the neutrino rest mass is nearly independent of the detailed mechanisms of neutrino emission.

2. TIME AND ENERGY CORRELATIONS OF SUPERNOVA NEUTRINOS

The detection of neutrinos from Supernova 1987a marked a new era in astrophysics, the birth of galactic neutrino astronomy.

First, the neutrinos were received preceding the observation of the optical eruption, as predicted in theory.[4-11] From the known distance to the Large Magellanic Cloud (LMC) one can compute the total energy dissipated in the form of neutrinos, which is remarkably close to theoretical predictions. Second, the detected neutrinos were created 5.10^{12} seconds ago. This free flight time is many orders of magnitude longer than what is achieved in laboratory experiments or in the observation of solar neutrinos. This allows certain properties of the neutrino to be deduced, including the life time and rest mass. Indeed, several upper limits on the rest mass of the neutrino have been obtained.[13,14,15]

The rest mass of the neutrino has been of great interest in particle physics. Various experimental techniques have been applied to ascertain the rest mass of the neutrino, if any. Most recent experiments yield an upper limit of 5-15 eV[16]. Since the neutrinos from Supernova 1987a have travelled a distance of some 160,000 light years, a small velocity difference among neutrinos of different energy, caused by the existence of a finite rest mass, will produce time dispersion in the arrival times. We will now discuss the conditions under which this time dispersion is detectable and can be used to determine the rest mass of the neutrino.[17,18,19,20]

The dispersion Δt of the arrival times, of two neutrinos emitted at the same instant (with nonzero mass energy m_ν and energies E_1, E_2 where

$E_2 > E_1$) after travelling over a time t, is

$$\Delta t/t = \frac{1}{2}\left[\frac{m_\nu}{E_1}\right]^2 [1 - (E_1/E_2)^2] . \qquad (7)$$

Consider first the case when all neutrinos are emitted simultaneously. An exact relationship between the energy E and the time of arrival t_a (with respect to the arrival time of the first neutrino) then exists as follows:

$$E = K_1(K_2 + t_a)^{-1/2} \qquad (8)$$

As demonstrated below, if the neutrino emission extends over a finite time interval, under certain circumstances it is still possible to observe an energy-arrival time correlation.

Now we may apply Eq. (7) to determine the mass of the neutrino via the time-energy correlation. Obviously a mass can be deduced from Eq. (7) only for a pair of neutrinos. In addition, a time correlation expressed by Eq. (7) is meaningful in terms of a rest mass via Eq. (7) only if the causality condition

$$(E_1 - E_2) \cdot (t_2 - t_1) > 0 \qquad (9)$$

is satisfied, which simply requires that neutrinos with higher energies arrive prior to lower energy ones. Using the experimentally determined times of arrival and energies for each pair of neutrinos, a mass m_ν is defined via Eq. (7). Recognizing that there are many factors (particularly the true time spread during emission) affecting the time of arrival, we designate the mass thus obtained as the *correlation mass*. If there are enough neutrinos which can be regarded as 'simultaneously emitted', in the correlation mass space one should find a sharp maximum in the distribution around the true mass (as shown by Eq. (8)), while the remaining time-uncorrelated pairs will be more or less randomly distributed in the correlation mass space. While this assertion is mathematically difficult to prove, it can be convincingly demonstrated by numerical simulations, as shown below.

In order to apply Eq. (7) to supernova neutrino data, it is necessary to know the energies of the neutrinos. In both published data sets (KII and IMB) only the energies of the secondary electrons are reported. One should note that both the production process and the detector system favor antineutrinos. Indeed, all detected neutrinos appear to be antineutrinos as indicated by the large angular dispersion. The antineutrino is detected via the Reines' reaction:

$$\bar{\nu} + p \rightarrow n + e^+ . \qquad (10)$$

In the reaction (10), e^+ will carry almost all energy if $E(\bar{\nu}) << m_p c^2 \simeq$ 980 MeV. This condition is fulfilled for all KII and IMB data.

Figure 1 shows the energy vs. time of arrival relationship of the KII and IMB data. While the timing accuracy of the data is extremely good (to better than 10^{-3} sec), the energies are only determined to around 20% accuracy. It is seen that most neutrinos are emitted within the first second or two. Within the group of initial neutrinos it appears that there is a trend for high energy neutrinos to precede low energy neutrinos, indicating a possible existence of a correlation described by Eq. (7). However, as the data set is small, it must be convincingly demonstrated that any conclusion drawn is not the result of random coincidences.

3. ANALYSIS OF NEUTRINO DATA AND NUMERICAL SIMULATIONS

As a first approach, we will ignore the experimental uncertainties. Next we will include them.

Applying Eq. (7) to all neutrino pairs satisfying the causality condition (9), one obtains correlation masses for all observed neutrinos. Figure 2 shows the histogram of population vs. correlation mass of neutrino pairs for the KII data (A) and for the IMB data (B). In the KII data two apparent peaks exist, one at around 4 eV and the other at 16 eV. In the IMB data one apparent peak exists at around 32 eV.

3.1. KII Data

The major peak around 4 eV consists of 6 correlation pairs within a correlation mass range of 0.5 eV. These 6 correlation pairs are produced by neutrinos 1, 2, 4, and 6 (in the time sequence of detection). From these 6 correlation pairs, the average correlation mass is 3.6 eV with a deviation from the mean of 0.2 eV. We suggest that this peak is due to the existence of a rest mass for the electron neutrino. A cross-examination of this proposition appears in Section 4.

3.2. IMB Data

No peak shows up at 3.6 eV in the IMB data. The absence of a 3.6 eV peak in these data is due to two factors: Firstly, in the IMB setup the experimental 'dead time' per event is 0.1 seconds (0.8 seconds of total 'dead time' among 8 events), so that the lowest correlation mass that could be obtained in this experiment is 5 eV for the given energy range, from Eq. (1). This experimental limitation eliminates the capability to detect a 3.6 eV mass in the IMB experiment as configured on Feb. 23, 1987. Secondly, this experiment is insensitive to low energy neutrinos. The significance of the low energy neutrinos for detecting a low neutrino mass is shown by the scaling of m_ν and E_1 (the lower energy) in Eq. (7).

3.3. Numerical Simulations

To illustrate the process of how a mass peak (due to a true mass) in the correlation pairs may emerge from the arrival time dispersion of supernova neutrinos, we performed a numerical simulation as follows: We studied a set of 50 neutrinos with randomly distributed energies within a given range, and randomly distributed in time within a certain pulse width. If the rest mass is zero, the time-energy relationship of the original data set is not modified after travelling and will be referred to as Data Set 1. Assuming the existence of a rest mass of 3.6 eV, and a distance of travel of 160,000 light years, application of Eq. (7) to Data Set 1 will produce a data set containing time dispersion, and this data set will be referred to as Data Set 2. Using the correlation mass method, the pair distributions of these two data sets can be compared.

Assuming a rest mass of 4 eV and the energy range of the KII data ($\sim$ 7 to 35 MeV) we find that a 4 eV peak in Data Set 2 becomes clearly discernable from the background with a peak to background ratio of roughly 3:1 when the pulse width of the simulated neutrino emission is chosen to be 0.35 second. A comparison of Data Sets 1 and 2 for these simulated KII events is shown in Figure 3A. Using the same rest mass of 4 eV and the energy range of the IMB data ($\sim$ 20 to 40 MeV), no peak could be found in a similar simulation with a pulse width of 0.35 seconds (Figure 3B). Therefore, the higher energy range of the IMB experiment can obscure the presence of a peak due to a low but nonzero neutrino mass.

The simulations above also illustrate what is meant by "emitted almost simultaneously" at the source. For the KII data set two neutrinos emitted farther apart than 0.35 seconds can be considered as too widely separated in time for correlation to be recognizable. For the IMB data set the criterion is only 0.08 second. The concept of 'simultaneity' of neutrino emission is thus closely related to the energy range of the detecting experiment.

3.4. KII Peak at 16 eV and IMB Peak at 32 eV

These apparent peaks may be coincidental. According to the statistical test described in Section 4, the probability for random clustering of 4 pairs to within 0.5 eV is 39%.

4. RANDOM COINCIDENCES AND PROBABILITY ANALYSIS

We now present a probability analysis to study chance coincidence for the existence of a mass peak. We will show that the peak described in Section 3 represents a statistically significant signal which indicates a nonzero neutrino rest mass.

4.1. Chance Coincidence at Nominal Energies

First we make a probability analysis, studying the probability that 12 neutrinos with energies specified by the KII data (and randomized arrival times) are distributed in a pattern that will produce 6 correlation mass results (of any values) lying within 0.5 eV of each other. The probability for a random coincidence is found to be less than 3%. Note that this analysis, involves *no assumption* regarding the emission process and is immune to detailed structures of emission models.

4.2. Mass Signature from Special Emission Processes

The next question is, can the strong apparent mass-correlation among the four neutrinos 1, 2, 4, and 6 observed by KII be generated by a special emission process?

To answer this question, it should be stressed that the mere fact that high energy neutrinos are emitted earlier does not insure the existence of a correlation mass. In order to observe a correlation mass the energy-time relationship must be the same as that shown in Eq. (8), i.e., there must exist an energy-time relationship as in Eq. (8), i.e.,

$$E = K_1(K_2 + t)^{-1/2} . \tag{8}$$

It now remains to verify that a physical process can exist causing a collapsing neutron star to emit a time-energy correlated neutrino burst precisely as given in Eq (8). There are two difficulties for such an emission process to exist: (a) neutrino emission takes place in a thermal medium and it is hard to imagine that a precise relationship Eq. (8) can result from random thermal processes; (b) even if such a process could exist, since at the density of the neutron star, the mean free path of neutrinos of energies in the range 7-50 MeV, is at most 1 km, roughly 1/10 of the radius of the neutron star, an average neutrino will suffer at least 100 scatterings within the neutron star before emerging. Thermalization of neutrinos is unavoidable and even if it were possible to produce neutrinos according to the precise energy-time relationship Eq. (8), such a precise energy time relationship certainly cannot survive random scatterings as the neutrinos diffuse outwards from the neutron star. At best one can only expect a precise relationship of the form Eq. (8) to be valid for the *average* energy of the neutrinos. This case will be discussed next.

4.3. Special Emission Process Models

We have demonstrated that at best one may expect a neutrino emission process to yield an average energy vs. time relationship as in Eq. (8).

We now investigate if this possibility may create a mass signature in the observed data. Let us concentrate our attention on the first 6 neutrinos in the KII data set, which play the key role in the mass interpretation. Assuming that an emission process whose temperature follows the E vs. t relationship Eq. (8) does exist, we made another probability analysis similar to the one performed in (4.1), as follows:

(i) Six neutrinos are randomly selected in time (within a time frame of 0.7 seconds);

(ii) Each value of time yields a temperature according to Eq. (8);

(iii) Using this temperature, the energy of the neutrino is randomly selected according to the Maxwellian distribution, requiring that the energy of the neutrinos be greater than the detection threshold (say, 6 MeV);

(iv) The time-energy of the 6 neutrinos are correlated according to Eq. (7). The criterion for a successful mass search is the existence of 6 correlation pairs with mass (of any value) within 0.5 eV of each other, as adopted in Sections 3.1 and 4.1. (Note that out of 6 neutrinos a maximum of 15 possible correlation pairs can exist.)

Out of 10,000 cases studied, only 159 cases show successful correlation giving rise to a peak similar to that shown in Fig. 1. We conclude that the probability for a random coincidence is less than 2%[20] when both energy and time are allowed to vary in a manner most likely to produce a mass signature that is due to chance.

We find that a simulation using 5 neutrinos that satisfy criteria (i)-(iv) above yields an even lower probability. Increasing the number of neutrinos to 12 and distributing them over 12 seconds is not meaningful; neutrinos arriving later than 1 second will most likely have energies below the detection threshold (at 7.5 MeV the detection efficiency has dropped to 50%) and those with higher energies will be out of the range of Eq. (7) to yield a correlation mass with earliest arrival neutrinos. Further, inclusion of detection efficiencies (the cross section for detection σ is proportional to E^2) will tend to bias towards an even smaller coincidence rate.

4.4. Inclusion of Experimental Errors

Experimental errors represent uncertainties in the measured values. The most probable value of a measurement is represented by a distribution function whose half width at half maximum is the quoted error. Thus, to take account of experimental errors, it is necessary to consider combinations of values of the experimental data consistent with the distribution function for the error, which usually is taken to be a Gaussian. This may be studied by a Monte Carlo simulation. The one we made consists of 10,000 randomly selected energies according to a Gaussian distribution of the experimental errors; the randomly selected energy sets are analysed according to our correlation mass method. Figure 4 shows our result.[19] The 3.6 eV peak stands out prominently while all other spikes are smoothed out. As it turns out, event 3, whose nominal energy is a trifle too large to be included in the peak analysed in Section 3.1, also contributes in a statistically significant manner to the peak in this Monte Carlo computation, so that there are actually 5 mass-correlated neutrinos in the KII data (events 1, 2, 3, 4, and 6).

A statistical estimate of the rest mass of the neutrino may be obtained from Figure 4. The peak of this curve is at 2.5 eV and the probability falls to 50% of the peak value at 1.5 and 4.5 eV respectively. The probability for $m_\nu < 0.55$ eV versus that within the range 1.5 eV $< m_\nu <$ 4.5 eV is $< 10^{-3}$.

4.5. Event 6, is it a Real Neutrino Event?

We have received numerous comments regarding the validity of Event 6 in the KII data. It has the lowest energy (6.3 MeV) and the detection efficiency is already down to 50% at 7.5 MeV^2. This event was recorded by less than 20 photomultipliers and was excluded from the initial search of burst signals, but it was included in the final published data set and should be regarded as a valid detection.[2] It is true that in a data set as small as the KII Supernova 1987a neutrino detections, the removal of one single data point may alter statistics considerably. However, in a recent analysis,[20] Event 6 was purposely excluded and a peak around 3.6 eV still stands out prominently, though, by nature of the smallness of the data set, of a lesser magnitude. This is largely due to the participation of Event 3, as discussed earlier. Thus our conclusion of a mass signature does not depend critically on Event 6, although, because of the unique relationship of Event 6 with respect to other events, we believe that Event 6 is a real detection.

Figure 5 shows the energy-time of arrival relationship for the first six KII neutrinos. Ideal energy-time of arrival relationships for simultaneously emitted neutrinos, with neutrino rest masses at 2.5 and 3.6 eV are shown for comparison. As we have shown in previous analysis, the close energy-time of arrival relationship for the neutrinos 1, 2, 3, 4, and 6 is subject to a chance uncertainty of less than 3%.

4.6. Time Scale for Collapse

In the original analysis 3.1, the four neutrinos 1, 2, 4, and 6 showed a remarkably close correlation indicating a possible nonzero mass of 3.6 eV with a deviation from the mean at 0.2 eV (see also Figure 5). This deviation *does not* represent an error estimate. Rather, it scales the simultaneity of the events during emission. Interpreting the deviation from the mean as the neutrino pulse width, a value of 0.09 seconds is obtained. If the mass interpretation we discussed is correct, conceivably the five events 1, 2, 3, 4, and 6 were emitted by a short strong burst at the source. Theory allows strong pulse(s) to occur in the millisecond scale during stellar collapse.[4-11] Figure 6 shows the energy-time of emission relationship of KII neutrinos at the source, assuming a neutrino rest mass of 3.6 eV. The energy-time of arrival relationship is also shown for comparison.

4.7. Life Time of the Neutrino

If the neutrino has a rest mass, its life time can be inferred from the neutrino detections. A lower limit of the life time of the neutrino is t_{tr}/γ where t_{tr} is the time of travel and γ is the time dilation factor:

$$\gamma = E_\nu m_\nu c^2 . \qquad (11)$$

Since the average value of γ is 3.10^7, from the travel time we infer the lower limit of the life time of the neutrino *at rest* to be 2.10^5 seconds.

5. COSMOLOGICAL IMPLICATIONS

5.1. Is Our Universe Closed?

We now apply the rest mass of the neutrino to the Big Bang cosmological model. According to this model, our universe was created at a high temperature state. As the universe expanded its temperature T decreased as $1/R$ where R is a scaling factor. Various physical processes take place at different temperatures. Roughly speaking helium was formed at $10^9 K$, and ion recombination took place at $10^4 K$. After recombination, radiation became decoupled from matter and as the universe further expanded, the radiation continued to cool via the $T \propto R^{-1}$ law until $T \simeq 3K$ today. By the same argument, at a temperature of, say, $10^{10} K$ (or greater), ν and $\bar{\nu}$ also can exist in equilibrium with matter. The same argument leading to the existence of a background microwave radiation also demands a background neutrino population, except that decoupling of neutrinos with baryonic matter took place at a much higher temperature ($> 10^9 K$).

If the rest mass of the neutrino is zero, then the neutrino energy density today is comparable to that of radiation ($\simeq 10^{-34}$ g cm^{-3}). However, a nonzero neutrino rest mass leads to a substantial background neutrino mass-energy, which may account for the missing matter. Using the value 3.6 eV as we deduced from the neutrino data, a background neutrino energy density $\rho_\nu \sim 3.10^{-30}$ g cm^{-3} is obtained. This energy density is substantially higher than the estimated matter energy density $\rho_{baryons} \sim$

5.10^{-31} g cm^{-3}, but still falls short of the generally discussed density for closure, which could be as high as 3.10^{-29} g cm^{-3}. On the other hand, if we include the energy density of neutrinos of other flavors (τ and μ) the total energy density may approach closure. In any case, the uncertainties in all cosmological quantities are too large to draw a definitive conclusion.

5.2. A New Way of Astronomical Distance Measurement?

The problem of the rest mass of the neutrino may soon be resolved in laboratory experiments. Two experimental groups studying the decay of tritium (^{3}H) report upper limits of the neutrino rest mass in the range of 17 eV[21,22] and it appears that the techniques used can be refined to reduce the upper limit to 10 eV in the next few years. The rare event of a supernova in our galaxy would also provide the required data.

If the rest mass of the neutrino can be established accurately in laboratories, and if our analysis of the mass signature of the Supernova 1987A neutrino detections is correct, the fundamental distance scale of the Universe, the absolute magnitude-period relationship of the classical Cepheids in LMC, may be determined terrestrially, *without the use of telescopes* (beyond the original photograph of the outbursts, which reveals its location and approximate distance). The experimentally determined neutrino rest mass may be used to re-examine the Supernova 1987a neutrino data from which the distance (in terms of the neutrino travel time) to the Large Magellanic Cloud may be accurately determined, to an accuracy say, 10%. This distance may be used to calibrate the classical Cepheids in LMC, the results of calibration can then be applied to nearby galaxies to obtain more accurate value of Hubble's constant.

6. ACKNOWLEDGEMENTS

I thank Drs. M. Goldhaber, P. Morrison, E. E. Salpeter, and M. M. Shapiro for their comments and discussions and my colleagues Drs. Kwing L. Chan and Yoji Kondo for their collaboration on work on which this review is based, and Dr. Stephen Maran for reading this manuscript. I acknowledge the generous support and encouragement of the Laboratory for Atmospheres at NASA/GSFC, especially of Drs. Marvin Geller and Richard Hartle, without which this and other related works could not have been done.

References
1. W. Kunke, B. Madore, A. Jones, IAU Circ. 4316, (1987).
2. M. Koshiba, IAU circular 4338, (1987), K. Hirata *et al.* Phys. Rev. Ltrs. 58, 1490 (1987).
3. R. Svoboda, IAU circular 4340, (1987), R. M. Bionta *et al.*, 58, 1494 (1987).

4. H. Y. Chiu, *Annals of Phys.* **26**, 364 (1964).
5. R. A. Saenz and S. L. Shapiro, *Astrophys. J.*, **229**, 1107 (1979).
6. A. Burrows, *Astrophys. J.*, **283**, 848 (1984).
7. W. Hildebrandt and E. Müller, in "Neutrino Physics and Astrophysics", ed. K. Kleinknecht and E. A. Paschos, World Scientific (Singapore) 1984. Additional references can be found in this comprehensive volume.
8. Y. Kondo, G. E. McCluskey and S. Sofia, *Astrophysics and Space Science*, **51**, 187 (1977).
9. J. Cooperstein, L. J. Van den Horn, E. A. Baron, *Astrophysics J.*, **309**, 653-666 (1986).
10. H. A. Bethe, A. Yahil, and G. E. Brown, *Astrophysics J.*, **262**, L7 (1982).
11. D. N. Schramm, R. Mayle, J. R. Wilson, *Nuovo Cimento C*, **9C**, 443 (1986).
12. W. D. Arnett, *Can. J. Phys.* **45**, 1621 (1967).
13. J. N. Bachall and S. L. Glashow, *Nature*, **326**, 476 (1987).
14. W. Hilderbrandt et al, *Astronomy and Astrophysics*, **177**, L41 (1987).
15. K. Sato and H. Suzuki, "Analysis of Neutrino Burst from the Supernova in LMC", preprint (1987).
16. F. Böhm and P. Vogel, *Ann. Rev. Nucl. Part. Sci.*, **34**, 125 (1984).
17. H. Y. Chiu, K. L. Chan and Y. Kondo, to be published (1987).
18. H. Y. Chiu and K. L. Chan, to be published (1987).
19. H. Y. Chiu, Y. Kondo and K. L. Chan, to be published (1987).
20. K. L. Chan, H. Y. Chiu and Y. Kondo, to be published (1987).
21. S. Boris, et al., *Phys. Rev. Ltrs.*, **58**, 2019 (1987)
22. J. F. Wilkerson, et al., *Phys. Rev. Ltrs.*, **58**, 2023 (1987).

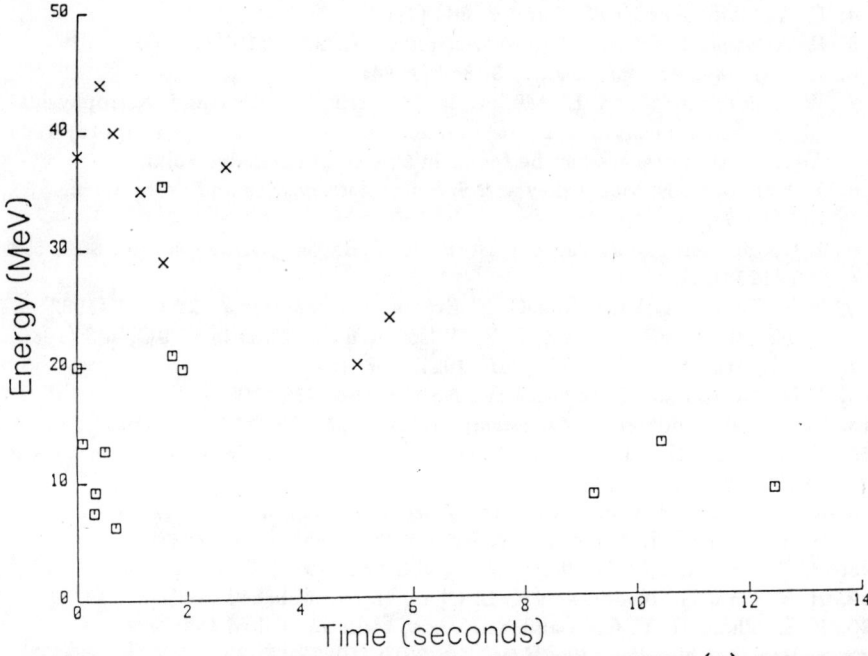

Figure 1. Energy-timing relationship from the Kamiokande II data ($\square$), and the IMB data ($\times$). Horizontal and vertical axes represent time (in seconds) and energy (in MeV) respectively. The arrival times are relative to the first event in each series.

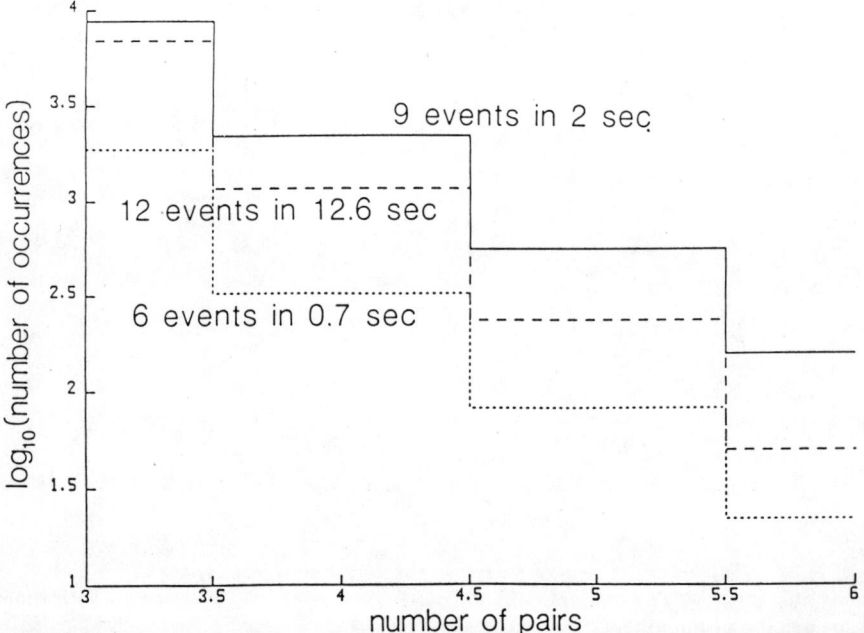

Figure 2. Distribution of all neutrino pairs in the correlation, mass space satisfying the causality condition Eq. (2). Curve A is the KII data and curve B is the IMB data.

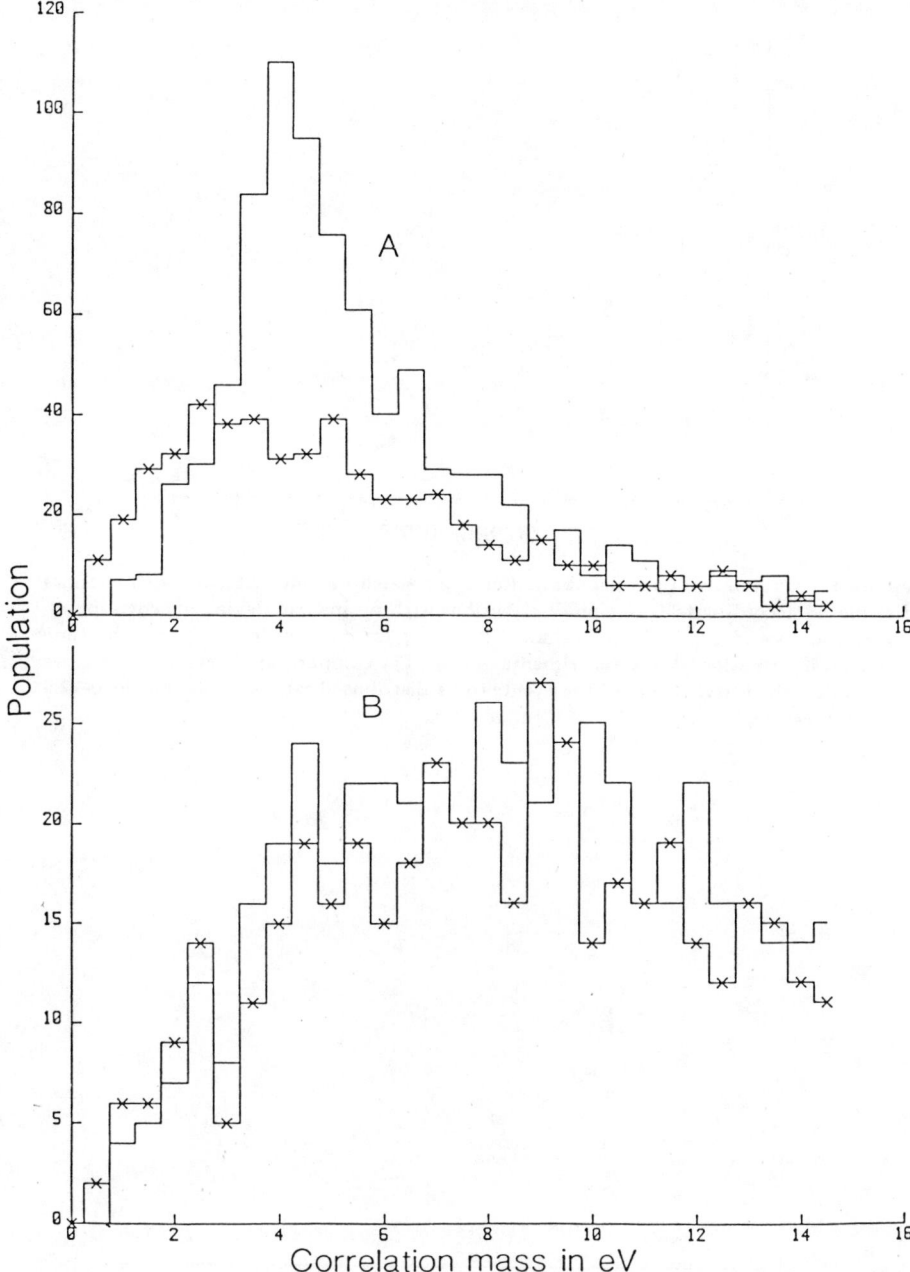

Figure 3. Distribution of neutrino pairs in the correlation mass space in a simulation using 50 neutrinos. The original random data (marked with ×) is compared with time dispersed data (unmarked). Case A uses the KII energy range and case B uses the IMB energy range.

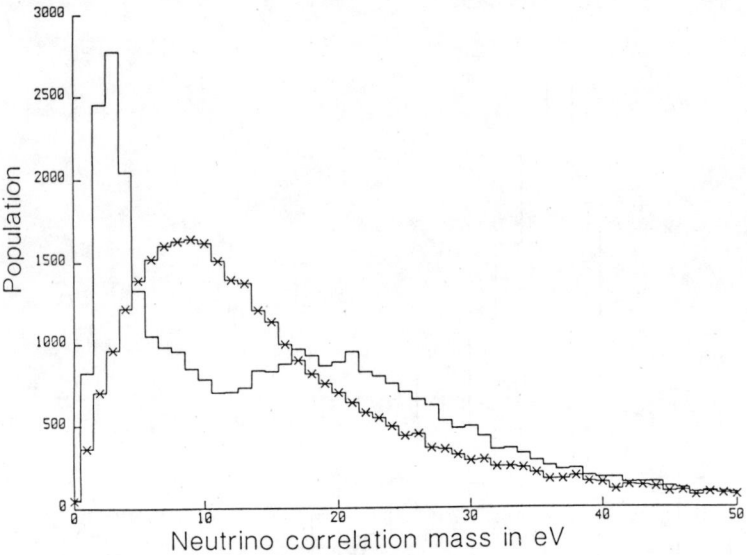

Figure 4. Cumulative pair distribution of cases which use the KII data with energies distributed according to Gaussian distributions with means and deviations corresponding to the quoted experimental values. Curve 1 is for cases which maintain the time of arrival information at the experimental values. For comparison, Curve 2 shows cases which allow the arrival times of the events to be distributed randomly within the 0-12.5 second interval.

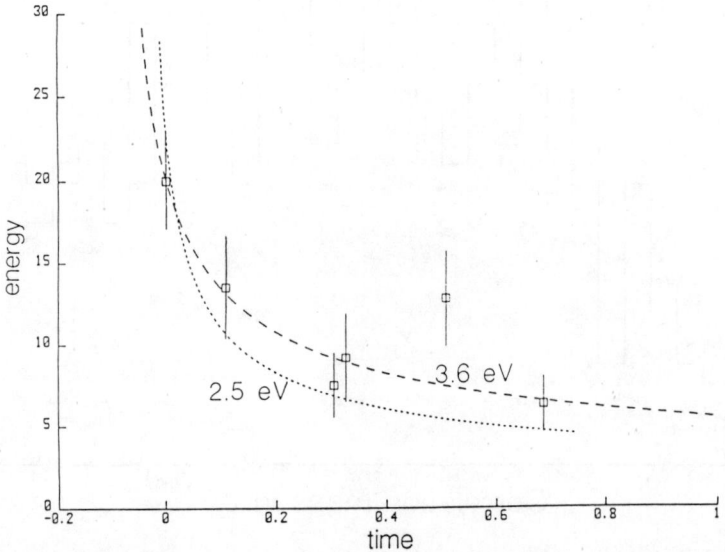

Figure 5. Enlarged view of the energy-time of arrival of the first KII Supernova 1987a neutrinos. The energy-time of arrival relationships of simultaneously emitted neutrinos with masses at 2.5 and 3.6 eV are shown for comparison.

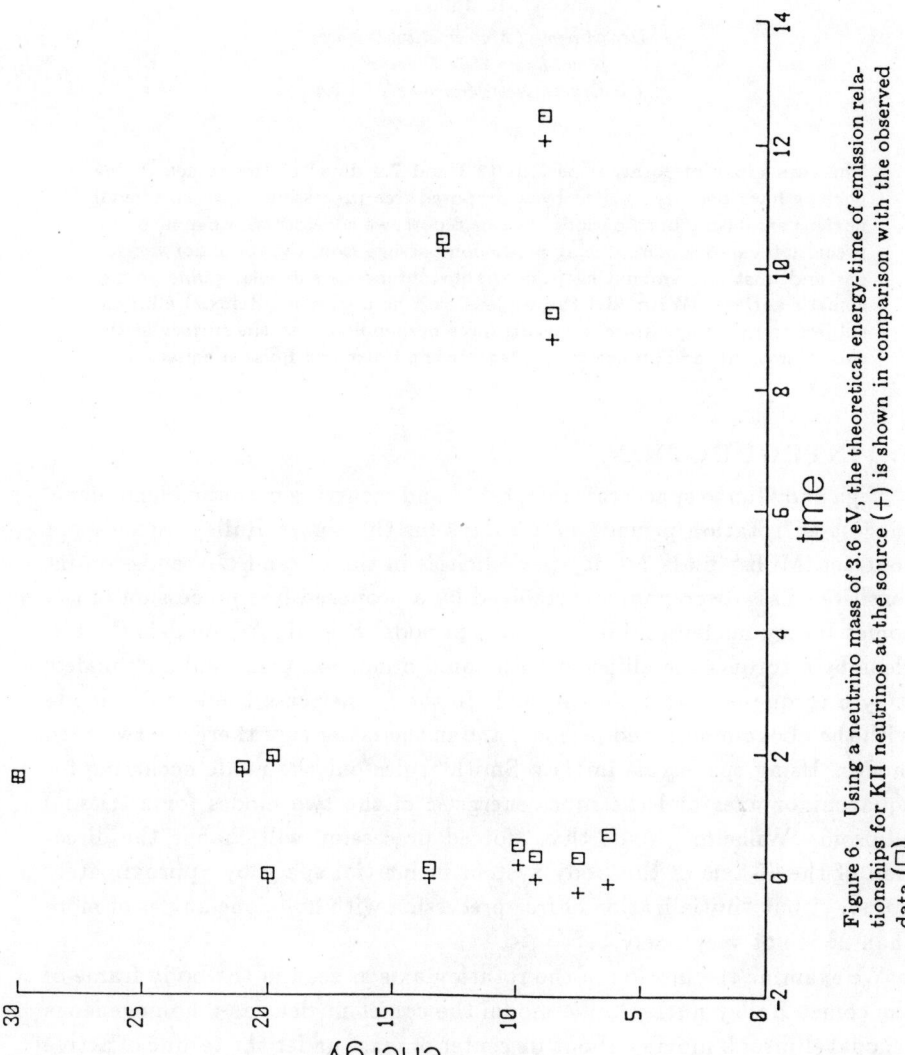

Figure 6. Using a neutrino mass of 3.6 eV, the theoretical energy-time of emission relationships for KII neutrinos at the source (+) is shown in comparison with the observed data (□).

FORCED PRECESSION OF THE COMET HALLEY NUCLEUS

W. H. Julian

Department of Mathematical Sciences
New Mexico State University
Las Cruces, New Mexico 88003 USA

The two apparent rotation periods (2.2 and 7.4 days) of the comet Halley nucleus have been reconciled by a proposed free precession of an ellipsoidal nucleus exhibiting both periods. In this report, we consider whether such free precession can originate during perihelion passage from the torque of varying gas and dust jets emanating from the instantaneous sub-solar point on the comet's surface. We model the nucleus as a homogeneous, triaxial ellipsoid subject to the torque from a varying force perpendicular to the surface at the sub-solar point, and numerically integrate the Euler and Poisson equations.

1. INTRODUCTION

Vega and Giotto spacecraft images[1-3] and recurring periodic phenomena[4] suggest a "rotation period" of 2.2 days for the comet Halley nucleus. In contrast, Millis[5] finds 7.4-day periodicities in the C_2 and CN emission intensities. This discrepancy is resolved by a proposed free precession of the comet Halley nucleus exhibiting both periods: Sekanina[6,7] models the nucleus by a torque-free ellipsoid with equal minor axes, and Julian[8] models it by a torque-free, triaxial ellipsoid. In the former case there is one mode with the above-mentioned periods, and in the latter case there are two such modes. Using spacecraft images, Smith[9] rules out the mode occurring for equal minor axes and the more energetic of the two modes for a triaxial ellipsoid. Wilhelm[10] finds that "forced precession will change the direction of the Z-axis of the body system in inertial space by approximately 30°...," but "initialization of free precession with half-cone angles of more than 2° is not very likely..."

We examine the motion of the rotation axis as seen in the body frame of the comet Halley nucleus. We model the comet nucleus as a homogeneous triaxial ellipsoid, moving about its center of mass under the torque of active gas and dust jets that emanate from the instantaneous sub-solar point. We numerically integrate the Euler and Poisson equations for three proposed variations of the jet strength: In the first, we crudely simulate perihelion passage by modulating the strength of the jet at the sub-solar point. In the second, we add a three-to-one jet strength asymmetry that depends on which hemi-ellipsoid happens to point toward the sun. In the third, we readjust the jet strength randomly at fixed time intervals.

2. EULER AND POISSON EQUATIONS

We model the nucleus of comet Halley by a homogeneous ellipsoid with principal axes of inertia of lengths $2a > 2b > 2c$. We denote by Ω_i the component of the angular velocity Ω along the moving i-th principal axis of the nucleus, by h_i the i-th component of the angular momentum, and by f_i the i-th component of a unit vector f pointing in the direction of the sun. We neglect the effect of the orbital motion on this direction. At the instantaneous *sub-solar* point the line perpendicular to the surface passes through the sun. We assume that a varying force $F(t)$ is perpendicular to the surface of the ellipsoid at this point, and calculate the motion of the nucleus due to the resulting torque. Contrast this with Wilhem's[10] specification "that the force be effective at the point in the X-Y plane that was closest to the sun." We denote by $|\Omega|$ the magnitude of the *initial* angular velocity of the nucleus, and scale time t and the angular velocity Ω in units of $T = 1/|\Omega|$ and $|\Omega|$, respectively. Thus, time is measured in radians of rotation. We denote by $A < B < C$ the moments of inertia of the ellipsoid about its principal axes; we have $A : B : C \equiv b^2 + c^2 : a^2 + c^2 : a^2 + b^2$.

The Euler equations for $\Omega_1, \Omega_2,$ and Ω_3 are

$$\frac{d}{dt}h_1 = A\frac{d}{dt}\Omega_1 = (B - C)(\Omega_2\Omega_3 + \varepsilon\lambda f_2 f_3)$$
$$\frac{d}{dt}h_2 = B\frac{d}{dt}\Omega_2 = (C - A)(\Omega_1\Omega_3 + \varepsilon\lambda f_1 f_3) \qquad (1)$$
$$\frac{d}{dt}h_3 = C\frac{d}{dt}\Omega_3 = (A - B)(\Omega_1\Omega_2 + \varepsilon\lambda f_1 f_2)$$

where $\lambda(t) \equiv L(a^2 f_1^2 + b^2 f_2^2 + c^2 f_3^2)^{-1/2}$, and L is the length scale $(a^2 + b^2 + c^2)^{1/2}$. The mass M of the nucleus enters through the dimensionless parameter $\varepsilon(t) \equiv 5F(t)/ML|\Omega|^2$. The products $\lambda f_i f_j$ are geometric factors governing the torque lever arm. Finally, we complete the system with the Poisson kinematical equations for $f_1, f_2,$ and f_3:

$$\frac{d}{dt}f_1 = f_2\Omega_3 - f_3\Omega_2$$
$$\frac{d}{dt}f_2 = f_3\Omega_1 - f_1\Omega_3 \qquad (2)$$
$$\frac{d}{dt}f_3 = f_1\Omega_2 - f_2\Omega_1$$

Small oscillations in system (1) near $\Omega_3 \sim 1$ (for $\varepsilon = 0$) have a natural period $T_1 = 2\pi[AB/(C-B)(C-A)]^{1/2}$. The same oscillations in system (2) have a natural period $T_2 = 2\pi$. We shall call motion with the period T_1 (7.4 days) a *free wobble* and motion with the period T_2 (2.2 days) a *forced wobble*.

3. NUMERICAL EXPERIMENTS
3.1. Basic Parameters

We integrate these equations using a standard Runge-Kutta scheme. We let the principal axis lengths $2a$, $2b$, and $2c$ be 16, 8, and 7 km, respectively.[8] Thus $T_1/T_2 = 3.33 \sim 7.4/2.2$. We set the ellipsoid spinning around its axis of greatest moment of inertia at an inclination of 35° to the orbital pole. In each experiment, we simulate perihelion passage in a crude manner by including an off-on-off factor in the parameter $\varepsilon(t)$.

A principal uncertainty is the perihelion value of the parameter $\varepsilon(t)$. The jets are volatilized material ejected with a velocity v of 900 m/s; the energy source is solar heating of the comet surface. Assuming that all of the solar energy intercepted by the nucleus is converted into gas and dust at the ejection velocity, we obtain an upper bound of $F \sim 4.3 \times 10^{13}$ dyne on the force of the jets at perihelion. The overall efficiency of this process is reduced by albedo, heat of vaporization, re-radiation, and divergence of the flow. With an overall efficiency of 20 per cent, we have a force F of 9×10^{12} dyne at perihelion. Then, assuming a nucleus density of 0.3 g cm^{-3}, we have $\varepsilon \sim 0.3$ at perihelion. Although we shall use this value here, it is an upper bound for the actual value.

3.2. Sub-solar point jet

We carry out the first computations with a sub-solar point jet whose strength is independent of its point of application on the surface of the comet. Figure 1 shows a typical computed polhode (or phase plot). The force-free ($\varepsilon = 0$) Euler equations (1) have a first integral

$$A(C-A)\Omega_1^2 + B(C-B)\Omega_2^2 . \tag{3}$$

Thus, we have the force-free ratio

$$\frac{\max(\Omega_2)}{\max(\Omega_1)} = \left|\frac{A(C-A)}{B(C-B)}\right|^{1/2} \sim 2.261 . \tag{4}$$

In contrast, this experiment yields a forced ratio $\max(\Omega_2)/\max(\Omega_1)$ of about seven. This forced wobble has a 2.2-day period.

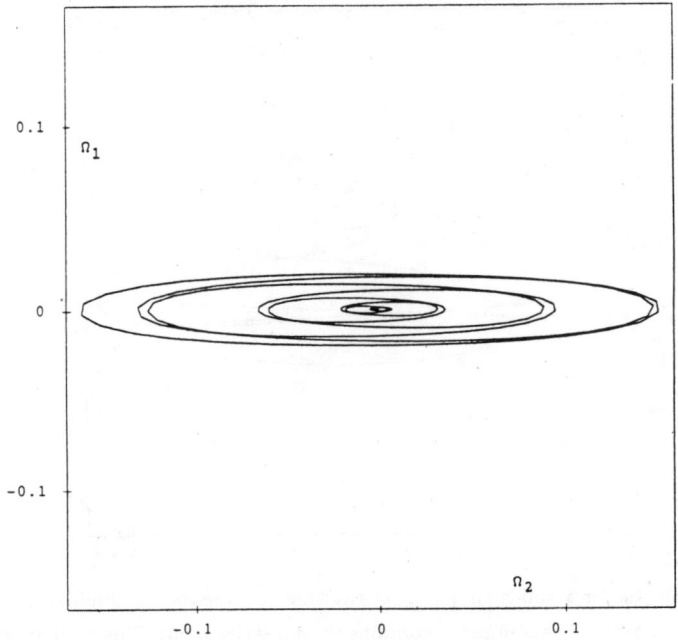

Fig. 1 Polhode for a sub-solar point jet of slowly varying amplitude.

The polhode starts and ends with $\Omega_1 = \Omega_2 = 0$; no free wobble is incited by the smooth forcing. This supports Wilhem's conclusion[10] that it is difficult to generate free wobble. Indeed, the system of equations (1) and (2) describe the motion of an ellipsoid subject to a slowly modulated disturbance. Goldreich and Toomre[11] find that the motion of a free ellipsoid with slowly changing moments of inertia is limited by an adiabatic invariant, the area enclosed by the tip of the angular momentum vector. Although their ellipsoid (the earth) is disturbed differently, nevertheless we suggest that there is similar limitation in the present situation. Indeed, the numerical calculations show that the corresponding area on the angular velocity ellipsoid starts and ends at zero. Thus, we conjecture that some adiabatic invariant[12–14] prevents the establishment of free wobbling by a slowly modulated force.

3.3. Sub-solar point jet with hemi-ellipsoid asymmetry

The comet Halley nucleus has an asymmetrical profile; the jets are stronger when the big end points at the sun. Thus, in a second numerical experiment, we introduce a three-to-one asymmetry in the jet amplitude depending on which hemi-ellipsoid contains the sub-solar point. A typical computed polhode is shown in Fig. 2 where the oval indicated by the arrow represents the induced, 7.4-day period, free wobble.

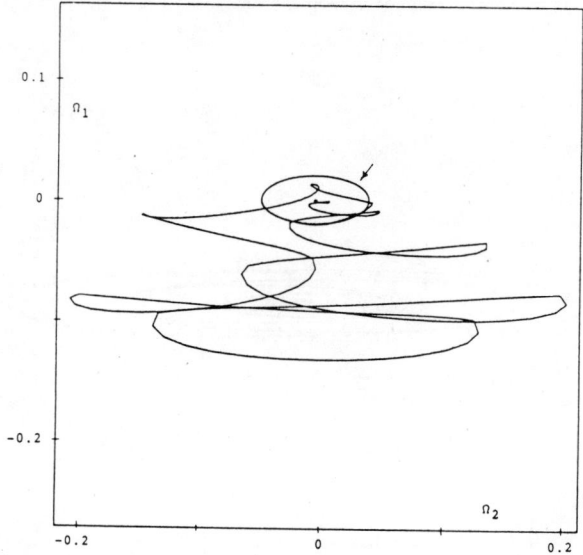

Fig. 2 Polhode for a sub-solar point jet that has a three-to-one amplitude asymmetry depending on which hemi-ellipsoid contains the sub-solar point. The polhode eventually oscillates on the small oval indicated by the arrow.

3.4. Random amplitude sub-solar point jet

We next model random variations in the strengths of the jets. At fixed 0.7-day time intervals, we select a random factor $r(t)$ uniformly distributed on the interval [0.5, 1.5) and include this factor in the parameter $\varepsilon(t)$. Figure 3 shows the resulting polhode. The terminus of the angular velocity vector eventually oscillates (with a 7.4-day period) around the oval indicated by the arrow.

4. CONCLUSIONS

Sub-solar point jets cause wobble of the rotation axis as seen in the body frame of the comet Halley nucleus. Forced and free wobbles are distinguished by their periods, i.e., 2.2 days for the forced and 7.4 days for the free. Sub-solar point jets with slowly modulated amplitude induce some forced wobble, but no free wobble. Sub-solar point jets having a hemi-ellipsoid dependent strength or a random strength induce both forced and free wobble. The induced mode of free wobble is the less energetic[8] of the two modes of a triaxial ellipsoid. The results are in accord with those of Wilhelm.[10]

I thank N. Julian, P. Goldreich, and A. Toomre.

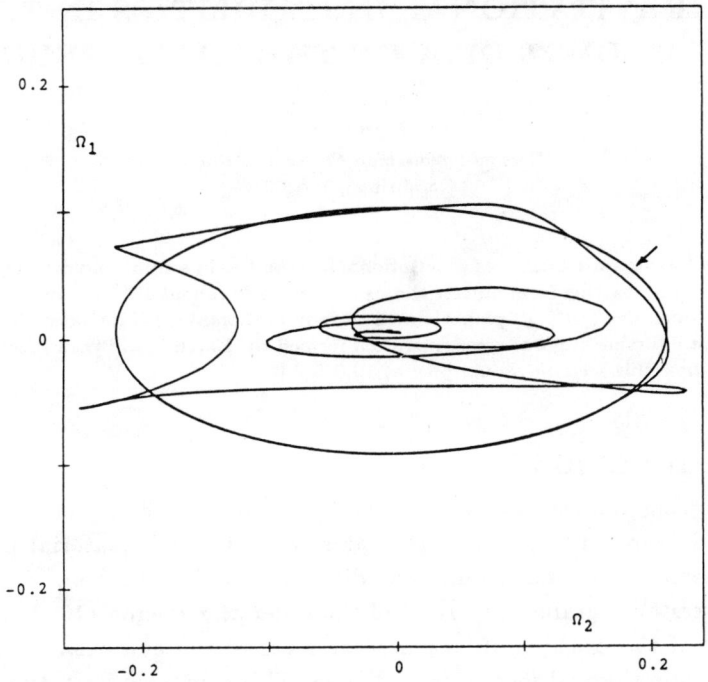

Fig. 3   Polhode for a sub-solar jet of random amplitude.

References
1. Sagdeev, R. Z. *et al. Nature* **321**, 262-266 (1986).
2. Sekanina, Z. & Larson, S. M. *Nature* **321**, 357-361 (1986).
3. Vaisberg, O. L. *et al. Nature* **321**, 274-276 (1986).
4. Sekanina, Z. *Adv. Space Res.* **5**, No. 12, 307-316 (1986).
5. Millis, R. L. & Schleicher, D. G. *Nature* **324**, 646-649 (1986).
6. Sekanina, Z. *Nature* **325**, 326-328 (1987).
7. Sekanina, Z. *I.A.U. Circ.* No. 4273 (17 November, 1986).
8. Julian, W. H. *Nature* **326**, 57-58 (1987).
9. Smith, B. A. *et al. Nature* **326**, 573-574 (1987).
10. Wilhelm, K. *Nature* **327**, 27-30 (1987).
11. Goldreich, P. & Toomre, A. *J. Geophys. Res.* **74**, 2555-2567 (1969).
12. Lenard, A. *Ann. Phys.* **6**, 261-276 (1959).
13. Kulsrud, R. M. *Phys. Rev.* **106**, 205-207 (1957).
14. Gardner, C. S. *Phys. Rev.* **115**, 791-794 (1959).

GRAVITATIONAL INSTABILITIES IN THE THIN DUST DISK OF THE SOLAR NEBULA

Myron Lecar
Harvard-Smithsonian Center for Astrophysics
Cambridge, MA 02138

I will review a sequence of gravitational instabilities in a thin, cold disk around the Sun, starting from objects the size of asteroids (about 10^{14} g), and ending in lunar size (10^{26} g) planetesimals. Many authors, myself included, believe that a further stage of conglomeration formed the Earth. I will then speculate on how this scenario would play at 10,000 AU.

INTRODUCTION

Following the scenario envisioned by Safronov (1969) and by Goldreich and Ward (1973), the dust grains settle to the equatorial plane, in the relatively low mass, quiescent disk surrounding the Sun. Since even the interstellar grains with radii of the order of a micron (10^{-4} cm) have masses of the order of 10^{12} times the mass of the hydrogen atom, their equilibrium thermal motion is negligible. This thin, cold dust disk is gravitationally unstable, on various scales. The smallest scale applies to bodies that collapse directly to solid body density. At 1 AU, their mass is about 10^{14} g. The next largest scale is the critical wavelength that just balances the gravitational and centrifugal accelerations. The critical mass at 1 AU is about 3×10^{21} g. Bodies of critical mass, at 1 AU, can further combine, in tens of thousands of years (Greenburg *et al.* 1978), to bodies of mass 4×10^{25} g (or about half the mass of the moon). Finally, a number of authors (Safronov 1969, Wetherill 1980, Hayashi 1986, Lecar and Aarseth 1986) have studied scenarios where the lunar-sized planetesimals collide and stick together to form the terrestrial planets, on a time scale of tens to hundreds of millions of years. I will review these processes that are thought to occur near 1 AU, and ask what would happen at 10^4 AU, if the nebula stretched that far.

SURFACE DENSITY OF GAS AND DUST

Weidenschilling (1977) and Hayashi (1981) derived the surface density of the gas and dust when the planets were formed, and found that it decreased as the 3/2 power of the distance from the Sun. This result was obtained by augmenting the planetary masses to solar composition, and then smearing them out. Augmenting the planetary masses to solar composition involved multiplying the present planetary masses by factors of a few hundred, which introduced considerable uncertainty into the result. Weidenschilling and Hayashi differ by about a factor of two. I use the surface densities proposed by Hayashi. They are, for the gas,

$$\sigma_g = 1700 X^{-3/2} \text{g/cm}^2 , \qquad (1)$$

where X is the distance from the Sun in AU, and for the dust,

$$\sigma_d = 7.1 X^{-3/2} \quad \text{for} \quad X < 2.7 \text{ AU} \qquad (2a)$$

$$\sigma_d = 30 X^{-3/2} \quad \text{for} \quad X > 2.7 \text{ AU} . \qquad (2b)$$

The discontinuity comes about because at distances greater than 2.7 AU, the disk is cold enough for ices to condense.

THE DUST GRAINS SETTLE TO THE PLANE

The half-thickness of the gas disk is c/Ω, where c is the thermal velocity of the gas and Ω is the angular velocity. Following Hayashi, I take the mean molecular weight of the gas to be 2.34, and the temperature of the gas to be $280\ X^{-1/2}$°k whence $c = 1.7 \times 10^5 X^{-1/4}$ cm/s. Dividing the surface density by the thickness of the disk, he obtains the mass density of the gas, ρ_g to be $1.4 \times 10^{-9} X^{-11/4}$ g/cm^3, and the number density of the gas, $n_g = 5.0 \times 10^{14} X^{-11/4}$ cm^{-3}. The mean free path of the gas molecules, l, assuming the collision cross-section for neutrals to be 10^{-15} cm^2, is given by $l = 2.0 X^{11/4}$ cm. If the acceleration in the Z-direction (perpendicular to the plane) is dominated by the Z-component of the Suns' attraction, the equation of motion for a dust grain is that of a dampled harmonic oscillator, viz;

$$\ddot{Z} = -\Omega^2 Z - 2\nu \dot{Z} , \qquad (3)$$

where $2\nu = \rho_g c_g / \rho_s r$, ρ_s is the solid-body density (which I take as unity) and r is the radius of the grain.

Neglecting the growth of the grain, the time for the grain to settle to the plane, t_s, is given by

$$t_s = 2\nu/\Omega^2 = (1/\Omega^2)(\rho_g c_g / \rho_s r) \qquad (4)$$

or, using $\rho_g(2c_g/\Omega) = \sigma_g$ and setting r equal to its initial value r_o, we have

$$t_o = (1/\Omega)(\sigma_g/2\rho_s r_o) = 1.35 \times 10^6 \text{ yrs}/r_o(\mu) \tag{5}$$

where $r_o(\mu)$ is the initial value of r in microns. The grains grow at a rate given by (Whipple 1971)

$$\dot{r} = (\rho_d c_d)/(4\rho_s) = (\sigma_d \Omega)/8\rho_s . \tag{6}$$

Letting $r = r_o + \dot{r}t_s$ in equation 4, we obtain an equation for the settling time allowing for grain growth. It is

$$\begin{aligned}\tau^2 + b\tau - 1 &= 0, \quad \text{where} \\ t &= (2/\Omega)\sqrt{(\sigma_g/\sigma_d)}\tau \quad \text{and} \\ b &= \sqrt{(r_o/\dot{r}t_o)} .\end{aligned} \tag{7}$$

For $X = 1, b = 3.64 \times 10^{-6}, t_s$ is about 5 years, and the grains grow to a few centimeters. For $X = 10^4, t = 1.1 \times 10^6$ years, and the grains do not grow appreciably. For $X > 5 \times 10^4, t_s$ reaches its asymptotic value of 1.4×10^6 years.

This scenario could be applicable to the formation of comets at 10^4 AU if the solar nebula (or at least the dust and ice disk) lasted for 10^6 years before being removed by the solar wind.

GRAVITATIONAL INSTABILITIES IN THE THIN, COLD DUST DISK

Axisymmetric instabilities in thin disks have been studied by Toomre (1964). Qualitatively, for a cold disk, it can be viewed as a competition between gravity and rotation. The equation of motion for r, the radius of a patch, is

$$\ddot{r} = r\Omega^2 - GM/r^2 . \tag{8}$$

In a thin disk, $M = \sigma \pi r^2$, so

$$\ddot{r} = r\Omega^2 - \pi G\sigma . \tag{9}$$

Let $r_c = \pi G\sigma/\Omega^2$. For comparison, Toomre's critical wavelength is $\lambda_c = 4\pi^2 G\sigma/\Omega^2$.

When $t = 0$, let $r = r_o = x_o r_c$. As r_o contracts to $xr_o (x < 1)$, the angular momentum, $r^2\Omega$ is conserved so $\Omega = \Omega_o/x^2$, and the mass, $\sigma\pi r^2$ is conserved, so $\sigma = \sigma_o/x^2$.

Therefore, the equation of motion of x is

$$\ddot{x} = \Omega_o^2[(1/x^3) - (1/x_o x^2)] . \tag{10}$$

The acceleration becomes positive when $x = x_o$, so if $r_o = x_o r_c$, r will contract to and oscillate around the value $r = x_o r_o = x_o^2 r_c$.

Let the mass that contracts to a solid body be given by

$$\pi \sigma r_o^2 = (4\pi/3)(\rho_s r_s^3) \tag{11}$$

where $r_o = x_o r_c$ and $r_s = x_o^2 r_c$. Then $x_o^4 = (3/4)\sigma/\rho_s r_c$.

Using the more exact formula for the critical wavelength,

$$M_c = \pi\sigma(\lambda_c/2)^2 = (2\pi)^2 (\pi\sigma_o R_o^2/M_o)^3 M_o X^{3/2} \tag{12}$$

and

$$x_o^4 = (3/8\pi^2)(M_o/\rho_s R_o^3)(1/X^3) \ . \tag{13}$$

where $R_o = 1$ AU and M_o is the mass of the Sun.

The mass of a body that contracts directly to solid-body densities is

$$M_s = x_o^2 M_c \ . \tag{14}$$

For $X < 2.7$, $M_c = 1.24 \times 10^{18} X^{3/2} g$ and $M_s = 1.86 \times 10^{14} g$. For $x > 2.7$, $M_c = 9.37 \times 10^{19} X^{3/2} g$ and $M_s = 1.41 \times 10^{16}$ g.

Note that except for the discontinuity at 2.7, M_s is constant, independent of X.

The next largest mass scale comes from the following construction. Assume that all the mass in the annulus between $R - \Delta R$ and $R + \Delta R$ is collected into a body at R. The mass of that body is

$$M_L = 4\pi R^2 (\Delta R/R) \tag{15}$$

Now let $\Delta R/R$ be at the L_2 point of M_L, i.e.,

$$(\Delta R/R)^3 = (1/3)(M_L/M_o) \ . \tag{16}$$

Combining equations 15 and 16, we obtain

$$M_L = \sigma 4\pi R^2 (\sigma \pi R^2/3 M_o)^{1/2} \ . \tag{17}$$

At 1 AU, $M_L = 3.65 \times 10^{25} g$ or about half the mass of the moon.

1 AU vs 10,000 AU

At 1 AU, Goldreich and Ward (*op. cit.*) argued that the approximately 10^4 solid bodies in a patch the size of the critical wavelength, would collect into a single body by dissipative processes; either gas drag or inelastic collisions of the solid bodies with each other. Both processes, at 1 AU have time scales less than 1,000 years.

The argument that takes bodies of the critical mass ($10^{18}g$) to a lunar mass ($10^{26}g$) is less direct. Imagine that the bodies of mass M are equidistant in R, and that $\Delta R = b$. A fly-by at a distance b with relative velocity v, perturbs the relative velocity (in the radial direction) by

$$\Delta v = 2GM/bv . \qquad (18)$$

The induced eccentricity Δe is $\Delta v/V$, where V is the circular velocity.

For adjacent orbits,

$$v = (1/2)(b/R)V \quad \text{so} \qquad (19)$$

$$\Delta e = 4MR^2/(b^2 Mo) \qquad (20)$$

where I have used the relation $RV^2 = GM_o$. We now ask how closely do the bodies have to be spaced so that the eccentricity induced in one fly-by causes adjacent orbits to overlap; i.e., that $R\Delta e$ be greater than b. The answer is that

$$(b/R)^3 < 4M/M_o . \qquad (21)$$

The requirement is that (b/R) must be closer to M than its L_2 point (to within a factor of 2.3) and the derivation follows the derivation of M_L. The growth of M_c to M_L depends linearly on the time. Collisions between such bodies were frequent, and assuming that the parties to a collision stuck, bodies of mass M_L were formed rapidly. This argument is, at best, only suggestive, but Greenberg et. al (1978) studied this process with a numerical simulation and obtained final bodies of lunar mass, in about 20,000 years, although most of the mass remained in small bodies.

Once this process has finished, the spacing is such that more than one fly-by was required, for the orbits to overlap. Since successive fly-bys are uncorrelated, the eccentricity random walked and grew only as the square-root of the time. A number of authors (see Wetherill, 1980 and Hayashi, 1985 for reviews) and Lecar and Aarseth (1986) studied this phase of the evolution, and with a number of caveats, formed an Earth.

At large distances from the Sun, say at 1,000 or 10,000 AU, this evolutionary sequence is aborted after the solid bodies condense. I can find no mechanism to cause the solid body masses to grow. The dissipative processes that operated on a time scale of 1,000 years at 1 AU, operate on a time scale of 10^{14} years at 10,000 AU. I conclude that solid bodies with masses a few times $10^{16}g$, will condense but will grow no further. As this is about the average comet mass, perhaps that is how the comets were made. It might have been easier to form comets at 1-10,000 AU than to form them at 40 AU and have planetary perturbations push them out to 1,000 AU where stellar perturbations could have been effective in swelling the cloud further.

References
1. Goldreich, P. and Ward, R., 1973, *Ap. J.*, **183**, 1051.
2. Greenberg, R., Wacker, J.F., Hartmann, W.K. and Chapman, C.R., 1978, *Icarus*, **35**, 1.
3. Hayashi, C., 1981, *Supplement of the Progress of Theoretical Physics* (Japan), **70**, 35.
3. Hayashi, C., Nakazawa, K. and Nakagawa, Y., 1985, in *Protostars and Planets II*, Black, D.C. and Matthews, M.S., ed., Univ. of Arizona Press, Tuscon, p. 1100.
4. Lecar, M. and Aarseth, S.J., 1986, *Ap. J.*, **305**, 564.
5. Safronov, V.S., 1969, *Evolution of the Protoplanetary Cloud and the Formation of the Earth and the Planets* (English transl., Jerusalem, Israel Program for Scientific Translations, 1972).
6. Toomre, A., 1964, *Ap. J.*, **139**, 1217.
7. Weidenschilling, S.J., 1977, *Ap. and Space Science*, **51**, 153.
8. Wetherill, G.W., 1980, *Ann. Rev. Astr. and Ap.*, **18**, 77.
9. Whipple, F.L., 1971, in From Plasma to Planet, New York, Wiley and Sons.

A PROBABLE MECHANISM FOR BIPOLAR OUTFLOWS NEAR YOUNG STELLAR OBJECTS

Z. Y. Yue[*], B. Zhang,[*] and G. Winnewisser

1. Physikalisches Institut
Universität zu Köln
D-5000 Köln 41

The present paper gives a short summary of a theoretical model which we would like to propose in an attempt to understand the major observational facts of bipolar outflows near young stellar objects. Here we will concentrate in explaining the major ideas on which the model rests, show some of the most important mathematical steps involved and a numerical example will be discussed and compared with the observational data. The details of this model will be described in a forthcoming paper of the same title.

The initial scenario required by the model is that of a central stellar object embedded in a parental cloud of gas and dust forming a halo whose shape deviates slightly from spherical symmetry, an assumption which should be met in practically all realistic cases.

Different gravitational attraction of the nonsymmetrical halo leads to a disk-like structure and initiates a mass inflow towards the central object. This radial disk-like inflow draws the main bulk of its mass from a basically non-rotating halo surrounding a rotating disk structure. On its way towards the central stellar object the inflowing matter traverses with free-fall velocity a mass-gap which exists between the inner rim of the rotating disk and the outer fringes of the central object. Upon reaching the neighborhood of the equator of the stellar object, the inflow divides itself into two parts: the inner part with mass flux $\dot{M}_1$ penetrates a small ring-like strong shock region and mixes finally with the surface layer of the stellar object, while the outer part of the mass flux $\dot{M}_2$, which is small compared to $\dot{M}_1$, is pushed away and bypasses the shock region moving in two directions towards the poles as schematically shown in Fig. 1.

---

[*]Permanent address: Department of Geophysics, Beijing University, Beijing, P. R. China

The kinetic energy carried by $\dot{M}_1$ is released in the form of radiation at a zone around the equator. A small part of the radiation energy is absorbed above the "radiation zone" (cf. Fig. 2) by the circumventing material $\dot{M}_2$, which turns at the poles into the outflowing material.

The energy carried by the material on the inflowing path and outflowing gas can be considered in terms of the appropriate Bernoulli equations. Before the outer-layer-flow $\dot{M}_2$ reaches the radiation zone, the flow $\dot{M}_2$ satisfies the Bernoulli equation:

$$V^2/2 + i + \psi = C_1$$

because the viscosity is negligibly small. v is the velocity of the inflowing material, i its enthalpy, and ψ the gravitational potential. When the flow travels above the radiation zone, the quantity $v^2/2 + i + \psi$ does not remain constant, but increases due to absorption of a small fraction of the energy radiated from the radiation zone. After the flow $\dot{M}_2$ leaves the radiation zone, the Bernoulli equation holds again but now with a larger constant C_2:

$$v^2/2 + i + \psi = C_2$$

If the fraction of the radiation energy which is absorbed by the outflow $\dot{M}_2$ is denoted by ε, we have for the energy difference:

$$C_2 - C_1 = \varepsilon \frac{\sigma T_e^4 S}{\dot{M}_2} = \varepsilon \frac{\frac{1}{2}\dot{M}_1 V_1^2}{\dot{M}_2} = \Delta Q$$

where $S = 2\pi R_* \cdot 2l_0$ represents the area of the radiation zone, T_e is the effective temperature corresponding to the black-body radiation, σ is the Stefan-Boltzmann constant, and l_0 is the half width of the radiation zone. ΔQ is the extra thermal energy input absorbed by the unit mass of the flow $\dot{M}_2$, and is responsible for the high velocity of the bipolar jets at very large distances. Calculations show that a very small fraction ($\varepsilon \sim 10^{-4} - 10^{-3}$) of the radiated energy is sufficient to drive the bipolar outflow so as to maintain its high velocity of order of 10^2 km/s at very large distance (10^3 - 10^4 AU) after overcoming the gravitational pull of the central stellar object. Detailed consideration of the absorption mechanism has shown that the flow $\dot{M}_2$ is indeed able to absorb such a fraction.

The lower limit of the jet velocity at very large distance is 118 km/s as calculated in a numerical example, where we have assumed a central stellar object of $2M_\odot$ and a ratio of the mass-flux of the outflow to be 5% of the total inflowing material.

The collimation of the bipolar jets is also easily explained by the model as follows: The outgoing flows $\dot{M}_2$ meet in the regions above the two poles and produce a pair of cone-like strong shocks, one above each polar region. Behind this shock region the bipolar jets are formed and moderately collimated. As the jet continues on its path away from the young stellar object the collimation is further enhanced by the decrease of the temperature within the jet and the increase of the Mach number. In the numerical example considered, the initial opening angle of the jet is 31°, decreasing to 8° at 5 AU and to 6° at 50 AU.

The radial inflow is a transient phenomenon lasting only about $(2\text{-}3) \times 10^4$ years if we assume a typical radial scale of 10^4 AU for the non-rotating halo which supplies the radial disk flow. When the non-rotating halo is exhausted, the radial inflow and the well-collimated bipolar outflow stops, leaving behind a rotating faint disk around the stellar object.

Thus, the four major observational features associated with bipolar outflows,

(i) its bipolar geometry,
(ii) the high velocity of the outflowing material of order of 10^2 km/s at very large distances ($10^3 - 10^4$ AU),
(iii) the high degree of collimation and
(iv) the time scale of about 10^4 years

can all be explained by the proposed dynamical model. The interaction with the local environment into which the bipolar flow is finally ejected determines its observed geometrical shape but will not be discussed here.

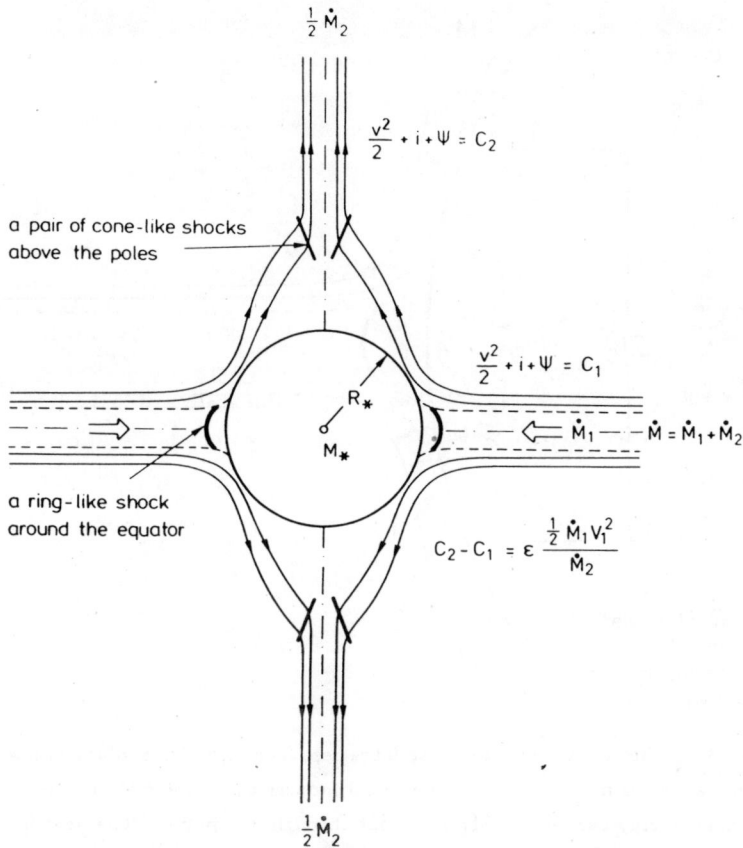

Fig. 1 Bipolar jets are induced by the radial disk-like inflow surrounding the equatorial region of the central stellar object. The radial inflow of mass flux $\dot{M}$ is divided into two parts: (i) the central major part of mass flux $\dot{M}_1$ flows along the plane of the surrounding disk through a strong ring-like shock region and eventually mixes with the surface material of the stellar object; (ii) the outer mass flow $\dot{M}_2$ is hindered from falling onto the central object by an increased vertical pressure gradient which redirects this flow towards the polar regions. There it forms collimated bipolar jets after passing through a pair of cone-like shocks above each of the poles. A small fraction of the released energy by $\dot{M}_1$ is absorbed by $\dot{M}_2$ in the form of radiation, producing the difference between the two constants C_2 and C_1 in the Bernoulli equations for the bipolar jets and for the radial inflow. This difference is responsible for the high velocity of bipolar jets even at very large distances (after overcoming the gravitational pull). This schematic picture is not drawn to scale: the thickness of the disk and the width of the ring-like shock around the equator should be only a few percent of the radius of the stellar object.

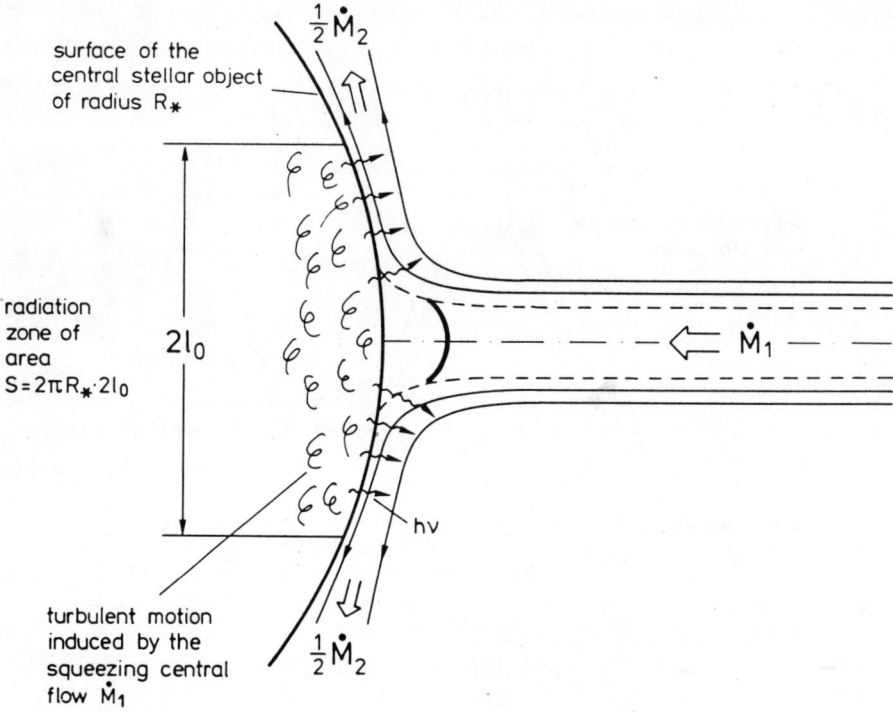

Fig. 2 A detailed picture of the equatorial region. The central part of the inflow of mass flux $\dot{M}_1$ squeezes into the stellar object and induces turbulent motion which converts the kinetic energy carried by $\dot{M}_1$ into heat through dissipation. The heat is radiated away from the radiation zone and partly absorbed by the outflow $\dot{M}_2$. A very small fraction of the radiated energy is shown to be sufficient for the bipolar jets to gain high velocities of 10^2 km/s even at very large distances after overcoming the gravitational pull of the stellar object.